Genetic Engineering

Principles and Methods

Volume 12

GENETIC ENGINEERING
Principles and Methods

Advisory Board

A Continuation Order Plan is available for this series. A continuation order will bring delivery of each new volume immediately upon publication. Volumes are billed only upon actual shipment. For further information please contact the publisher.

Genetic Engineering

Principles and Methods

Volume 12

Edited by

Jane K. Setlow

Brookhaven National Laboratory
Upton, New York

Plenum Press · New York and London

The Library of Congress cataloged the first volume of this title as follows:

Genetic engineering: principles and methods, v. 1-
    New York, Plenum Press [1979–
    v. ill. 26 cm.
    Editors: 1979–   J. K. Setlow and A. Hollaender.
    Key title: Genetic engineering, ISSN 0196-3716.

    1. Genetic engineering—Collected works. I. Setlow, Jane K. II. Hollaender,
Alexander, date.
QH442.G454   575.1   79-644807
                                                       MARC-S

ISBN 978-0-306-43616-1

CONTENTS OF EARLIER VOLUMES

VOLUME 1 (1979)
Introduction and Historical Background ● *Maxine F. Singer*
Cloning of Double-Stranded cDNA ● *Argiris Efstratiadis and Lydia Villa-Komaroff*
Gene Enrichment ● *M. H. Edgell, S. Weaver, Nancy Haigwood and C. A. Hutchison III*
Transformation of Mammalian Cells ● *M. Wigler, A. Pellicer, R. Axel and S. Silverstein*
Constructed Mutants of Simian Virus 40 ● *D. Shortle, J. Pipas, Sondra Lazarowitz, D. DiMaio and D. Nathans*
Structure of Cloned Genes from *Xenopus:* A Review ● *R. H. Reeder*
Transformationof Yeast ● *Christine Ilgen, P. J. Farabaugh, A. Hinnen, Jean M. Walsh and G. R. Fink*
The Use of Site-Directed Mutagenesis in Reversed Genetics ● *C. Weissmann, S. Nagata, T. Taniguchi, H. Weber and F. Meyer*
Agrobacterium Tumor Inducing Plasmids: Potential Vectors for the Genetic Engineering of Plants ● *P.J.J. Hooykaas, R. A. Schilperoort and A. Rörsch*
The Chloroplast, Its Genome and Possibilities for Genetically Manipulating Plants ● *L. Bogorad*
Mitochondrial DNA of Higher Plants and Genetic Engineering ● *C. S. Levings III and D. R. Pring*
Host–Vector Systems for Genetic Engineering of Higher Plant Cells ● *C. I. Kado*
Soybean Urease—Potential Genetic Manipulation of Agronomic Importance ● *J. C. Polacco, R. B. Sparks Jr. and E. A. Havir*

VOLUME 2 (1980)
Cloning of Repeated Sequence DNA from Cereal Plants ● *J. R. Bedbrook and W. L. Gerlach*
The Use of Recombinant DNA Methodology in Approaches to Crop Improvement: The Case of Zein ● *Benjamin Burr*
Production of Monoclonal Antibodies ● *Sau-Ping Kwan, Dale E. Yelton and Matthew D. Scharff*
Measurement of Messenger RNA Concentration ● *S. J. Flint*
DNA Cloning in Mammalian Cells with SV40 Vectors ● *D. H. Hamer*
Adenovirus–SV40 Hybrids: A Model System for Expression of Foreign Sequences in an Animal Virus Vector ● *Joseph Sambrook and Terri Grodzicker*
Molecular Cloning in *Bacillus subtilis* ● *D. Dubnau, T. Gryczan, S. Contente and A. G. Shivakumar*
Bacterial Plasmid Cloning Vehicles ● *H. U. Bernard and D. R. Helinski*
Cloning with Cosmids in *E. coli* and Yeast ● *Barbara Hohn and A. Hinnen*
DNA Cloning with Single-Stranded Phage Vectors ● *W. M. Barnes*
Bacteriophage Lambda Vectors for DNA Cloning ● *Bill G. Williams and Frederick R. Blattner*

VOLUME 3 (1981)
Constructed Mutants Using Synthetic Oligodeoxyribonucleotides as Site-Specific Mutagens ● *M. Smith and S. Gillam*

Evolution of the Insertion Element IS1 That Causes Genetic Engineering of Bacterial
    Genomes *In Vivo* ● *E. Ohtsubo, K. Nyman, K. Nakamura and H. Ohtsubo*
Applications of Molecular Cloning to *Saccharomyces* ● *M. V. Olson*
Cloning Retroviruses: Retrovirus Cloning? ● *W. L. McClements and G. F. Vande Woude*
Repeated DNA Sequences in *Drosophila* ● *M. W. Young*
Microbial Surface Elements: The Case of Variant Surface Glycoprotein (VSG) Genes of
    African Trypanosomes ● *K. B. Marcu and R. O. Williams*
Mouse Immunoglobulin Genes ● *P. Early and L. Hood*
The Use of Cloned DNA Fragments to Study Human Disease ● *S. H. Orkin*
Physical Mapping of Plant Chromosomes by *In Situ* Hybridization ● *J. Hutchinson,
    R. B. Flavell and J. Jones*
Mutants and Variants of the Alcohol Dehydrogenase-1 Gene in Maize ● *M. Freeling and
    J. A. Birchler*
Developmentally Regulated Multigene Families in *Dictyostelium discoideum* ● *R. A. Firtel,
    M. McKeown, S. Poole, A. R. Kimmel, J. Brandis and W. Rowekamp*
Computer Assisted Methods for Nucleic Acid Sequencing ● *T. R. Gingeras and R. J. Roberts*

VOLUME 4 (1982)
New Methods for Synthesizing Deoxyoligonucleotides ● *M. H. Caruthers, S. L. Beaucage,
    C. Becker, W. Efcavitch, E. F. Fisher, G. Galluppi, R. Goldman, P. deHaseth, F. Martin,
    M. Matteucci and Y. Stabinsky*
An Integrative Strategy of DNA Sequencing and Experiments Beyond ● *J. Messing*
Transcription of Mammalian Genes *In Vitro* ● *J. L. Manley*
Transcription of Eukaryotic Genes in Soluble Cell-Free Systems ● *N. Heintz and R. G. Roeder*
Attachment of Nucleic Acids to Nitrocellulose and Diazonium-Substituted Supports ●
    B. Seed
Determination of the Organization and Identity of Eukaryotic Genes Utilizing Cell-Free
    Translation Systems ● *J. S. Miller, B. E. Roberts and B. M. Paterson*
Cloning in Streptomyces: Systems and Strategies ● *D. A. Hopwood and K. F. Chater*
Partial Sequence Determination of Metabolically Labeled Radioactive Proteins and Peptides
    ● *C. W. Anderson*
Molecular Cloning of Nitrogen Fixation Genes from *Klebsiella pneumoniae* and *Rhizobium
    meliloti* ● *F. M. Ausubel, S. E. Brown, F. J. deBruijn, D. W. Ow, G. E. Riedel,
    G. B. Ruvkun and V. Sandaresan*
The Cloning and Expression of Human Interferon Genes ● *R. M. Lawn*
Cloning by Complementation in Yeast: The Mating Type Genes ● *J. B. Hicks, J. N. Strathern,
    A.J.S. Klar and S. L. Dellaporta*
Construction and Screening of Recombinant DNA Libraries with Charon Vector Phages ●
    B. A. Zehnbauer and F. R. Blattner

VOLUME 5 (1983)
Microcloning of Microdissected Chromosome Fragments ● *V. Pirrotta, H. Jackle and
    J. E. Edstrom*
Transient Expression of Cloned Genes in Mammalian Cells ● *J. Banerji and W. Schaffner*
Transposable Elements in Archaebacteria ● *W. F. Doolittle, C. Sapienza, J. D. Hofman,
    R. M. Mackay, A. Cohen and W.-L. Xu*
The Application of Restriction Fragment Length Polymorphism to Plant Breeding ● *B. Burr,
    S. V. Evola, F. A. Burr and J. S. Beckmann*
Antibodies against Synthetic Peptides ● *G. Walter and R. F. Doolittle*
Wheat α-Amylase Genes: Cloning of a Developmentally Regulated Gene Family ● *D. Baulcombe*
Yeast DNA Replication ● *J. L. Campbell*
Chromosome Engineering in Wheat Breeding and Its Implications for Molecular Genetic
    Engineering ● *C. N. Law*

Bovine Papillomavirus Shuttle Vectors ● *N. Sarver, S. Mitrani-Rosenbaum, M.-F. Law,*
    W. T. McAllister, J.C. Byrne and P. M. Howley
Chemical Synthesis of Oligodeoxyribonucleotides: A Simplified Procedure ● *R. L. Letsinger*

VOLUME 6 (1984)
Cloning of the Adeno-Associated Virus ● *K. I. Berns*
Transformation in the Green Alga *Chlamydomonas reinhardii* ● *J..-D. Rochaix*
Vectors for Expressing Open Reading Frame DNA in *Escherichia coli* Using *lacZ* Gene
    Fusions ● *G. M. Weinstock*
An Enigma of the Leghemoglobin Genes ● *J. S. Lee and D.P.S. Verma*
Yeast Transposons ● *G. S. Roeder*
Rearrangement and Activation of C-MYC Oncogene by Chromosome Translocation in B Cell
    Neoplasias ● *K. B. Marcu, L. W. Stanton, L. J. Harris, R. Watt, J. Yang, L. Eckhardt,*
    B. Birshtein, E. Remmers, R. Greenberg and P. Fahrlander
Screening for and Characterizing Restriction Endonucleases ● *I. Schildkraut*
Molecular Studies of Mouse Chromosome 17 and the T Complex ● *L. M. Silver, J. I. Garrels*
    and H. Lehrach
Use of Synthetic Oligonucleotide Hybridization Probes for the Characterization and Isolation
    of Cloned DNAs ● *A. A. Reyes and R. B. Wallace*
Hybridization of Somatic Plant Cells: Genetic Analysis ● *Yu. Yu. Gleba and D. A. Evans*
Genetic Analysis of Cytoskeletal Protein Function in Yeast ● *P. Novick, J. H. Thomas and*
    D. Botstein
Use of Gene Fusions to Study Biological Problems ● *L. Guarente*
The Use of the Ti Plasmid of *Agrobacterium* to Study the Transfer and Expression of Foreign
    DNA in Plant Cells: New Vectors and Methods ● *P. Zambryski, L. Herrera-Estrella,*
    M. De Block, M. Van Montagu and J. Schell
Analysis of Eukaryotic Control Proteins at Their Recognition Sequences by Scanning
    Transmission Electron Microscopy ● *P.V.C. Hough, M. N. Simon and I. A. Mastrangelo*
The Mass Culture of a Thermophilic Spirulina in the Desert ● *K. Qian, G. H. Sato, V. Zhao*
    and K. Shinohara
DNA-Mediated Gene Transfer in Mammalian Gene Cloning ● *F. H. Ruddle, M. E. Kamarck,*
    A. McClelland and L. C. Kühn

VOLUME 7 (1985)
Biochemical and Genetic Analysis of Adenovirus DNA Replication *In Vitro* ● *B. W. Stillman*
Immunoscreening λGT11 Recombinant DNA Expression Libraries ● *R. A. Young and R. W. Davis*
In Situ Hybridization to Cellular RNAs ● *R. C. Angerer, K. H. Cox and L. M. Angerer*
Computer Methods to Locate Genes and Signals in Nucleic Acid Sequences ● *R. Staden*
Biochemical and Molecular Techniques in Maize Research ● *N. Fedoroff*
Analysis of Chromosome Replication with Eggs of *Xenopus laevis* ● *R. A. Laskey,*
    S. E. Kearsey and M. Mechali
Molecular Genetic Approaches to Bacterial Pathogenicity to Plants ● *M. J. Daniels and*
    P. C. Turner
Synthesis of Hybridization Probes and RNA Substrates with SP6 RNA Polymerase ● *P. A. Krieg,*
    M. R. Rebagliati, M. R. Green and D. A. Melton
Identification and Isolation of Clones by Immunological Screening of cDNA Expression Libraries
    ● *D. M. Helfman, J. R. Feramisco, J. C. Fiddes, G. P. Thomas and S. H. Hughes*
Molecular Studies on the Cytomegaloviruses of Mice and Men ● *D. H. Spector*
Gene Transfer with Retrovirus Vectors ● *A. Bernstein, S. Berger, D. Huszar and J. Dick*
HPRT Gene Transfer as a Model for Gene Therapy ● *T. Friedmann*
Catabolic Plasmids: Their Analysis and Utilization in the Manipulation of Bacterial Metabolic
    Activities ● *S. Harayama and R. H. Don*

Transcription of Cloned Eukaryotic Ribosomal RNA Genes ● *B. Sollner-Webb, J. Tower, V. Culotta and J. Windle*
DNA Markers in Huntington's Disease ● *J. F. Gusella*

VOLUME 8 (1986)
Regulation of Gene Activity During Conidiophore Development in *Aspergillus nidulans* ● *W. E. Timberlake and J. E. Hamer*
Regulation of Expression of Bacterial Genes for Bioluminescence ● *J. Engebrecht and M. Silverman*
Analysis of Genome Organization and Rearrangements by Pulse Field Gradient Gel Electro-phoresis ● *C. L. Smith, P. E. Warburton, A. Gaal and C. R. Cantor*
Structural Instability of *Bacillus subtilis* Plasmids ● *S. D. Ehrlich, Ph. Noirot, M.A. Petit, L. Jannière, B. Michel and H. te Riele*
Geminiviruses, The Plant Viruses with Single-Stranded DNA Genomes ● *A. J. Howarth*
The Use of Bacterial Plasmids in the Investigation of Genetic Recombination ● *A. Cohen*
Shuttle Mutagenesis: A Method of Introducing Transposons into Transformable Organisms ● *H. S. Seifert, M. So and F. Heffron*
Genetic Advances in the Study of *Rhizobium* Nodulation ● *S. R. Long*
Galactokinase Gene Fusion in the Study of Gene Regulation in *E. coli, Streptomyces*, Yeast and Higher Cell Systems ● *M. Rosenberg, M. Brawner, J. Gorman and M. Reff*
Structure and Function of the Signal Recognition Particle ● *V. Siegel and P. Walter*
Alteration of the Structure and Catalytic Properties of Rubisco by Genetic Manipulation ● *S. Gutteridge*
Electrophoresis of DNA in Denaturing Gradient Gels ● *L. S. Lerman*
Caulimoviruses as Potential Gene Vectors for Higher Plants ● *R. J. Shepherd*
An Insect Baculovirus Host–Vector System for High-Level Expression of Foreign Genes ● *D. W. Miller, P. Safer and L. K. Miller*
Preparation of cDNA Libraries and the Detection of Specific Gene Sequences ● *J. Brandis, D. Larocca and J. Monahan*
Construction of Human Chromosome Specific DNA Libraries: The National Laboratory Gene Library Project ● *L. L. Deaven, C. E. Hildebrand, J. C. Fuscoe and M. A. Van Dilla*
New Approaches to the Expression and Isolation of a Regulatory Protein ● *D. Bastia, J. Germino, S. Mukherjee and T. Vanaman*

VOLUME 9 (1987)
Gene Transfer in the Sea Urchin ● *B. R. Hough-Evans and E. H. Davidson*
Properties and Uses of Heat Shock Promoters ● *H. Pelham*
The Expression of Introduced Genes in Regenerated Plants ● *D. Dunsmuir, J. Bedbrook, D. Bond-Nutter, C. Dean, D. Gidoni and J. Jones*
Control of Maize Zein Gene Expression ● *R. S. Boston and B. A. Larkins*
DNase I Footprinting as an Assay for Mammalian Gene Regulatory Proteins ● *W. S. Dynan*
Use of Gene Transfer in the Isolation of Cell Surface Receptor Genes ● *D. R. Littman and M. V. Chao*
A New Method for Synthesizing RNA on Silica Supports ● *D. J. Dellinger and M. H. Caruthers*
Activity Gels: Reformation of Functional Proteins from SDS-Polyacrylamide Gels ● *R. P. Dottin, B. Haribabu, C. W. Schweinfest and R. E. Manrow*
Plasmid Vectors Carrying the Replication Origin of Filamentous Single-Stranded Phages, ● *G. Cesareni and J.A.H. Murray*
High Level Production of Proteins in Mammalian Cells ● *R. J. Kaufman*
Plant Microinjection Techniques ● *R. J. Mathias*
Genetic Transformation to Confer Resistance to Plant Virus Disease ● *R. N. Beachy, S. G. Rogers and R. T. Fraley*
Alternative Splicing: Mechanistic and Biological Implications of Generating Multiple Proteins from a Single Gene ● *B. Nadal-Ginard, M. E. Gallego and A. Andreadis*

VOLUME 10 (1988)

Genomic Footprinting ● *P. B. Becker and G. Schütz*

Theoretical and Computer Analysis of Protein Primary Sequences: Structure Comparison and Prediction ● *P. Argos and P. McCaldon*

Affinity Chromatography of Sequence-Specific DNA-Binding Proteins ● *C. Wu, C. Tsai and S. Wilson*

Applications of the Firefly Luciferase as a Reporter Gene ● *S. Subramani and M. DeLuca*

Fluorescence-Based Automated DNA Sequence Analysis ● *L. M. Smith*

Phosphorothioate-Based Oligonucleotide-Directed Mutagenesis ● *J. R. Sayers and F. Eckstein*

Design and Use of *Agrobacterium* Transformation Vectors ● *M. Bevan and A. Goldsbrough*

Cell Commitment and Determination in Plants ● *F. Meins, Jr.*

Plasmids Derived from Epstein-Barr Virus: Mechanisms of Plasmid Maintenance and Applications in Molecular Biology ● *J. L. Yates*

Chromosome Jumping: A Long Range Cloning Technique ● *A. Poustka and H. Lehrach*

Isolation of Intact MRNA and Construction of Full-Length cDNA Libraries: Use of a New Vector, λgt22, and Primer-Adapters for Directional cDNA Cloning ● *J. H. Han and W. J. Rutter*

The Use of Transgenic Animal Techniques for Livestock Improvement ● *R. M. Strojek and T. E. Wagner*

Plant Reporter Genes: The GUS Gene Fusion System ● *R. A. Jefferson*

Structure of the Genes Encoding Proteins Involved in Blood Clotting ● *R.T.A. MacGillivray, D. E. Cool, M. R. Fung, E. R. Guinto, M. L. Koschinsky and B. A. Van Oost*

VOLUME 11 (1989)

DNA Methylases ● *A. Razin*

Advances in Direct Gene Transfer Into Cereals ● *T. M. Klein, B. A. Roth and M. E. Fromm*

The Copy Number Control System of the 2μm Circle Plasmid of *Saccharomyces cerevisiae* ● *B. Futcher*

The Application of Antisense RNA Technology to Plants ● *W. R. Hiatt, M. Kramer and R. E. Sheehy*

The Pathogenesis-Related Proteins of Plants ● *J. P. Carr and D. F. Klessig*

The Molecular Genetics of Plasmid Partition: Special Vector Systems for the Analysis of Plasmid Partition ● *A. L. Abeles and S. J. Austin*

DNA-Mediated Transformation of Phytopathogenic Fungi ● *J. Wang and S. A. Leong*

Fate of Foreign DNA Introduced to Plant Cells ● *J. Paszkowski*

Generation of cDNA Probes by Reverse Translation of Amino Acid Sequence ● *C. C. Lee and C. T. Caskey*

Molecular Genetics of Self-Incompatibility in Flowering Plants ● *P. R. Ebert, M. Altschuler and A. E. Clarke*

Pulsed-Field Gel Electrophoresis ● *M. V. Olson*

PREFACE TO VOLUME 1

This volume is the first of a series concerning a new technology which is revolutionizing the study of biology, perhaps as profoundly as the discovery of the gene. As pointed out in the introductory chapter, we look forward to the future impact of the technology, but we cannot see where it might take us. The purpose of these volumes is to follow closely the explosion of new techniques and information that is occurring as a result of the newly-acquired ability to make particular kinds of precise cuts in DNA molecules. Thus we are particularly committed to rapid publication.

Jane K. Setlow

ACKNOWLEDGMENT

Again June Martino has done a careful and intelligent job processing the manuscripts, for which the Editor is enormously grateful.

CONTENTS

FOLDING OF EUKARYOTIC PROTEINS PRODUCED IN ESCHERICHIA COLI.. 1
     R.F. Kelley and M.E. Winkler

HUMAN RETINOBLASTOMA SUSCEPTIBILITY GENE.................... 21
     C.-C. Lai and W.-H. Lee

α-OLIGODEOXYNUCLEOTIDES (α-DNA): A NEW CHIMERIC NUCLEIC
ACID ANALOG... 37
     F. Morvan, B. Rayner and J.-L. Imbach

THE UTILITY OF STREPTOMYCETES AS HOSTS FOR GENE CLONING...... 53
     P.K. Tomich and Y. Yagi

FROM FOOTPRINT TO FUNCTION: AN APPROACH TO STUDY GENE
EXPRESSION AND REGULATORY FACTORS IN TRANSGENIC PLANTS....... 73
     E. Lam

PURIFICATION OF RECOMBINANT PROTEINS WITH METAL CHELATE
ADSORBENT... 87
     E. Hochuli

DETERMINANTS OF TRANSLATION EFFICIENCY OF SPECIFIC mRNAs
IN MAMMALIAN CELLS.. 99
     D.S. Peabody

THE POLYMERASE CHAIN REACTION............................... 115
     N. Arnheim

REGULATION OF ALTERNATIVE SPLICING.......................... 139
     M. McKeown

STRUCTURE AND FUNCTION OF THE NUCLEAR RECEPTOR SUPERFAMILY
FOR STEROID, THYROID HORMONE AND RETINOIC ACID.............. 183
     V. Giguère

IDENTIFICATION AND FUNCTIONAL ANALYSIS OF MAMMALIAN
SPLICING FACTORS.. 201
     A. Bindereif and M.R. Green

THE GENES ENCODING WHEAT STORAGE PROTEINS: TOWARDS A
MOLECULAR UNDERSTANDING OF BREAD-MAKING QUALITY AND ITS
GENETIC MANIPULATION... 225
     V. Colot

CONTROL OF TRANSLATION INITIATION IN MAMMALIAN CELLS......... 243
     R.J. Kaufman

ELECTROPORATION OF BACTERIA: A GENERAL APPROACH TO
GENETIC TRANSFORMATION....................................... 275
     W.J. Dower

THE ISOLATION AND IDENTIFICATION OF cDNA GENES BY THEIR
HETEROLOGOUS EXPRESSION AND FUNCTION......................... 297
     G.G. Wong

MOLECULAR CLONING OF GENES ENCODING TRANSCRIPTION
FACTORS WITH THE USE OF RECOGNITION SITE PROBES.............. 317
     H. Singh

INDEX.. 331

FOLDING OF EUKARYOTIC PROTEINS PRODUCED IN ESCHERICHIA COLI

Robert F. Kelley[1] and Marjorie E. Winkler[2]

Departments of Biomolecular Chemistry[1] and
Recovery Process R&D[2]
Genentech, Inc.
460 Pt. San Bruno Blvd.
South San Francisco, CA 94080

INTRODUCTION

The development of recombinant DNA technology has allowed production in prokaryotic hosts of proteins derived from eukaryotic organisms. The use of heterologous expression has had a large impact on the pharmaceutical industry since it enables large scale production of proteins which may have been difficult to isolate from a natural source. This approach also avoids the potential for contamination with disease agents associated with the isolation of a protein from human tissues. An example of this approach is the production in Escherichia coli of human growth hormone (1), used in the treatment of pituitary dwarfism. Studies of protein structure-function are also facilitated if the target protein can be produced in a prokaryote. For these studies, heterologous expression enables rapid production of variant proteins, either by directed or random mutagenesis, in sufficient quantities for detailed characterization using biochemical and biophysical methods.

A critical issue for heterologous expression of any eukaryotic protein is how to obtain correct folding of the recombinant protein. High level intracellular expression of recombinant proteins in prokaryotes is often accompanied by the formation of inclusion bodies (also referred to as refractile bodies) in the cytoplasm (2,3). Inclusion bodies usually contain large amounts of the recombinant protein in an insoluble, aggregated form. The mechanism of inclusion body formation is unclear but may reflect incorrect or incomplete folding of the protein, as well as poor solubility of the native protein (4). Nonetheless, active, properly folded protein can often be recovered from inclusion bodies with the use of denaturant solubilization followed by refolding.

The focus of this review is on practical methods for obtaining properly folded eukaryotic protein from a prokaryotic host. We will concentrate on the recovery of eukaryotic protein from E. coli since this prokaryote is widely used for heterologous expression work in the biotechnology industry. Criteria for judging correct folding will be discussed as well as procedures for refolding a protein if it is found to be in a non-native conformation. Results of experiments with the kringle-2 domain of tissue plasminogen activator (t-PA) will be used to illustrate these methods for expression of proteins in E. coli. Although some description of the relative merits of various expression systems is required in our discussion of protein folding, our coverage of this topic is limited and the reader is directed to more extensive reviews on the molecular biology of protein expression in prokaryotes (5).

THEORY OF PROTEIN FOLDING

A tenet of molecular biology is that all the information necessary to specify the native conformation of a protein is contained in the amino acid sequence. This principle has been elucidated from the study of the reversible denaturation of small globular proteins in vitro (for review see refs. 6-8). Anfinsen and colleagues (9,10) performed some of the seminal experiments in this field by showing that unfolded ribonuclease, prepared by reduction of its four disulfide bonds, could be refolded to yield active protein, indistinguishable from native ribonuclease, by simple air oxidation in vitro. Most small globular proteins, after denaturation produced by a change in environmental conditions, i.e. pH, temperature, etc., can be reversibly folded in vitro by a return to the original conditions. Instances of irreversible protein folding can usually be explained by secondary reactions upon denaturation such as aggregation or covalent modification of the polypeptide chain. Protein folding in aqueous solution is thus a spontaneous process and is presumed to be driven by the favorable free energy change resulting from burial of hydrophobic groups upon folding (11).

Denaturation of small globular proteins is almost always a highly cooperative process such that folding intermediates are not detected at equilibrium. Most proteins fold rapidly ($t_{1/2} < 10s$) and any slow steps are usually attributable to formation of disulfide bonds or to the cis-trans isomerization of proline peptide bonds (12,13). Protein folding occurs much more rapidly than expected if folding were to result from a random conformational search (14). Therefore, most models of protein folding postulate that folding intermediates, although having low equilibrium stability relative to native or unfolded protein, are at least populated transiently and serve to direct the folding process. Indeed, transient intermediates have been observed for refolding of several proteins (cf. 15-17) using rapid kinetics methods.

PROTEIN FOLDING IN E. COLI

If all the information for folding is specified in the primary sequence, and proteins fold rapidly, then one would expect that any eukaryotic protein could be produced in a prokaryote simply by transforming the host with a plasmid having the protein coding sequence and the appropriate expression elements. A number of factors, however, can impact the folding of a eukaryotic protein in a prokaryotic host. 1) Many extracellular proteins of eukaryotes contain disulfide bonds. Proteins having multiple disulfide bonds may form non-native disulfide bonds during folding from the reduced species. Further folding is then blocked unless the incorrect disulfide bond is cleaved by reduction with an external thiol or by attack from a protein thiol. Thus, eukaryotic organisms that secrete disulfide-containing proteins also have machinery for ensuring proper disulfide bond formation. The components of this machinery are not well defined at present. One of the components may be the enzyme protein disulfide isomerase (PDI) (18), which is often found in eukaryotic cells that secrete disulfide-containing proteins, and has been shown to catalyze disulfide shuffling in vitro. PDI by itself, however, is not capable of catalyzing refolding of reduced proteins. In contrast to eukaryotic organisms, E. coli produce very few proteins containing disulfide bonds and do not appear to have machinery for ensuring a proper disulfide bond. Thus, intracellular expression in E. coli of proteins having multiple disulfide bonds is often unsuccessful. Although E. coli contain large amounts of the protein thioredoxin, which has been shown to have PDI-like activity in vitro (19), this may be insufficient for disulfide bond formation in vivo. 2) Eukaryotic proteins are often made as larger precursors or "pro" forms and then processed to yield the mature protein. As observed for insulin (20), only the unprocessed form of the protein may be capable of reversible folding and thus expression of the mature protein would be unsuccessful. 3) Eukaryotic proteins often have sequences specifying carbohydrate attachment which are not utilized by the prokaryote. Only a few studies of the role of attached carbohydrate in protein folding have been done. For ribonuclease (21) and invertase (22), removal of the attached carbohydrate does not have a direct effect on either the rate or equilibria of folding. In the case of invertase, however, deglycosylation results in poor solubility of the unfolded protein such that the refolding process is inhibited at high protein concentration. Effects of deglycosylation on protein solubility might lead to precipitation in vivo, resulting in improper folding of a eukaryotic protein when expressed in E. coli. 4) Some proteins have prosthetic groups, either attached covalently or non-covalently, that influence the stability or solubility of the protein. These prosthetic groups may not be synthesized by the prokaryote or are synthesized in amounts insufficient for over-expression of the recombinant protein. For example, removal of the non-covalently attached heme group from myoglobin results in a conformational change of the protein and decreased solubility. Nonetheless, Varadarajan et al. (23) were able to obtain expression of apomyoglobin in E. coli by expressing it as a fusion with a phage protein. The fusion protein product was contained in inclusion bodies and recovered by denaturant extraction and refolding in the presence of heme.

EXPRESSION OF EUKARYOTIC PROTEIN IN E. COLI

Three general strategies for expression of a foreign protein in E. coli will be discussed: 1) direct intracellular expression, 2) intracellular expression as a fusion with a stable host protein and 3) secretion of the protein into the periplasmic space of E. coli. In the direct expression strategy, a plasmid is constructed having a prokaryotic promoter and ribosome binding site fused to the N-terminus of the eukaryotic protein coding sequence. Direct expression often fails due to degradation of the foreign protein by the proteolytic enzymes of E. coli. Degradation may sometimes be circumvented by expressing the protein as a fusion with a protein that is normally stable in E. coli. The fusion protein is usually constructed so that the protein of interest is attached to the C-terminus of the stable protein with the use of a linker that can be specifically cleaved chemically or enzymatically. For example, many proteins have been expressed in E. coli as fusions with the S. aureus protein A (24). This strategy has the distinct advantage in that the immunoglobulin binding capacity of protein A can be used in the purification of the fusion protein. The disadvantage of this approach, however, is that the cleavage of the linker between the two proteins often results in unwanted secondary cleavages. In addition, because of the high level expression of fusion proteins, these molecules are often deposited in inclusion bodies, especially if the protein has disulfide bonds.

For some proteins, these expression problems can be eliminated if the protein is secreted into the periplasmic space. Since the intracellular redox potential of E. coli favors reduction of disulfide bonds, secretion into the more oxidative environment of the periplasmic space may enhance the chances for correct disulfide formation in proteins having multiple disulfide bonds. The periplasmic space also has fewer proteolytic enzymes than the cytoplasm. Recovery of secreted proteins is simplified by the compartmentalization of the proteins in the periplasmic space. Secreted proteins may often be extracted with the use of simple osmotic shock protocols that result in selective release of periplasmic contents.

The minimal requirement for secretion is that an E. coli secretion signal, such as the STII signal, must be attached to the amino terminus of the recombinant protein. Upon secretion of the product, the signal should be removed by the E. coli signal peptidase. One example of a successful secretion strategy in E. coli is the use of either the alkaline phosphatase or STII signal sequence to produce correctly folded bovine pancreatic trypsin inhibitor (BPTI) (25), which has 3 disulfide bonds. BPTI could be released from the cells with an osmotic shock procedure and was found to have the N-terminal sequence as expected for the proper removal of the signal peptide. X-ray crystallographic measurements indicate that the BPTI produced by secretion in E. coli has a structure identical to that determined for the non-recombinant protein (C. Eigenbrot and A. Kossiakoff, unpublished results).

One disadvantage of the secretion strategy is that the expression levels are generally much lower than for fusion proteins expressed intracellularly. Also, it is difficult to predict whether secretion of a eukaryotic protein into the periplasmic space of E. coli will be successful since the secretion pathway in E. coli is not well understood. Although a signal sequence is obligatory for secretion, other sequences of the mature protein may also play a role in secretion (26). There is evidence that secretion must occur prior to the completion of folding and that E. coli possess soluble factors which help to maintain the nascent polypeptide chain in a translocation-competent conformation (27,28). It is unknown how eukaryotic protein might interact with these factors.

GUIDELINES FOR RECOVERY OF INTRACELLULARLY EXPRESSED PROTEIN

A protein which is expressed intracellularly in E. coli may often accumulate in inclusion bodies which appear as dense cytoplasmic granules when the cells are observed under a light microscope. A general strategy for recovering active protein from inclusion bodies is outlined in Figure 1. Inclusion bodies will sediment at low g forces and thus can be separated from many of the other intracellular E. coli proteins. A typical protocol for inclusion body isolation involves mechanical disruption of the cells, e.g., sonication, followed by centrifugation for 30 min at 4700 g. Further purification can be done by washing the pellet with the buffer used during the cell disruption, or by centrifuging the resuspended pellet through a cushion of 40 to 50% glycerol (29).

Dissolution of the pelleted recombinant protein usually requires the use of denaturants. 7 M guanidine hydrochloride or

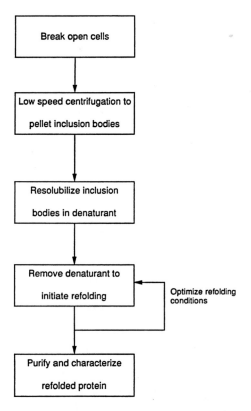

Figure 1.   Outline of a general strategy for recovering properly folded protein from inclusion bodies.

8 M urea (the urea should be freshly deionized to prevent carbamylation reactions) are usually used, although sometimes alkaline buffers (pH>10) can also be successful. The choice of denaturant depends on the behavior of each particular recombinant protein, and several conditions should be tested, in order to get the highest yield. If the solubilized protein contains cysteine residues, the protein will usually be in the form of disulfide-bonded dimers, trimers and higher order polymers, as observed on SDS-PAGE gels in the absence of reducing agent. The amount of aggregation may continue to increase with time if the protein is allowed to remain in the denaturant.

Removal of the denaturant from the solubilized inclusion bodies by dialysis or desalting columns will cause the protein to precipitate under conditions where the native protein is known to remain in solution if the protein needs to be refolded. A misfolded protein solution can also have a very low specific activity in biological assays. Each protein requires somewhat different conditions to get optimum yields of native protein, and these conditions must be determined empirically. Therefore it is extremely important to have a good biological or immunological assay that is sensitive only to properly folded protein in order to monitor refolding yields. Without such an assay, the chances of finding the best refolding conditions for a protein are minimal.

In some cases, contaminating proteins will interfere with the refolding process. Here, the best yields can be obtained if the protein is purified further before refolding. This can be done with ion exchange chromatography in the presence of urea, or size exclusion chromatography in the presence of urea or guanidine. In other cases, better yields can be obtained by refolding the protein as soon as possible after solubilization in denaturant since further aggregation of the protein may be detrimental to refolding. Problems with proteolysis will also be minimized by folding the protein into its native state since many of the susceptible sites will become buried in the interior of the protein.

Refolding of disulfide bonded aggregates of recombinant proteins can occur through disulfide shuffling. In this technique, both reduced and oxidized thiol reagents are added to the solution in a ratio which sets the redox potential of the solution such that disulfide bonds can easily be made and broken. As the protein's disulfide bonds shuffle, the protein will come to rest in low energy states, which, in non-denaturing buffers, will often include the native state of the protein. This reaction is best done in dilute solutions of protein to avoid polymer formation, and again should be optimized for each particular protein. Typical redox buffers include either reduced and oxidized glutathione, cysteamine and cystamine, reduced and oxidized dithiothreitol, or reduced and oxidized β-mercaptoethanol (reduced dithiothreitol or β-mercaptoethanol in the presence of air will generate enough oxidized thiol to promote disulfide shuffling). It may be necessary to compare the effects of several thiol reagents on the yield of refolding since these compounds differ in redox potential. For example, native ribonuclease can be refolded from the reduced form using oxidized glutathione but not oxidized dithiothreitol (30).

The choice of buffer solution for refolding is critical for high yields of refolded native protein. Theoretically, the highest yield of properly folded protein should occur in a non-denaturing buffer. Often, however, the recombinant protein is not soluble in non-denaturing buffers before it has been refolded. Therefore, it is necessary to determine the conditions which are LEAST denaturing for the protein, while still keeping it in solution. Acceptable yields of native protein can sometimes be obtained after refolding in 1 M guanidine hydrochloride or 8 M urea. The addition of other reagents such as glycerol, ethylene glycol, polyethylene glycol or sucrose to the refolding solution can aid in the solubilization of the denatured protein and thereby increase the final yield of native protein. Other reagents such as salts, metals, detergents or amino acids have also been shown sometimes to give increased yields of properly folded protein (31). The inclusion of a ligand, substrate, pseudosubstrate or inhibitor in the refolding buffer can also help stabilize native-like folding intermediates and thus increase both the rate and yield of refolding (32). There is no way to predict beforehand which of these reagents will be useful in the refolding of a particular recombinant protein. The pH should be kept above 7.0 so that the formation of disulfide bonds is not in competition with protonation of free thiols. Generally, optimum yields are found around pH 8.0 to pH 9.0. Temperature can also affect refolding yields, and should be optimized for each set of conditions.

A typical set of refolding conditions which were optimized for the refolding of transforming growth factor-α(TGFα), expressed as a fusion protein (see Figure 2), are as follows (33). Inclusion bodies from E. coli containing the amino terminal trpLE-fusion protein of TGFα, are solubilized in 7 M guanidine hydrochloride and the trpLE-TGFα fusion is partially purified on a G-75 Sephadex column in 4 M guanidine hydrochloride. The TGFα pool is diluted to 1 M guanidine hydrochloride, with a buffer which contains 50 mM Tris, pH 9.0, 1.25 mM reduced glutathione and 0.25 mM oxidized glutathione. The protein concentration is about 150 μg/ml. The disulfides are allowed to reshuffle at 4°C for 24 hr, with stirring. These conditions give a yield of 37% active trpLE-TGFα fusion protein as calculated by comparing the EGF receptor binding activity with the total amount of trpLE-TGFα protein in solution, determined by amino acid

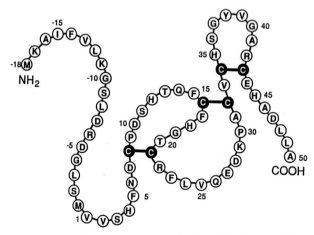

Figure 2. Amino acid sequence and disulfide pairing of trpLE-TGFα fusion protein. The first amino acid of TGFα is numbered "1".

composition. The fusion protein can then be cleaved with cyanogen bromide to release native TGFα.

The solubilization of cysteine-containing proteins from inclusion bodies may be aided by the addition of reducing agents. However, caution should be used with this approach. When the trpLE-TGFα fusion protein is reduced with β-mercaptoethanol during the solubilization of the inclusion bodies in 7 M guanidine hydrochloride, the final refolding yield drops from 37 to 6.5% under optimal refolding conditions (34). The reason for the decreased yield is due to the fact that one of the three disulfide bonds necessary for the activity of TGFα has formed either intracellularly, or during the disruption of the cells and solubilization of the protein (34). If this disulfide bond is reduced in a denaturing solution, it does not form easily during the disulfide shuffling process due to interference by the amino terminal trpLE fusion. (Reduced TGFα can be refolded to give 65% yield if the trpLE fusion is not attached.)

Since the extra amino acids contained in a fusion protein can interfere with the folding of that protein, the question arises as to whether the fusion should be removed before refolding. One must consider the nature of the fusion protein and the conditions of the cleavage reaction before deciding when to do the cleavage reaction. The fusion protein may not only decrease the refolding yield by making the native form of the protein less stable, but it may also decrease the solubility of the recombinant protein. On the other hand, the fusion protein can also be designed to add to the solubility of the denatured protein, and may be a useful tool in the purification of the protein after the refolding. The conditions of the cleavage reaction, however, may cause cysteine residues to be irreversibly altered so that they are no longer able to form disulfide bonds. This a serious side reaction caused by cleavages done with cyanogen bromide. For the trpLE-TGFα fusion protein, which is designed to be cleaved with cyanogen bromide, similar yields are obtained by removing the fusion before refolding or by removing the trpLE peptide after refolding. Thus in this case, the refolding yield must be lowered equally by the interference of the fusion protein and by oxidative reactions of cyanogen bromide. Fusion proteins can be produced in which the linker has a sequence that is specifically cleaved by a proteolytic enzyme (35,36) thus avoiding the potential for unwanted side reactions associated with chemical cleavage.

If conditions which are only weakly denaturing cannot be found for disulfide shuffling, then it may become necessary to modify the protein reversibly to increase its solubility before refolding. One reaction which has found utility in protein folding is the sulfitolysis reaction. Here, the protein is reacted with sodium sulfite (20 mg/ml) and sodium tetrathionate (10 mg/ml) under denaturing conditions, at pH 8.0 for several hours. The reaction causes existing disulfides to be reduced, and free cysteine residues to be protected as the S-sulfonate. In this way, disulfide bonded aggregates cannot form as denaturant is removed or diluted out. The protein can then be diluted into buffers which contain less denaturant and the cysteine residues can be restored to their unprotected state during disulfide shuffling.

Another reaction which may be useful for the solubilization of denatured proteins under weakly denaturing conditions is the reaction with citraconic anhydride at pH 8.5. This reaction will modify the exposed lysine residues of the protein, changing their positively charged amino function to the negatively charged carboxylate. The change in total charge may help to prevent aggregation of the denatured protein. The reaction is fully reversible so that after refolding, the pH can be lowered and the primary amino function of the lysine residues will be regenerated to yield the native protein.

Once the optimal yields for refolding a particular recombinant protein have been established, it is necessary to find purification steps which will separate misfolded protein forms from properly folded, native protein. This can be the most difficult part of the process since misfolded protein molecules will not differ from the native protein in terms of overall size or charge. If an active site analog or inhibitor to the protein is known, it can be used to make an affinity resin which may be very powerful in resolving native protein from misfolded protein. In this way, benzamidine Sepharose was used to purify native urokinase after refolding from E. coli supernatants (37). Immunoaffinity chromatography can be useful if a monoclonal antibody can be identified which recognizes a nonlinear epitope in the protein and thus is specific for the native form of the protein. Other techniques which are useful include hydrophobic interaction chromatography and reverse phase HPLC, which take advantage of the fact that most of the misfolded protein structures are more hydrophobic than the native protein structure. Native TGFα can be purified from misfolded forms of the protein by chromatography on a reverse phase C18 HPLC column with the use of an acetonitrile gradient in 0.1% trifluoroacetic acid (31). Care must be taken when these techniques are used, however, because some misfolded protein structures may co-purify with the native protein.

CRITERIA FOR CORRECT FOLDING OF PROTEINS EXPRESSED IN E. COLI

Once the recombinant protein has been purified, it is very important to establish that the protein recovered from E. coli is properly folded. A comparison of the biological activity of the recombinant protein to its non-recombinant form is the most straightforward method for assessing the folding of the polypeptide. If enzymatic activity measurements cannot be made, then reaction with antibodies raised against the native, non-recombinant protein or measurements of ligand binding may be used initially to examine folding. In some cases, however, the non-recombinant protein may not be available to use as a standard in the assay or to use as an immunogen for antibody production. Anti-peptide antibodies may be used for detecting expression but they will probably not be specific for the folded conformation.

Although the best comparison of the folding of a recombinant and non-recombinant protein would be to determine the x-ray structure, this often is not possible due to lack of suitable crystals or the time and expense required. Instead, a variety of spectroscopic methods including absorption, fluorescence, circular dichroic (CD), and nuclear magnetic resonance spectroscopy (NMR) may be used to characterize the recombinant protein (see ref. 38 for review of these techniques). The method of choice for preliminary characterization is CD since these measurements are the easiest to make, require only small amounts of protein, and provide information on secondary structure content and global conformation. A much more detailed picture of the conformation may be constructed using NMR. This technique requires much higher protein concentrations but can be used to determine the three-dimensional structure of proteins of less than 10 to 20 kD if resonances can be assigned.

The folding of the recombinant and non-recombinant proteins may also be compared by measurements of thermal stability. If the native conformations are identical, then unfolding should occur at the same temperature with the same values for ΔH, the enthalpy change for unfolding, and ΔC_p, the change in heat capacity accompanying unfolding. The most reliable method for determination of these parameters is differential scanning calorimetry (DSC) (39). In this technique, two solutions, one containing buffer and the other containing the protein, are heated in a calorimeter. The excess heat required to maintain the protein solution at the same temperature as the buffer solution is recorded. As the protein unfolds, heat must be put into the protein solution and thus a transition is observed.

PRODUCTION OF t-PA KRINGLE-2 DOMAIN IN E. COLI

Human tissue-type plasminogen activator (t-PA) is a large (527 amino acids) glycoprotein (40) used as a thrombolytic agent in the treatment of myocardial infarction. As shown in Figure 3, sequence homology studies suggest that t-PA is composed of several domains defined by the pattern of intramolecular disulfide bonds. In addition to a C-terminal proteolytic domain, t-PA has a multi-domain N-terminal extension consisting of a fibronectin type I "finger" domain, a domain homologous to epidermal growth factor and two "kringle" domains, homologous to those first observed in prothrombin. Efforts toward understand-

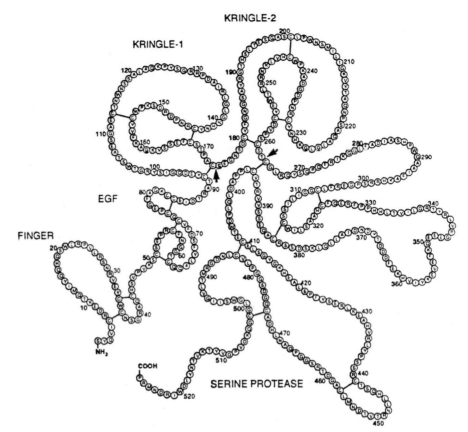

Figure 3. Amino acid sequence and domain structure proposed for human tissue plasminogen activator (t-PA). Disulfide bonds are shown as solid lines joining cysteine residues and were assigned on the basis of sequence homology to other proteins. Arrows indicate the endpoints of the fragment secreted in E. coli.

ing the role of these domains in the mechanism of t-PA action have been hindered by the complexity of the t-PA protein. Our approach has been to simplify the problem by using recombinant DNA technology to express the individual domains in E. coli (41,42).

Previous work with isolated kringles, obtained by proteolytic digestion of plasminogen, had indicated that these domains were capable of folding independently of the rest of the polypeptide (43,44). Based on these results, we chose to isolate t-PA kringle-2 by expression of the domain in E. coli. Three different strategies were tested for expression of t-PA kringle-2. 1) Direct expression with the use of the trp promoter. 2)

Expression as an N-terminal fusion with the trp leader (trpLE) segment. 3) Secretion of kringle-2 into the periplasmic space with the use of a construct having the alkaline phosphatase promoter and STII signal sequence joined to the coding sequence of kringle-2 (1,41). Expression of kringle-2 was initially tested by Western analysis of crude extracts run on SDS-polyacrylamide gels. A polyclonal antibody raised against unfolded t-PA (prepared by reduction of disulfide bonds followed by alkylation of the cysteines generated) was used in the Western analysis. Proper folding of the expressed protein was then assayed by ELISA with a polyclonal antibody raised against native t-PA.

In the direct expression system no kringle-2 was detected by Western analysis. Large amounts of antigen were observed in extracts of cells expressing kringle-2 as a fusion with the trp leader segment. Examination of these cells under a light microscope showed the presence of inclusion bodies and the fusion protein was found to sediment with these particles. Although the fusion protein was expressed at high levels and could be partially purified by sedimenting the inclusion bodies, attempts to recover kringle-2 proved unsuccessful due to poor refolding of the fusion protein and destruction of the protein under conditions necessary to remove the fusion.

In contrast to the two other expression strategies, secretion of kringle-2 yielded protein that appeared to be properly folded. Western analysis detected a protein having a molecular weight appropriate for kringle-2 with the signal peptide removed. At least a portion of the expressed kringle-2 was extracted from the cells by an osmotic shock procedure designed to cause release of the contents of the periplasmic space. Kringle-2 in this fraction was detected in the ELISA for native t-PA.

For purification of kringle-2, E. coli transformed with the secretion plasmid was fermented at 37°C in 10 liters of low phosphate-containing media yielding about 1 kg of cell paste. Preliminary experiments indicated that some of the secreted kringle-2 could be released from the cells by an osmotic shock protocol involving a freeze-thaw cycle. Western analysis, however, indicated that most of the kringle-2 remained cell associated after this treatment. An additional amount of kringle-2 equal to that obtained by osmotic shock was released from the cells by sonication. Kringle-2 in these two fractions was detected in the ELISA for native t-PA. The remaining kringle-2, 60 to 80% of the total, pelleted with the cell debris and required both denaturant (6 M guanidine hydrochloride) and reducing agent (5 mM dithiothreitol) for solubilization. By using the ELISA to monitor formation of native kringle-2, a refolding procedure was developed. We found that refolding could be accomplished simply by diluting the solubilized kringle-2 with 4 volumes of 50 mM Tris-HCl, pH 8, containing 1.2 mM reduced and 1.2 mM oxidized glutathione followed by stirring at room temperature for 2 hr. Longer incubation periods did not increase the yield whereas shorter refolding periods decreased the yield. While lower concentrations of glutathione decreased the yield, higher concentrations resulted in only marginal increases in the yield. Significantly greater or smaller volumes of dilution also impaired refolding. As revealed in the experiments described below, some denaturant must be present in order to solubilize the reduced form of kringle-2. The refolded fraction was centrifuged to remove insoluble material, concentrated by ammonium sulfate precipitation, dialyzed and further purified as described below.

Since kringle-2 was presumed to have a binding site for L-lysine, affinity chromatography on a column of lysine-Sepharose was tested as a method of purification. Kringle-2 in all fractions was retained on the column and could be specifically eluted with a lysine analog. Kringle-2 purified from each of the three fractions was found to be identical by amino acid composition, N-terminal sequence (indicating correct processing of the signal peptide), apparent molecular weight by SDS-PAGE, and affinity for L-lysine. Results of in vitro refolding experiments indicate that the reduced form of kringle-2 is quite insoluble and precipitates if refolding is initiated in concentrations of guanidine hydrochloride below 1 M. This result suggests the following explanation for the association of kringle-2 with the cell debris fraction. If disulfide bond formation is slow relative to the rate of translocation and processing, then non-covalent aggregation of reduced kringle-2 in the periplasm could compete with folding. Intermolecular disulfide formation may then cross-link the aggregates as suggested from the requirement of a reducing agent for solubilization. Thus, aggregate formation can be a problem for secretion as well as intracellular expression.

To characterize the t-PA kringle-2, enzymatic assays were not appropriate so an ELISA based on a polyclonal antibody raised against native t-PA and also lysine binding measurements were used in the development of the refolding protocol. Purified, recombinant kringle-2 was found to bind L-lysine with a dissociation constant of 100 μM. t-PA has very poor solubility under the conditions in which quantitative ligand binding experiments must be performed, so that a direct comparison of ligand affinity was not possible. However the pattern of elution of the two proteins from lysine-Sepharose with different lysine analogs indicated that the geometry of the binding pocket was maintained in the isolated domain.

The ultraviolet CD spectrum of kringle-2 is shown in Figure 4. In the near ultraviolet region, CD results from the aromatic chromophores, tyrosine and tryptophan, and reflects the asymmetry of the environment of these residues. For kringle-2 which has been reduced, presumably an unfolded molecule, no CD band in this region is observed. In the far ultraviolet, CD results from absorption by the peptide bond and thus can indicate secondary structure. Most proteins have a negative CD band at 220 nm due

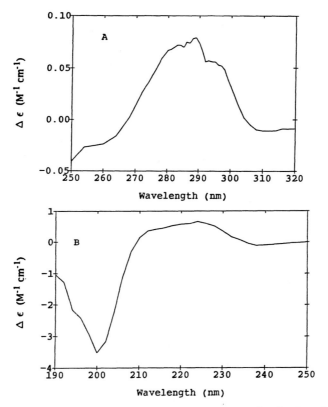

Figure 4. Circular dichroic spectra of kringle-2 domain in the near (A) and far (B) ultraviolet regions. $\Delta\epsilon$ is the difference in molar extinction coefficient between left and right circularly polarized light.

to the presence of α-helix and β-sheet. In contrast, kringle-2 has positive ellipticity in this region. The CD spectra is very similar to that observed for the kringle-4 domain of plasminogen isolated by elastase digestion of human plasminogen (45). These spectra are consistent with the x-ray structure of prothrombin kringle-1 (46) which has no α-helix, a small β-sheet and a high turn content.

A ^{1}H-NMR spectrum of kringle-2 is shown in Figure 5. This spectrum indicates that kringle-2 is folded since there is a spread of resonances over a wide chemical shift range. Similar spectra are observed for the plasminogen kringles produced by proteolytic digestion of non-recombinant plasminogen. In particular, all kringles examined thus far have a "fingerprint" upfield shifted (-1 ppm) methyl resonance arising from a conserved leucine residue. Although the size of t-PA precludes NMR measurements, the similarities between the plasminogen kringles

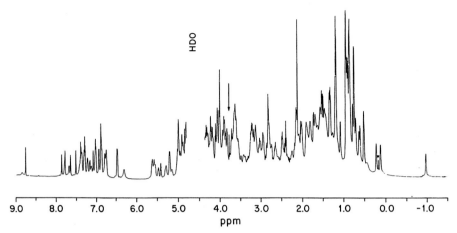

Figure 5. ¹H-NMR spectrum of kringle-2 recorded in D₂O at 620 MHz, 1.5 mM K2, pH 4.5, 37.8°C. The authors are grateful to M. Llinas and I.-J. Byeon (Carnegie-Mellon University) for supplying this figure.

and t-PA kringle-2 strongly suggest that the recombinant kringle-2 has the same conformation as found in full length t-PA. Despite the lack of glycosylation of the 184 site for kringle-2 expressed in E. coli, the domain folds correctly.

A typical DSC profile observed for recombinant kringle-2 is shown in Figure 6. This domain is quite heat stable and unfolds reversibly at pH values less than 5. Since t-PA is a multi-domain protein, a DSC curve reflects unfolding of several structures and thus a direct comparison to the isolated kringle-2 domain is difficult. However, both intact t-PA and the isolated kringle-2 domain denature in the same temperature range indicating that expression in E. coli does not lead to destabilization of the domain.

Our results with t-PA kringle-2 suggest that secretion is an effective strategy for heterologous expression of proteins having multiple disulfide bonds. However, expression by secretion does not ensure proper folding and protection from proteolytic degradation. For example, we have not been able to express the homologous kringle-1 domain of t-PA using the STII secretion system. Secretion of a larger fragment of t-PA consisting of the kringle-2 and proteolytic domains resulted in degradation of the proteolytic domain in the periplasmic space. In some cases, secretion in a strain lacking a periplasmic protease, such as the degP strains of E. coli (47), may improve expression levels by preventing degradation of the product. It may also be important to vary the temperature at which cells secreting a eukaryotic protein are grown. Lowering the growth temperature has been shown to result in improved secretion of properly folded subtili-sin in E. coli (48). Furthermore, as observed for intracellular expression, secretion of the eukaryotic protein as a fusion with

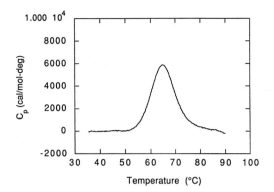

Figure 6. Thermal denaturation of recombinant kringle-2. Differential scanning calorimetric measurements were made on a 2 mg/ml solution of kringle-2 containing 0.1 M citrate pH 4.5. The excess heat capacity (Cp) required to maintain the sample and reference cells at the same temperature during the heating process is plotted as the solid line.

a protein that does fold correctly in E. coli may result in efficient expression. For example, a BPTI mutant that is not secreted using the STII signal sequence alone is secreted as a fusion with the S. aureus protein A (B. Nilsson, unpublished results).

SUMMARY

Although intracellular expression in E. coli may result in accumulation of the eukaryotic protein in inclusion bodies, the protein may often be recovered by first solubilizing with denaturant followed by refolding. Some general guidelines for developing a refolding procedure are apparent but the specific protocol must be empirically determined for each protein. Convenient and rapid assays for detecting native protein are critical for developing a refolding procedure. Maintaining solubility during refolding is a common feature of recovery processes. Proper folding should be assessed by a number of methods including activity, spectroscopic and stability measurements. For some proteins, properly folded protein may be obtained by secretion from E. coli; however, secretion does not ensure correct folding and protection from proteolytic degradation.

REFERENCES

1   Gray, G.L., Baldridge, J.S., McKeown, K.S., Heyneker, H.L. and Chang, C.N. (1985) Gene 39, 247-254.

2   Williams, D.C., Van Frank, R.M., Muth, W.L. and Burnett, J.P. (1982) Science 215, 687-689.

3   Schoner, R.G., Ellis, L.F. and Schoner, B.E. (1985) Bio/Technology 3, 151-154.

4   Mitraki, A. and King, J. (1989) Bio/Technology 7, 690-697.

5   Goeddel, D.V. (1990) Methods in Enzymology (in press).

6   Baldwin, R.L. (1975) Annu. Rev. Biochemistry 44, 453-475.

7   Kim, P.S. and Baldwin, R.L. (1982) Annu. Rev. Biochemistry 51, 459-489.

8   Goldenberg, D.P. (1988) Annu. Rev. Biophys. and Biophys. Chem. 17, 481-507.

9   Anfinsen, C.B., Haber, E., Sela, M. and White, F.H. (1961) Proc. Nat. Acad. Sci. U.S.A. 47, 1309-1314.

10 Anfinsen, C.B. (1973) Science 181, 223-230.

11 Kauzmann, W. (1959) Adv. Protein Chem. 14, 1-63.

12 Brandts, J.F. Halvorson, H.R. and Brennan, M. (1975) Biochemistry 14, 4953-4963.

13 Kelley, R.F. and Richards, F.M. (1987) Biochemistry 26, 6765-6774.

14 Levinthal, C. (1968) J. Chim. Phys. 65, 44-45.

15 Cook, K.H., Schmid, F.X. and Baldwin, R.L. (1979) Proc. Nat. Acad. Sci. U.S.A. 76, 6157-6161.

16 Schmid, F.X. and Baldwin, R.L. (1979) J. Mol. Biol. 135, 199-215.

17 Kelley, R.F., Wilson, J., Bryant, C. and Stellwagen, E. (1986) Biochemistry 25, 728-732.

18 Creighton, T.E., Hillson, D.A. and Freedman, R.B. (1980) J. Mol. Biol. 142, 43-62.

19 Pigiet, V.P. and Schuster, B.J. (1986) Proc. Nat. Acad. U.S.A. 83, 7643-7647.

20 Steiner, D.F. and Clark, J.L. (1968) Proc. Nat. Acad. Sci. U.S.A. 60, 622-629.

21 Grafl, R., Lang, K., Vogl, H. and Schmid, F.X. (1987) J. Biol. Chem. 262, 10624-10629.

22 Shulke, N. and Schmid, F.X. (1988) J. Biol. Chem. 263, 8832-8837.

23 Varadarajan, R., Szabo, A. and Boxer, S.G. (1985) Proc. Nat. Acad. Sci. U.S.A. 82, 5681-5684.

24 Nilsson, B., Abrahamsen, L. and Uhlen, M. (1985) EMBO J. 4, 1075-1080.

25 Marks, C., Vasser, M., Ng, P., Henzel, W. and Anderson, S. (1986) J. Biol. Chem. 261, 7115-7118.

26 Dalbey, R.E. and Wickner, W. (1986) J. Biol. Chem. 261, 13844-13849.

27 Crooke, E. and Wickner, W. (1987) Proc. Nat. Acad. Sci. U.S.A. 84, 5216-5220.

28 Randall, L.L., Hardy, S.J.S. and Thom, J.R. (1987) Annu. Rev. Microbiology 41, 507-541.

29 Winkler, M.E. and Blaber, M. (1986) Biochemistry 25, 4041-4045.

30   Creighton, T.E. (1977) J. Mol. Biol. 113, 329-341.

31   Schaffer, S.W., Ahmed, A. and Wetlaufer, D.B. (1975) J. Biol. Chem. 250, 8483-8486.

32   Teipel, J.W. and Koshland, D.E. (1971) Biochemistry 10, 792-805.

33   Winkler, M.E., Bringman, T. and Marks, B.J. (1986) J. Biol. Chem. 261, 13838-13843.

34   Winkler, M.E. (1987) in Protein Structure, Folding and Design 2 (Oxender, D.L., ed.), pp. 363-372, Alan R. Liss, Inc., NY.

35   Nagai, K. and Thogersen, H.C. (1987) Methods in Enzymology 153, 461-481.

36   Carter, P., Nilsson, B., Burnier, J.P., Burdick, D. and Wells, J.A. (1989) Proteins: Structure, Function and Genetics (in press).

37   Winkler, M.E., Blaber, M., Bennett, G.L., Holmes, W. and Vehar, G. (1985) Bio/Technology 3, 990-1000.

38   Cantor, C.R. and Schimmel, P.R. (1980) Biophysical Chemistry, Part II, W.H. Freeman, San Francisco, CA.

39   Privalov, P. (1979) Adv. Protein Chem. 33, 167-241.

40   Pennica, D., Holmes, W.E., Kohr, W.J., Harkins, R.N., Vehar, G.A., Ward, C.A., Bennett, W.F., Yelverton, E., Seeburg, P.H., Heyneker, H.L. and Goeddel, D.V. (1983) Nature (London) 301, 214-220.

41   Cleary, S., Mulkerrin, M.G. and Kelley, R.F. (1989) Biochemistry 28, 1884-1891.

42   Kelley, R.F. and Cleary, S. (1989) Biochemistry 28, 4047-4054.

43   Castellino, F.J., Ploplis, V.A., Powell, J.R. and Strickland, D.K. (1981) J. Biol. Chem. 256, 4778-4782.

44   Trexler, M. and Patthy, L. (1983) Proc. Nat. Acad. Sci. U.S.A. 80, 2457-2461.

45   Castellino, F.J., DeSerrano, V.S., Powell, J.R., Johnson, W.R. and Beals, J.M. (1986) Arch. Biochem. Biophys. 247, 312-320.

46   Park, C.H. and Tulinksy, A. (1986) Biochemistry 25, 3977-3982.

47   Strauch, K.L. and Beckwith, J. (1988) Proc. Nat. Acad. Sci. U.S.A. 85, 1576-1580.

48   Takagi, H., Morinaga, Y., Tsuchiya, M., Ikemura, H. and Inouye, M. (1988) Bio/Technology 6, 948-950.

HUMAN RETINOBLASTOMA SUSCEPTIBILITY GENE

Chen-Ching Lai and Wen-Hwa Lee

Department of Pathology, M-012,
and Center for Molecular Genetics
University of California, San Diego
La Jolla, California 92093

GENETICS OF RETINOBLASTOMA (RB)

Genetic alterations or predisposition of human cancers have been studied extensively in the last few decades (1-3). These genetic changes, which may be responsible for the genesis of cancers, can either be inherited or occur as a result of somatic mutation. In many cases of heritable human cancers, such as retinoblastoma (RB), Wilms' tumor (nephroblastoma) and familial polyposis (4-6), the transmission of these genetic defects in affected families is suggested to render them at higher risk and/or higher susceptibility to certain types of cancer (7,8). One of the most well studied heritable cancers is retinoblastoma. Retinoblastoma occurs in young children at an incidence of about 1 in 20,000 live births and it is the most common intraocular tumor of the pediatric age group (9). Two forms of retinoblastoma are distinguished based on their genetic origins (10). Typically, bilateral and/or multifocal retinoblastoma occurs in hereditary cases which comprise 40% of all retinoblastomas. The hereditary form of retinoblastoma is transmitted as an autosomal-dominant trait with 90% penetrance (11,12). In addition to the occurrence of childhood retinoblastomas, those people carrying the trait often develop secondary primary tumors, such as osteosarcoma or fibrosarcoma, at higher incidence than the normal population (13,14). By contrast, monofocal and/or unilateral tumors are characteristics of nonhereditary retinoblastoma (6).

As a prototype of hereditary cancers, the mechanism of retinoblastoma formation has been studied extensively. Based on clinical observations and statistical data, Knudson postulated that retinoblastoma can result from as few as two mutational

events (15). It was then suggested that these two hits serve to
inactivate both alleles of a single gene, presumably the retinoblastoma susceptibility gene (RB), which may functionally
suppress retinoblastoma formation (16). An individual inheriting
a mutant RB allele in all somatic cells is predisposed to
retinoblastoma formation, and the cancer arises from an additional mutation of the residual normal RB allele in retinoblasts.
On the other hand, sporadic retinoblastoma occurs when both RB
alleles are inactivated by two independent somatic mutations.
This model can explain both the early onset and multiple tumors
in predisposed individuals. However, the validity of this
hypothesis remained to be demonstrated at the molecular level.

By karyotyping and molecular genetic studies, the putative
RB gene was localized to chromosome band 13q14. For much of the
past several years, this laboratory and others have been dedicated to unraveling the molecular basis of retinoblastoma
formation by identifying the RB gene, characterizing its protein
product, and studying its tumor suppression activity by reintro-
ducing the normal gene back into RB deficient tumors. It is now
known that the RB gene encodes a nuclear phosphoprotein of 110
kiloDalton (kD) with DNA binding activity. This suggests it may
regulate other genes. By studying the naturally occurring as
well as genetically engineered RB mutants, it is possible to
reveal the physiological role of the RB gene. Furthermore, upon
reintroduction into RB deficient tumor cells by retroviral
mediated gene transfer, this gene has pronounced effects in
suppressing the neoplastic phenotype of these tumor cells. This
evidence strongly suggests that the retinoblastoma gene is a
tumor suppressor gene and its inactivation results in the tumor
formation.

LOCALIZATION OF THE RETINOBLASTOMA GENE ON CHROMOSOME 13q14

Cytogenetic studies of patients with hereditary retinoblastoma revealed occasional deletions of the long arm of chromosome
13 (17,18). Similar deletions were also observed in established
retinoblastoma cell lines (19). Band 13q14 was common to all
deletions. In addition, Cavenee et al. (20) demonstrated
specific loss of heterozygosity of several markers on chromosome
13 in retinoblastomas compared to somatic cells from the same
patients. These results indicated that partial or complete loss
of one chromosome 13 is a common event during retinoblastoma
formation and this 13q14 region presumably contained a locus
determining susceptibility to hereditary retinoblastoma.

MOLECULAR CLONING OF GENES FROM CHROMOSOME REGION 13q14

By biochemical and genetic studies, a polymorphic marker
enzyme, esterase D, was found to be closely linked to retinoblas-

toma in this chromosome 13q14 region (21-23). Since there was no previous knowledge about the identity of the RB gene, candidate genes were to be identified solely based on the appropriate chromosomal location and presumed "recessive" behavior that an intact RB gene product expresses in normal retinal tissue but not in retinoblastoma. "Reverse genetic" cloning strategies require a collection of one or more DNA probes from the region of interest. These may include probes for other closely-linked genes, or randomly isolated DNA probes mapped to this chromosome 13 location. In order to clone the RB gene, several laboratories made major efforts to obtain probes for region 13q14. Esterase D was therefore a potential candidate as a start point in cloning the retinoblastoma susceptibility gene.

By generating specific antisera for screening a cDNA expression library, we and others have cloned the esterase D cDNA (24-26). Also available were several DNA probes mapping to 13q14, such as H3-8, H2-42 (27) and 7D2 (28) which were isolated by random selection from chromosome 13-specific libraries. These probes were used in several laboratories to attempt cloning of the putative RB gene.

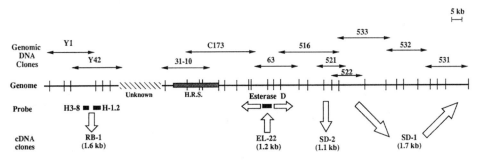

Figure 1. Summary of bidirectional chromosome walking and isolation of cDNA clones from chromosome 13q14 region. Starting points were the esterase D cDNA clone(EL-22), located at 13q14.11, and H3-8, a 13q14-specific DNA probe. These were used to screen Charon 30, Charon 4A, EMBL-3, and cosmid genomic libraries from DNA of human peripheral blood lymphocytes or the Y79 retinoblastoma cell line. Unique sequence subfragments of genomic clones were then used to screen λgt10 or gt11 cDNA libraries made from human fetal retinal or placental mRNA.   3 specific cDNA clones, SD-1, SD-2 and RB-1, were obtained. The RB-1 was further identified as the putative retinoblastoma gene. A 20 kb region 3' to the esterase D gene contained many highly repetitive sequences (H.R.S.). The exact distance between the RB-1 locus and the esterase D gene is still unknown. The EcoRI sites are indicated on the map.

We initiated bidirectional chromosome walking from the esterase D gene to create contigs (regions of overlapping genomic clones) of progressively larger size. At 20 kb intervals of walking distance, unique sequences thus identified were used as probes to isolate cDNA clones from fetal retina and placenta libraries. By alternately screening genomic and cDNA libraries, we established a contig covering 120 kb around the esterase D gene (Figure 1). Two cDNA clones, called SD-1 and SD-2, were isolated using probes 5' to the esterase D gene. Chromosome walking 3' to the esterase D gene was hampered by a 20 kb region containing highly repetitive sequences (H.R.S.). A second bidirectional chromosome walk extending over 30 kb was started from probe H3-8, which was found to be homozygously deleted in two retinoblastoma tumors (27). A unique DNA fragment in this region identified two overlapping cDNA clones of 1.6 kb (RB-1) and 0.9 kb (RB-2) in human cDNA libraries (Figure 1).

IDENTIFICATION OF THE RB GENE

It was expected that candidate RB genes would be expressed in fetal retinal cells but not in retinoblastoma cells. Therefore, cDNA clones isolated above were used as probes in RNA Northern blotting analysis to detect mRNA transcripts in fetal retina, normal human placenta, and cultured cells from six retinoblastomas, three neuroblastomas and two medulloblastomas. Esterase D transcripts were detected in all tumor and tissue samples, consistent with the known "constitutive" expression of esterase D (29). Although undetectable in retinoblastoma cell, neither SD-1 nor SD-2 appeared to be a candidate for the RB gene. Clone RB-1 detected a 4.7 kb mRNA transcript in fetal retina and placenta. Three retinoblastomas demonstrated abnormal mRNA transcripts measuring approximately 4.0 kb. In two retinoblastomas, mRNA transcripts were not observed, while faint bands of about 3.8 and 4.5 kb were visible only after prolonged exposure. Three neuroblastomas and two medulloblastomas displayed identical transcripts of 4.7 kb, equivalent to those in normal tissues. Alterations in gene expression were thus found in 6 of 6 retinoblastomas, but not in two normal tissues and two other related human tumors of neuroectodermal origin. In independent studies, both Friend et al. (29) and Fung et al. (30) also detected a putative RB gene with properties similar to those described above: ubiquitous expression in normal tissues but absent or altered transcription in retinoblastoma. This provided strong support that RB-1 represented part of the putative RB gene (31).

COMPLETE cDNA SEQUENCE AND GENOMIC ORGANIZATION OF THE RB GENE

Complete RB cDNA was obtained after rescreening several cDNA libraries with RB-1 as probe. Clone RB-5 (3.5 kb insert) overlapped RB-1 by restriction mapping and nucleotide sequence analysis. Another RB cDNA clone was isolated that extended an additional 242 base pairs beyond the 5' end of RB-1 (32). Together these clones defined a cDNA sequence of 4757 nucleotides (33) with a long reading frame. Several in-frame methionine codons were detected, and it was later concluded that the first in-frame Met (nucleotide 139) instead of the second (nucleotide 475) was the real translation start based on the evidence that in vitro translation yielded full-size product only when the first 200 nucleotides were present (unpublished data). With the use of the first methionine, the predicted RB protein had 928 amino acids and a molecular weight of 1.1×10^8. Features of the predicted amino acid sequence included one potential leucine zipper domain similar to those found in nucleic acid binding proteins (34). However, this predicted RB protein had no close relatives in current protein sequence databases.

Genomic clones of the RB gene were obtained by screening genomic libraries with cDNA probes. More than two dozen nonredundant, overlapping phage clones assorted into three overlapping groups (contigs) based on shared restriction fragments and exons. Total span of the RB gene exceeded 200 kb of genomic DNA with 27 exons. Exons were initially identified as minimal-length EcoRI and/or HindIII restriction fragments containing sequences hybridizing to RB cDNA clones. About 50 oligonucleotides were synthesized and located in the exon map by hybridization to DNA blots of genomic clones. These oligonucleotides were used as primers to define additional exon/intron junctions by sequencing genomic clones. The length of individual exons ranged from 31 to 1889 base pairs. The largest intron was >60 kb and the smallest had only 80 base pairs (35).

The 5' untranslated portion and first methionine were found in a single exon (exon 1) of the genomic map. The coincidence of cDNA length (4757 bp) and mRNA size (4.7 kb), as well as S1 mapping studies, indicated that our cDNA sequence was essentially complete, and that the first exon was correctly located. Restriction site and nucleotide sequence analysis demonstrated that the last exon was 1.9 kb in length and included the translation stop codon (nucleotides 2924 to 2926).

PRODUCTION OF ANTI-RB ANTIBODIES AND IDENTIFICATION OF RB PROTEIN

The cloned RB cDNA sequence contained a long open reading frame of 2784 base pairs, which when translated yielded a hypothetical protein of 928 amino acids. To identify and further characterize this predicted RB protein, rabbit antisera against this hypothetical RB protein were prepared (Figure 2).

With the use of the hypothetical protein sequence, three recombinant plasmids that express TrpE-RB fusion proteins were constructed. Three DNA fragments (representing amino acids: 301

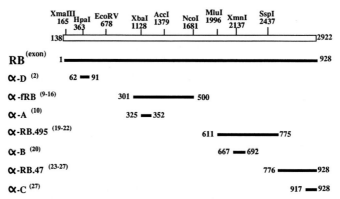

Figure 2.   Summary of anti-RB antibodies.   3 fusion proteins containing TrpE and RB proteins from amino acids 301 to 500, 611 to 775, 776 to 928 were prepared by inserting RB cDNA fragments in-frame into the pATH based vector. After induction, the fusion proteins were purified and used as antigen to produce rabbit antisera.   Oligopeptides (amino acids 62 to 91, 325 to 352, 667 to 692, and 917 to 928) were chosen based on the hydrophobicity plot, linked to rabbit serum albumin, and subsequently used to generate antisera.

to 500, 611 to 775, and 776 to 928) obtained from RB cDNA clones were fused in-frame with the trpE gene in pATH vectors (36) (Figures 2, 3A).   Upon induction, fusion proteins of 57 kD, 54 kD, and 50 kD (in which 37 kD was derived from TrpE protein) were expressed.   Large quantities of fusion protein were prepared and purified by preparative SDS-polyacrylamide gel electrophoresis (SDS-PAGE) and were used to produce rabbit antisera as well as monoclonal antibodies (Figure 2) (37).

In addition, 3 synthetic oligopeptides, peptide A (amino acids 325 to 352), B (amino acids 667 to 692) and C (amino acids 917 to 928) were selected as antigenic determinants on the basis of hydrophobicity and coupled to rabbit serum albumin to produce antibodies (Figure 2).

With these antibodies against the predicted RB protein, a protein of 110 kD was recognized by all of these antibodies (Figure 3B).   In general, cells were metabolically labeled with ^{35}S-methionine, immunoprecipitated with purified anti-RB IgG, resolved by SDS-PAGE and then autoradiographed.   This protein was present in all normal cell lines examined while absent in retinoblastoma as well as several other tumor cell lines.   The absence of antigenically detectable RB protein in retinoblastoma cells further supports the notion that oncogenicity by mutant RB genes is achieved through complete loss of gene product function even in those cell lines containing shortened RB mRNAs.

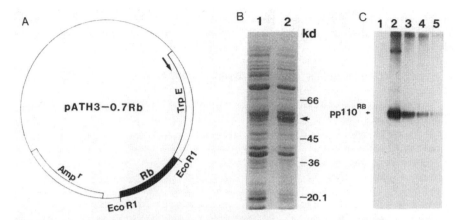

Figure 3. (A) Structure of pATH based vectors for fusion protein preparation. Subfragments of RB cDNA were inserted in-frame into this vector downstream from the trpE gene. (B) Purification of TrpE-RB fusion proteins by polyacrylamide gel electrophoresis (PAGE). (C) Immunoprecipitation of RB protein with anti-RB antibodies. The antibodies prepared (Figure 2) recognized a single protein species from the normal cell lysates. Lane 1: immunoprecipitation with pre-immunized sera; Lane 2-5: with anti-fRB, anti-A, anti-B, and anti-C antibodies, respectively. (Copy from ref. 37 with the permission of authors.)

Post-translational phosphorylation of this protein was shown by metabolically labelling the RB protein with ^{32}P-phosphoric acid. Phosphoamino acid analysis indicated that serine and threonine but not tyrosine residues were phosphorylated (37). Recent studies have shown that the phosphorylation and dephosphorylation of this RB protein were modulated during cell division and cell differentiation (38). Further experiment indicated that the RB protein can be phosphorylated by the cell division cycle kinase (cdc2 kinase) (B.T.-Y. Lin et al., unpublished data). Immunofluorescence studies indicated that the majority of the protein was located in the nucleus. This protein was retained by single- or double-stranded DNA-cellulose columns and was eluted in 0.3 to 0.5 M NaCl, suggesting that pp110^{RB} is associated with DNA binding activity. Together, these results imply that pp110^{RB} may function in regulating other genes during the cell growth and/or differentiation. All these studies have benefited from the availability of these polyclonal and monoclonal antibodies against RB protein and the development of immunoprecipitation and Western blotting analysis. Besides as an excellent tool to study the biological properties of pp110^{RB}, these antibodies could serve as specific probes for the diagnosis of RB protein defects in other types of cancer.

MUTATIONS OF THE RB GENE IN OTHER CANCERS AND THEIR ANALYSIS

From the Northern analysis of RB mRNA and immunoprecipitation analysis of RB protein, it was found that the RB gene is ubiquitously expressed in all normal tissues. From the structural analysis of the (5')-upstream sequence of the RB gene, the RB promotor had the characteristics associated with many other "housekeeping" genes, such as lack of TATA box and no obvious regulatory sequences. This suggests that the RB mutations may be involved in more types of human cancers than retinoblastoma. Indeed, studies from this laboratory and others have implicated RB gene defects in a wide variety of human cancers, such as osteosarcomas (37), synovial sarcomas (37), breast cancers (32,39), leukemia (Chien, C. et al., unpublished data), small cell lung carcinomas (Shew, J.Y. et al., unpublished data), and prostate carcinomas (Bookstein, R. et al., unpublished data).

Mutational analysis of the genomic structure of these RB mutants has revealed long range deletions in several tumor cell lines (40,41). However, small deletions or point mutations of the RB gene are probably sufficient to disrupt the RB function in tumor cells. With mRNA analysis and RB protein screening by immunoprecipitation, shortened RB proteins were observed in several cases, such as in the J82 bladder carcinoma cell line (42), the DU145 prostate carcinoma cell line (Bookstein, R. et al., unpublished data), and fresh small cell lung carcinoma cells (Shew, J.Y. et al., unpublished data).

Characterization of these shortened RB proteins was made possible by the development of reverse transcription-polymerase chain reaction (RT-PCR) analysis of the RB mRNA (43,44). Total RNA was purified from cell lysates and single-stranded cDNA was synthesized from this RNA by AMV reverse transcriptase. The cDNA was then PCR-amplified with several pairs of oligonucleotides chosen from the known RB sequence (Figure 4) and analyzed by polyacrylamide gel electrophoresis. Shortened fragments were observed in RNA obtained from the small cell lung carcinoma cells, and DU 145 prostate cancer cells (Shew, J.Y. et al., unpublished data, and Bookstein, R. et al., unpublished data). Detailed characterization to reveal mutational regions provides useful insight for our understanding of functional domains of RB protein.

TUMOR SUPPRESSION ACTIVITY OF THE RB GENE

Genetic identification of the RB gene must be further confirmed by assays for its expected tumor suppression activity

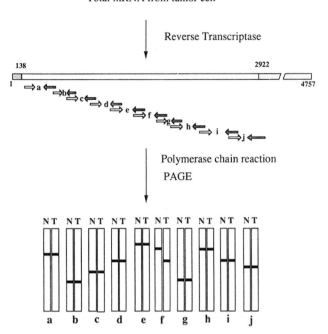

Figure 4. RT-PCR analysis of mutant RB. Total RNA was prepared from small cell lung carcinoma cells, and was reverse transcribed to single-stranded cDNA. The cDNA was then amplified with RB cDNA specific oligonucleotides spanning from amino acids 1-928. The amplified PCR mixtures were then resolved by PAGE. RNA from normal cells was included as control. (N: normal cells; T: tumor cells.)

(45). For this purpose, the RB gene was inserted into Moloney murine leukemia virus (MuLV) under the regulation of MuLV long terminal repeats (LTR) (Figure 5a). A selectable marker gene, neomycin phosphotransferase (Neo), under the regulation of the Rous sarcoma viral (RSV) promotor, was included for selection purposes. A control virus containing another reporter gene, luciferase (Lux), was also constructed (Figure 5a). These viral constructs were transfected into PA12 cells, which carry a packaging-deficient provirus and express all necessary components for the packaging of viral particles, to produce amphotropic virus (46). Infection of these amphotropic virus into ecotropic helper ψ2 cells produced an ecotropic retrovirus capable of infecting human cells (Figure 5b). The RB virus and Lux virus thus formed were used to infect the human retinoblastoma cell line, WERI-Rb27, and osteosarcoma cell line, Saos-2. Upon G418

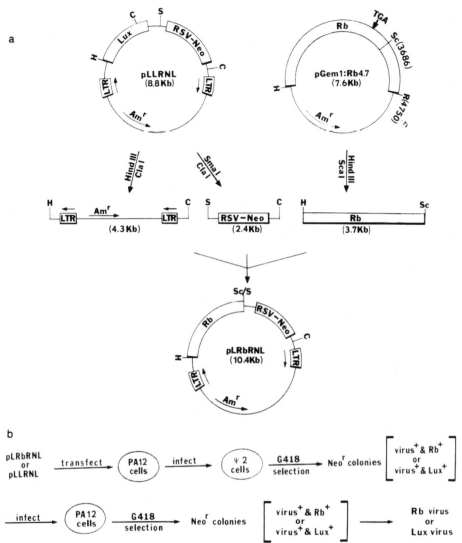

Figure 5 (a).   Construction of RB virus.   Plasmids pLLRNL and
pGem1:RB4.7 were digested with restriction endonucleases as
shown, and appropriate fragments were ligated to form pLRbRNL.
Only selected restriction sites are shown (H, HindIII; C, ClaI;
S, SmaI; Sc, ScaI; R, EcoRI).   LTR, long terminal repeat of
Moloney murine leukemia virus; Lux, luciferase gene; RSV, Rous
sarcomas viral promotor; Neo, Tn5 neomycin phosphotransferase
gene, AmR, ampicillin resistant gene; Rb, RB cDNA; TGA, stop
codon.   (b) Production of amphotropic RB and Lux viruses.
Plasmids pLRbNL or pLLRNL were transfected into amphotropic
packaging cell line PA12 by calcium phosphate co-precipitation,

(Figure 5 cont.) and the cells were grown in Dulbecco's modified essential medium with 10% fetal calf serum and 800 µg of G418/ml. Viral supernatants were harvested after 48 hr and used to infect ecotropic packaging cell line ψ2. After 3 to 4 weeks, resistant colonies were assayed for virus production by infecting 208F rat fibroblasts. RB or Lux gene expression in G418-resistant 208F colonies was determined by immunoprecipitation with anti-fRB antibody as described earlier or by assay of luciferase activity. The RB virus- or Lux virus-producing lines were used to produce viral particles to study the RB gene replacement in RB-deficient tumor cells. (Copy from ref. 45 with the permission of authors. Copyright 1988 by the AAAS.)

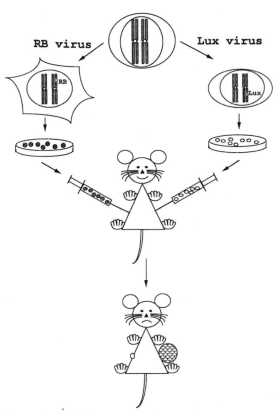

Figure 6.   Tumorigenicity assay in nude mice.   After G418 selection, the RB virus- or Lux virus-infected tumor cells were injected subcutaneously into each side of an individual nude mouse.   After 3 to 5 weeks, gross tumors were detectable on the side injected with Lux virus-infected tumor cells but not in the side injected with RB virus-infected tumor cells.

Table 1
Tumorigenicity of RB or Lux virus-infected WERI-Rb27 cells

Experiment	Number of mice with tumor/ Number injected		
	Parental	Lux	RB
1	5/5	5/5	0/5
2		2/2	0/2

Cells infected with RB or Lux viruses were selected with G418 for 3 weeks (experiment 1) or 5 weeks (experiment 2). After the cell viability was checked, 2×10^7 viable cells were injected subcutaneously into each flank of the same nude mice as shown in Figure 6. For control, the same amount of uninfected parental cells were injected into other mice. The formation of palpable tumor mass was examined one month after the injection.

selection, the RB virus-infected cells were found to express RB protein, pp110^{RB}, while Lux virus-infected cells expressed no RB protein. Those tumor cells which regained RB gene were found to have lost several malignant phenotypes, such as soft agar colony formation, and were morphologically different from the parental RB-deficient tumor cells in tissue culture.

Tumorigenicity of these cells in nude mice was also studied (45). Parental WERI-Rb27 and Saos-2 cells formed tumors in nude mice three weeks after injection of 2×10^7 cells. The effect of RB expression on the tumorigenicity of WERI-Rb27 cells was examined by injecting Lux virus-infected and RB virus-infected cells into separate flanks of a nude mouse (Figure 6). It was shown that Lux virus-infected cells formed palpable tumors 3 to 5 weeks after injection, while RB virus-infected cells formed no detectable tumors (Table 1).

SUMMARY

It is clear that the RB-deficient tumor cells lost their tumorigenicity in nude mice after regaining the RB gene expression. However, the mechanism of tumor suppression by the RB gene is still unknown. More studies on the biological activities of RB protein, pp110^{RB}, are necessary to answer this question. Recent studies have shown that several oncogenic viral proteins, such as SV40 large T antigen (47) and adenoviral E1A protein (48), bind to RB protein. The significance of these bindings remains unclear; nevertheless, they suggest that depletion of functional RB protein by viral proteins may provide another mechanism of RB inactivation. Continued study of naturally occurring as well as engineered RB mutants may give us some information on the biological activity of RB protein, and its roles in oncogenesis, differentiation, development and gene regulation.

Additionally, direct detection of RB gene mutations would have great clinical utility. Probes for the RB gene and gene product will be useful for genetic diagnosis of cancer susceptibility in affected families. Therefore, antibodies to the RB protein will be excellent tools for diagnostic and/or prognostic application in clinical medicine.

Acknowledgments: We thank our colleagues Jin-Yuh Shew, Phang-Lang Chen, Frank Hong, Nan-Ping Wang, Hoang To, Sue Huang, Robert Bookstein, Peter Scully and Eva Lee for their assistance and preliminary results. These studies were supported by grants from the National Institutes of Health, from March of Dimes and from the University of California Cancer Research Coordinating Committee. C.C. Lai is a recipient of fellowships from the Fight for Sight Co. (1988-89) and the University of California Cancer Research Coordinating Committee (1989-90).

REFERENCES

1   Klein, G. (1987) Science 238, 1539-1545.
2   Bishop, J. (1987) Science 235, 305-311.
3   Klein, G. and Klein, E. (1985) Nature 315, 190-195.
4   Schimke, R., Lowman, J.T. and Cowan, G.A.B. (1974) Cancer 34, 2077-2079.
5   Cohen, A.J., Li, F.P., Berg, S., Marchetto, J., Tsai, S., Jacobs, S.C. and Brown, R.S. (1979) New. Engl. J. Med. 301, 592-595.
6   Greene, M.H., Clark Jr., W.H., Tucker, M.A., Kraemer, K.H., Elder, D.E. and Fraser, M.C. (1985) Ann. Intern. Med. 102, 458-465.
7   Knudson, A.G. (1973) Adv. Cancer Res. 17, 317-352.
8   Knudson, A.G. (1975) Cancer 35, 1022-1026.
9   Shields, J.A. (1983) in Diagnosis and Management of Intraocular Tumors pp. 437-438, C.F. Mosby Co., St. Louis, MO.
10 Francois, J., Matton, M.T., DeBie, S., Tanaka, Y. and Vandenbulcke, D. (1976) Ophthalmologica 170, 405-425.
11 Matsunaga, E. (1978) Amer. J. Hum. Genet. 30, 406-424.
12 Vogel, F. (1979) Hum. Genet. 52, 1-54.
13 Jensen, R.D. and Miller, R.W. (1971) New Engl. J. Med. 285, 307-311.
14 Abramson, D.H., Ellsworth, R.M., Kitchin, D. and Tung, G. (1984) Ophthalmology 91, 1351-1355.

15 Knudson, A.G. (1971) Proc. Nat. Acad. Sci. U.S.A. 68, 820-823.

16 Comgins, D.E. (1973) Proc. Nat. Acad. Sci. U.S.A. 70, 3324-3328.

17 Francke, U. 91976) Birth Defects 12, 131-137.

18 Yunis, J.J. and Ramsay, N. (1978) Amer. J. Dis. Child. 132, 161-163.

19 Balaban, G., Gilbert, F., Nichols, W., Meadows, A.T. and Shields, J. (1982) Cancer Genet. and Cytogenet. 6, 213-221.

20 Cavenee, W.K., Dryja, T.P., Phillips, R.A., Benedict, W.F., Godbout, R., Gallie, B.L., Murphree, A.L., Strong, L.C. and White, R.L. (1983) Nature 305, 779-784.

21 Sparkes, R.S., Sparkes, M.C., Wilson, M.G., Towner, J.W., Benedict, W., Murphree, A.L. and Yunis, J.J. (1980) Science 208, 1042-1044.

22 Sparkes, R.S., Murphree, A.L., Lingua, R., Sparkes, M.C., Field, L.L., Funderburk, S.J. and Benedict, W.F. (1983) Science 219, 971-973.

23 Godbout, R., Dryja, T., Squire, J., Gallie, B.L. and Phillips, R.A. (1983) Nature 304, 451-453.

24 Lee, W.H., Wheatley, W., Benedict, W.F., Huang, C.M. and Lee, E.Y.-H.P. (1986) Proc. Nat. Acad. Sci. U.S.A. 83, 6790-6794.

25 Lee, E.Y.-H.P. and Lee, W.H. (1986) Proc. Nat. Acad. Sci. U.S.A. 83, 6337-6341.

26 Squire, J., Dryja, T.P., Dunn, J., Goddard, A., Hofmann, T., Musarella, M., Willard, H.F., Becker, A.J., Gallie, B.L. and Phillips, R.A. (1986) Proc. Nat. Acad. Sci. U.S.A. 83, 6573-6577.

27 Lalande, M., Dryja, T.P., Schreck, R.R., Shipley, J., Flint, A. and Latt, S.A. (1984) Cancer Genet. and Cytogenet. 13, 283-295.

28 Leppert, M., Callahan, P., Cavenee, W., Holm, W., O'Connell, P., Thompson, K., Lalouel, J.M. and White, R. (1985) Cytogenet. Cell Genet. 40, 679.

29 Friend, S.H., Bernards, R., Rogelj, S., Weinberg, R.A., Rapaport, J.M., Albert, D.M. and Dryja, T.P. (1986) Nature 323, 643-646.

30 Fung, Y.K.T., Murphree, A.L., T'Ang, A., Qian, J., Hinrichs, S.H. and Benedict, W.F. (1987) Science 236, 1657-1661.

31 Lee, W.H., Bookstein, R., Hong, F., Young, L.J., Shew, J.Y.and Lee, E.Y.-H.P. (1987) Science 235, 1394-1399.

32 Lee, E.Y.-H.P., Bookstein, R., Young, L.J., Lin, C.J., Rosenfeld, M.G. and Lee, W.H. (1988) Proc. Nat. Acad. Sci. U.S.A. 85, 6017-6021.

33 Lee, W.H., Shew, J.Y., Hong, F.D., Sery, T.W., Donoso, L.A., Young, L.J., Bookstein, R. and Lee, E.Y.-H.P. (1987) Nature 329, 642-645.

34 Landschulz, W.H., Johnson, P.F. and McKnight, S.L. (1988) Science 240, 1759-1764.

35   Frank, D.H., Huang, H.-J.S., To, H., Young, L.-J., Oro, A., Bookstein, R., Lee, E.Y.-H.P. and Lee, W.H. (1989) Proc. Nat. Acad. Sci. U.S.A. 86, 5502-5506.

36   Boyle, W.J., Lipsick, J.S. and Baluda, M.A. (1986) Proc. Nat. Acad. Sci. U.S.A. 83, 4685-4689.

37   Shew, J.Y., Ling, N., Yang, X., Fodstad, O. and Lee, W.H. (1989) Oncogene Res. 1, 205-214.

38   Chen, P.L., Scully, P.A., Shew, J.Y., Wang, J. and Lee, W.H. (1989) Cell 58, 1193-1198.

39   Lee, E.Y.-H.P., To, H., Bookstein, R., Scully, P. and Lee, W.H. (1988) Science 241, 218-221.

40   Bookstein, R., Lee, E.Y.-H.P., Peccei, A. and Lee, W.H. (1989) Mol. Cell. Biol. 9, 1628-1634.

41   Shew, J.Y., Lin, B.T.-Y., Chen, P.L., Tseng, B.Y., Yang-Feng, T.L. and Lee, W.H. (1989) Proc. Nat. Acad. Sci. U.S.A. (in press).

42   Horowitz, J.M., Yandell, D.W., Park, S.-H., Canning, S., Whyte, P., Buchkovich, K., Harlow, E., Weinberg, R.A. and Dryja, T.P. (1989) Science 243, 837-940.

43   Rappolee, D.A., Wang, A., Mark, D. and Werb, Z. (1988) J. Cell. Biochem. 39, 1-11.

44   Rapolee, D., Brenner C.A., Schultz, R. and Werb, Z. (1988) Science, 241, 1823-1826.

45   Huang, S.H.-J., Yee, J.K., Shew, J.Y., Chen, P.L., Bookstein, R., Friedmann, T., Lee, E.Y.-H.P. and Lee, W.H. (1988) Science 242, 1563-1566.

46   Miller, A.D., Law, M.-F. and Verma, I.M. (1985) Mol. Cell. Biol., 5, 431-437.

47   DeCaprio, J.A., Ludlow, J.W., Figge, J., Shew, J.-Y., Marsillo, E., Paucha, E. and Livingston, D.M. (1988) Cell, 54, 275-283.

48   Whyte, P., Buchkovich, K.J., Horowitz, J.M., Friend, S.H., Raybuck, M., Weinberg, R.A. and Harlow, E. (1988) Nature, 334, 124-129.

α-OLIGODEOXYNUCLEOTIDES (α-DNA):

A NEW CHIMERIC NUCLEIC ACID ANALOG

F. Morvan, B. Rayner and J.-L. Imbach

Laboratoire de Chimie Bio-Organique, UA 488 CNRS
Université des Sciences et Techniques du Languedoc
Place E. Bataillon
34060 Montpellier Cédex 1, France

INTRODUCTION

Nucleic acids have long been strategic targets for approaches to chemotherapy in view of their roles in replication, transcription and translation (1). The use of synthetic oligonucleotides, which bind specifically to complementary sequences of nucleic acids (RNA or DNA) through base pairing, is now under extensive investigation. In principle, relatively short oligomers (20 bases or less) can specifically hybridize with DNA or RNA and thus be used for drug design strategies involving targeted interference of genetic expression.

However, potential chemotherapeutic applications resulting from sequence-specific hybridization require analogs that are resistant to degradation by various nucleases. Oligonucleotide analogs presenting modifications on the phosphate backbone (i.e., methylphosphonates (2-5) and phosphorothioates (6)) have been introduced and have been shown to have good resistance to enzyme-mediated depolymerization. However, such modifications introduce asymmetric linkages and, as the synthesis is not stereo-controlled, lead to a mixture of 2^n stereoisomers (where n is the number of these linkages), each of them presenting different binding capacities. In addition, in the case of methylphosphonate the replacement of the negatively charged oxygen atom by a neutral group could interfere with eventual enzymatic interactions (ionic linkage) and also with the hydrophilic properties of these oligomers. These phosphate backbone modifications have always been considered starting from "natural" 2'-deoxy-β-D-ribofuranosyl nucleosides as constituting synthons. But another possibility is to consider sugar-modified

Genetic Engineering, Vol. 12
Edited by J.K. Setlow
Plenum Press, New York, 1990

nucleosides as building blocks and to synthesize the
corresponding oligomers with natural phosphodiester linkages.

Let us consider first the structure of a natural 2'-deoxy-β-
D-ribofuranosyl nucleoside (Figure 1). This molecule presents
three chiral carbon atoms in 1', 3' and 4' positions. In order
to maintain the geometry of the sugar phosphate backbone with a
3', 4'-transorientation, configurations of 3' and 4' carbon atoms
must be changed together. This modification leads to 2'-deoxy-L-
ribonucleosides. The oligomers of L-2'-dU have been shown to be
much more resistant to snake venom phosphodiesterase than the
oligomers of D-2'-dU, although no evidence was found for binding
of a 18-mer of L-2'-dU to poly(dA) (7). Inversion of the 1'
carbon atom leads to 2'-deoxy-α-D-ribofuranosyl nucleosides.
Only α-oligodeoxyribonucleotides (α-DNA), consisting exclusively
of α-anomeric deoxyribonucleotides, will be considered in this
chapter. In 1973, U. Sequin (8), using Dreiding models,
considered the possibility of α-oligonucleotides exhibiting
secondary structure similar to that of the natural nucleic acids,
featuring base pairing, base stacking and helix formation. This
study predicted that an α-strand may form a helix duplex with a
complementary β- or α-strand by base pairing, and the two strands
should exhibit parallel and antiparallel polarity, respectively.
Shortly thereafter, isomeric dithymidine monophosphates bearing
one or two α-dTs were found to exhibit nuclease resistance (9).

The aim of this work, initiated in 1985, was to set up
efficient synthesis of α-oligodeoxynucleotides and to study their
physical and biological properties.

SYNTHESIS

In 1986 α-oligonucleotides (hexamers) were first synthesized
in solution with the phosphotriester approach (10). However, in
this primary publication only pyrimidine nucleosides were used,
but the method was rapidly extended to the four usual bases (11)
and to a solid phase approach (12,13).

As the reactivity of an α-nucleoside is very close to that
of a β-anomer, the methodology used for synthesizing an α-
oligonucleotide is the same as for the usual β-one. The main
problem arises from the preparation of a sufficient amount of the
starting α-nucleoside synthons, which are not commercially
available except for α-dT.

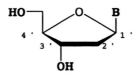

Figure 1. Structure of deoxy-2'-β-ribose.

The nucleoside α-dT can either be obtained in 16% overall yield following an anomerization procedure as described (14) or with a 47% yield through a direct condensation of 1-chloro-3,5-di-p-toluyl-2'-deoxyribofuranose with bistrimethyl silyl thymidine followed by base-catalyzed deprotection (15). Similarly, the N-6-benzoyl-α-dC can be obtained in a 26% yield through a self-anomerization procedure starting with 3',5'-diacetyl-N-6-benzoyl-β-dC (10). The α-2'-deoxy-adenosine derivative can be made as described (14) through a transglycosylation procedure or by a direct condensation method with 1-chloro-3,5-bis-p-nitrobenzoyl-2'-deoxyribofuranose. In this latter case, the yield of the N-6-benzoyl-α-dA is 34% (16). In any case, α-dG is more difficult to obtain. The literature reported only a very low yield of the desired α-nucleoside (17), but a transglycosylation procedure with the use of N-2 protected guanine and a protected cytosine did increase the yield to about 29% (18). At this stage one can see that α-nucleoside building blocks can be obtained in large amount with appropriate synthesis. The phosphotriester approach in solution was used first to synthesize "large" amounts (i.e., 20 mg) of pure α-oligomers for biophysical and NMR studies.

In two different papers (10,11), we have reported the physical data of all required phosphotriester synthons. In Figure 2 is drawn the structure of a phosphotriester intermediate that we use. We have noticed that no striking difference (reaction time, formation of side products) could be detected during α-oligonucleotide synthesis when compared to that of natural β-oligonucleotides (10,11).

The automated solid phase approach was designed using the appropriate α-nucleoside phosphoramidites. Two kinds of phosphate protecting groups were used, the cyanoethyl (12) and

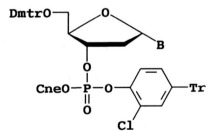

B : 6-N-benzoyladenine,
    4-N-benzoylcytosine,
    2-N-palmitoylguanine,
    thymine.

Cne, 2-cyanoethyl; Dmtr, 4,4'dimethoxytrityl; Tr, trityl.

Figure 2. Structure of an α-phosphotriester intermediate.

the methyl (13). The necessary synthons were synthesized
following the usual procedure and their physical data have been
described (13). Fractosil 500 (12) and long chain alkylamino-CPG
(13) were used as solid supports, and derivatized as described
for the β-anomers. The amount of loaded nucleoside was 26 to 27
μmol per gram in the latter case. It was possible to assemble α-
oligomers automatically with an automatic DNA synthesizer either
with the same cycle as for a β-anomer (but only α-
oligonucleotides which contain A, C and T have been described)
(12) or with a slightly modified one to take into account the
poor solubility of the α-guanosine phosphoramidites (13). Using
such methodology we have synthesized and described a 20-mer
containing the four usual bases (13); it was obtained in 29%
yield after purification with an average coupling yield of 98.3%
and its primary structure was confirmed by Maxam and Gilbert
sequence analysis.

These data show that α-oligonucleotides of any length can be
obtained with the usual α-oligonucleotide methodology and that
the starting α building blocks are readily available in multigram
amounts.

STRUCTURE AND PHYSICO-CHEMICAL PROPERTIES

Single-stranded α-Oligodeoxynucleotides

The first α sequence which was examined, i.e., α-[d(CCT-
TCC)], contained only pyrimidine residues (10) and was
investigated by NMR spectroscopy. It was shown that the cytosine
and thymine bases adopt an anti conformation at the glycosyl
bonds. In addition, the deoxyriboses of the thymidines adopt an
average conformation approximating C_3, -endo, while the cytidine
furanose groups are close to C_2, -exo. Further studies on
heterogeneous sequences α-[d(CATGCG)] and α-[d(CGCATG)] confirmed
the anti conformation of the bases, the sugar rings being C_3,-
exo (19). It was then shown, based upon hypochromicity
considerations, that the α-[d(CATGCG)] exhibits a higher level of
base-base interaction than the corresponding β-strand. But this
sequence is also able to self-anneal, thus complicating the
interpretation (20). For instance, in the case of an homogenous
α-oligothymidylate: α-[d(Tp)$_7$T], CD spectra indicate base-base
interactions whose thermodynamic parameters are of the same order
of magnitude as those of the β-[(pT)$_8$] even though this
interaction is formed with a different geometry (21).

NMR studies on α-[d(CATGCG)] and on α-[d(CGCATG)] give some
insight into the self-annealing of α-strands (20). We were able
to detect base pairing due to self-annealing and to conclude that
there was antiparallel self-recognition of these α-strands, whose
thermodynamic stability was compatible with Watson-Crick rather
than Hoogsteen base pairing. Once it was established that α-

oligonucleotides are able to undergo strong antiparallel self-annealing, the important question of annealing to a complementary β-strand remained to be answered.

Formation of α-β-Double Strands

Much work dealing with this question allows one to state unambiguously that α-oligonucleotides are able to form stable duplexes with their β-counterparts. Both hypochromicity in thermal denaturation (10,22) and detection of base-paired imino protons in ^{1}H -NMR studies provide evidence for the annealing of α-[d(CCTTCC)] with β-[d(GGAAGG)]. Melting experiments (22) lead to the same conclusion for α-[d(GGAAGG)] and β-[d)CCTTCC)]. Furthermore, the annealing of α-[d(CATGCG)] with β-[d(GTACGC)] (19) and of α-[d(TCTAAAC)] with β-[d(AGATTTG)] (23) were demonstrated and studied in detail with the use of NMR. Applying 1-D and 2-D strategies, the authors derived a right-helical structure for those complexes with the orientation of the bases at the glycosyl bond with respect to the sugar moiety being anti (19,23). Furthermore, the pucker of the sugar moiety has been found to be 2'-endo-3'-exo (19) or 3'-exo (23). As far as the polarity of the complex is concerned, the authors propose a parallel orientation of the two chains.

Other work dealing with the annealing of α-oligothymidylate and its β-complement, with techniques different from that of NMR: photo-crosslinking reactions (24), circular dichroism spectra (21) and fluorescence (25), lead to the same conclusion that double helices are formed with parallel strands. Furthermore, it has been shown (22) that an α-hexamer forms a complex with its complementary β-sequence that is more stable than the corresponding β,β-complex. This conclusion raises the question of the stability of the unnatural α,β-duplex compared to the natural β,β one.

Stability of α-β-Oligonucleotide Duplexes

The melting temperature of the duplex α-[(Tp)$_7$T]: β-[d(pA)$_8$] is reported to be 31°C compared to 24°C for the corresponding β-[(Tp)$_8$]: β-[d(Ap)$_8$] (21), suggesting that α,β-annealing is thermodynamically more stable than β,β-annealing. On the other hand the α strand containing a stretch of nine thymidines flanked by two guanosines at each end, α-[d(G$_2$T$_9$G$_2$)] hybridized with its complementary β-strand β-[d(C$_2$A$_9$C$_2$)] less readily (T$_m$ = 34°C) than the counterpart β-[d(G$_2$T$_9$G$_2$)] (T$_m$ = 40°C) (18). This result suggests that α-guanosine may destabilize the α,β duplexes. However, mixing of α-[d(TCTAAAC)] with β-[d(AGATTTG)] produces a duplex whose melting temperature is 33°C (2 mM/2 mM) compared to 36°C for the corresponding β-[d(CAAATCT)]: β-[d(AGATTTG)] duplex (2.5 mM/2.5 mM), suggesting that α,β-annealing has a stability close to that of the β,β duplex (23). Detailed thermodynamics studies on the duplexes α-[d(CCT-TCC)]: β-[d(GGAAGG)] and α-[d(GGAAGG)]:β-[d(CCTTCC)] lead to the conclusion that their stabilities depend upon the nature of the bases (purines or pyrimidines) involved in the building of the α-strand, at least for homogeneous purine or pyrimidine sequences (22). Comparison of the thermal denaturation of β-[d(CCTTCC)]: β-[d(GGAAGG)] and α-[d(CCTTCC)]: β[d(GGAAGG)] duplexes indicates that the α,β-duplex is more stable ($T_{m\alpha,\beta}$ = 28°C) than the β,β-duplex ($T_{m\beta,\beta}$ = 19°) in 1M NaCl. However, in the symmetrical situation where the purine strand is made of unnatural α-nucleotides, α-[d(GGAAGG)], the stability of the duplex is in favor of the β,β-system ($T_{m\alpha,\beta}$ = 14°C for α-[d(GGAAGG)]: β-[d(CCTTCC)]).

This result leads to a conclusion that the stability of the α,β-annealing may be dependent upon base composition of the α-sequence involved in the formation of the unnatural duplex. It also appears that base pairing formed with α-dG and β-dC may destabilize the α-β duplexes. The stability of the α,β duplexes raises the question of the annealing selectivity of such duplexes.

Annealing Selectivity of α-Oligodeoxynucleotides

Preliminary results obtained through melting experiments showed the influence of the nature of a single central mismatch on α,β heteroduplex stability, in comparison with that observed in the corresponding β,β duplexes (Table 1) (18). The influence

Table 1

A)

β-d(C_2 AAAAXAAAAC$_2$)	α-d(G_2 TTTTTTTTTG$_2$)		β-d(G_2 TTTTTTTTTG$_2$)	
X = A	33.9°C	ΔTm	40.0°C	ΔTm
= C	24.2°C	-9.7	27.5°C	-12.5
= G	25.9°C	-8.0	30.8°C	-9.2
= T	25.3°C	-8.6	28.1°C	-11.9

B)

α-d(TTTTTTXTTTTTT)	poly rA	
X = T	43.5°C	ΔTm
= C	36.6°C	-6.9
= G	31.6°C	-11.9
= A	33.6°C	-9.9

Melting temperature of A): α-DNA;β-DNA and B) α-DNA;β-RNA duplexes. The concentration of each oligonucleotide is 3μM in buffer: 0.01M sodium cacodylate, 0.1M NaCl, pH 7. ΔT_m represents the variation in T_m between a duplex bearing a single mismatch and the corresponding perfectly matched duplex.

of a β-dX, α-dT mismatch was first examined, X being C,G or T
(Table 1A). In all cases, a sharp decrease of T_m was observed
ranging from 8 to 9.7°C, the C and T mismatches inducing the most
important effect in both α,β and β,β duplexes. Although the
decreases of T_m are smaller in α,β duplexes than they are in the
corresponding β,β duplexes, they allow a sharp discrimination
between a perfectly matched α,β heteroduplex and an analogous
duplex exhibiting a single mismatch. Higher decreases in T_m,
ranging from 6.9 to 11.9°C, were observed in the case of the α-
dX, β-rA mismatches (Table 1B) with a maximum for a G,A mismatch.
These results are consistent with a high sequence-specificity of
α-oligonucleotides for both DNA and RNA targets. The study of
the stability and the selectivity of the α,β complexes, also
raised the question of the stoichiometry of such complexes.

Stoichiometry of α,β-complexes

Results obtained by circular dichroism on α-oligothymidylate
(21), UV absorbance and fluorescence on α-[d(CCTTCC)] and α-
[d(GGAAGG)] (22) suggest that unnatural α-sequences form 1α, 1β-
duplexes with complementary β-oligodeoxynucleotide sequences.
However, experiments performed at high salt concentration with a
α-octathymidylate covalently linked to a photo-crosslinking
reagent and a 27-mer duplex containing a $(dA \cdot dT)_8$ sequence
indicated that a local triple helix was formed in which the α-
octathymidylate was located in the major groove of the initial
double helix and was oriented parallel to the adenine-containing
strand. No evidence was obtained that local formation of a
triple helix can be observed with sequences different from
oligopurine-oligopyrimidine sequences in a DNA double helix (26).
More recently it has been shown that the stoichiometry of the
α,β-complex is similar for an oligonucleotide and for a
polynucleotide and does not depend on the nature of the sugar
(ribose or deoxyribose) in the β-strand (21). This last point
led us to consider the formation of duplexes between α-
oligodeoxynucleotides and complementary β-oligoribo- or
polyribonucleotides.

Formation of Double-stranded Structures by Association of α-Strands with Complementary β-Oligoribo- or Polyribonucleotides

Most of the results concern the interaction between an α-
oligothymidylate and oligo or poly(rA). A detailed study of such
interaction has been carried out with circular dichroism (21).
It appears that a duplex is formed between the α- and the β-
strands with a 1:1 stoichiometry. This duplex is more stable
than the corresponding β,β-duplex: T_m = 29°C in 0.1 M NaCl for α-
[d(Tp)₇T] associated with poly(rA) or r(Ap)₇rA and 14 or 15°C for
β-[d(Tp)₈] associated with poly(rA) or r(Ap)₇rA as already shown
(27). A stabilization of the α-DNA:β-RNA complex has also been

demonstrated when α-d$[G_2T_{12}G_2)]$ is associated with poly(rA): T_m = 24.4°C for β-oligo:RNA and 52.8 for α-oligo:RNA (28). Furthermore there is great similarity between the complexes formed by α-$[d(Tp)_7T]$ with β-r(Ap)$_7$rA and those formed with poly(rA) (21). On the other hand, energy minimization calculations have led to the conclusion that the more stable structure for the α-oligo dT:β-oligo rA complex, as long as the hybrid has an A conformation, should be the antiparallel orientation, in contrast to what is expected and demonstrated for the α-oligo dT:β-oligo dA complex (25). Actually an antiparallel orientation of the α-DNA:β-RNA hybrid is experimentally suggested in the case of oligo α-thymidylate (29), in contrast with previous Northern blot experiments which showed that a 20-mer α-oligonucleotide complementary, in parallel orientation of the codon initiation region of a mRNA, did hybridize with its target (28). However the A type of the α-oligo dT:β-oligo dA double helix was not established and it remains to be determined whether the antiparallel orientation observed for the complex α-(dT)$_8$ with β-(rA)$_8$ is a special property due to the formation of adjacent A-T base pairs involving, e.g., Hoogsteen or reverse Watson-Crick (30,31), rather than the classical Watson-Crick hydrogen bonds. More recently, a thorough investigation clearly demonstrates the parallel annealing of α-oligonucleotide with complementary RNA sequences including the four bases (32); ps- and aps-α-oligonucleotides were designed to recognize in parallel (ps) or antiparallel (aps) orientation two different sites of a 1000 base-long mRNA. Northern blot experiments indicate that only ps-α-oligonucleotides were able to hybridize to the mRNA target. Furthermore, competition experiments confirm the parallel annealing of α-oligonucleotides containing the four bases with a natural mRNA (32).

BIOLOGICAL PROPERTIES

Recognition by Enzymes

Enzymes which recognize single-stranded DNA. α-Oligos can be 5'-labeled by T4 polynucleotide kinase in the presence of [γ-^{32}P]-ATP. After 1 hr incubation at 37° the yield is 75% when compared to β-oligonucleotides with the same sequence for a homopolymeric oligonucleotide (13-mer) and with identical length and base composition for an oligonucleotide with four bases (20-mer) (28). Alkaline phosphatase activity was measured (33) and found to be identical for α- or β-nucleotide with a phosphate in the 5'-position. Although phosphatase activity was not measured on α-oligonucleotides it can be expected to be in the same range as kinase activity when compared to β-oligonucleotides.

Concerning the behavior of various nucleases, it has been shown that α-oligos are poor substrates at best. The lyophilized venom of <u>Crotalus adamenteus</u>, which is rich in 5'-nucleotidase, has no action on a 5'-phosphate α-pyridone after 60 min at 37°C and 40 hr at 23°C, whereas the 5-phosphate β-pyridone is completely cleaved within 10 min at 37°C (33). The rate constant for the degradation of a β-oligonucleotide, as measured by UV absorbance (6-mer), is 30 times higher than for the analogous α-oligonucleotide (11). HPLC and UV experiments show that the 5'-exonuclease, calf spleen phosphodiesterase, has no action on α-oligonucleotides whereas, under the same experimental conditions (10 min at 37°C), 90% of their β-counterparts are degraded (11). The α-structure therefore lowers to a variable extent the action of enzymes which recognize the extremities of the oligomer. The inhibition ratio with the α-structure is likely related to the role of the bases in the enzymatic sites for the usual nucleic acids. α-Oligonucleotides are also poor substrates for S1 endonuclease, an enzyme specific for single-stranded DNA. After 10 min incubation at 37°C, β-[d(CATGCG)] was completely digested whereas 93% of α-[d(CATGCG)] was intact (11), the amount of oligonucleotide being measured by HPLC. Digestion kinetics performed at 260 nm fully confirm this result (11).

<u>Enzymes which recognize double-strand nucleic acids.</u> Hybrids between α-[d($G_2T_{12}G_2$)] and r(A_{12}) were not degraded by <u>E. coli</u> RNase H, at variance with the results obtained on duplexes containing the corresponding β anomer (28). In a more detailed study, the α-DNA:β-RNA duplexes have been shown to be competitive inhibitors of RNase H. Different reports indicate that α-DNA:β-RNA duplexes have a higher melting temperature than that of the corresponding β-DNA:β-RNA, although no precise structure is yet available for these duplexes (27,28).

It has been shown (33) that <u>E. coli</u> and <u>Drosophila</u> embryo RNases H are displaced from their substrate by a competitive α-DNA:β-RNA hybrid which is resistant to attack by RNases. However, strand exchanges could occur between the substrate and the competitive inhibitor. With [^{3}H]poly(rA): poly(dT)$_n$ as a substrate for E. coli RNase H, the inhibitory properties of a poly(rI):α-[d(C_{20})] have been shown (Figure 3) (33). It is assumed that the affinity of RNase H for [^{3}H]poly(rA): poly(dT) substrate decreases according to the digestion of [^{3}H]poly(rA). When the reaction proceeds, competitive binding becomes more and more in favor of the inhibitor. This explains the occurrence of a lower plateau for digestion of the substrate in the presence of the inhibitor when compared to the control (Figure 3). The available data suggest that the affinities of the RNase H for the inhibitors and for their natural substrates are of the same order of magnitude if they are similar in length. The ability of RNase H, for which DNA:RNA duplexes are its natural substrates, to bind to the unnatural α-DNA:β-RNA hybrids could likely be extended to other enzymes which recognize nucleic acids: i.e., reverse transcriptases.

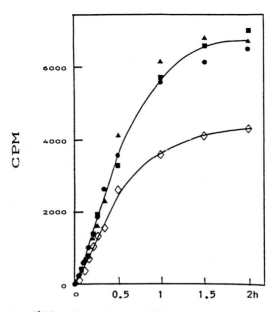

(■) [³H] poly (rA):poly(dT)

(◇) [³H] poly (rA):poly(dT) + poly(rI):α-d(C₂₀)

(•) [³H] poly (rA):poly(dT) + poly(rI)

(▲) [³H] poly (rA):poly(dT) + α-d(C₂₀)

Figure 3.   Digestion kinetics of poly(rA):poly(dT) by E. coli RNase H with or without poly(rI):α-d(C$_{20}$).   [³H̅] poly (rA):poly(dT) was prehybridized before being incubated with RNase H.

Reverse transcriptase. An inhibitory effect was observed on M-MLV reverse transcriptase polymerizing activity with poly(rC) as a template and r(A$_{12}$): α-d(G$_2$T$_{12}$G$_2$)] (150 µm) as a competitor (75% inhibition after 20 min incubation at 20°C) (34). Retroviral reverse transcriptases are also endowed with a nucleolytic activity. It would therefore be important to have inhibitors able to be efficient with respect to both polymerizing and RNase H activities. Accordingly, it has been shown (35) that the RNase H associated activity of avian myeloblastosis virus reverse transcriptase is inhibited by a poly(I): α-[d(C$_{20}$)] hybrid (50% inhibition after 120 min incubation at 20°C, with both substrate and inhibitor at 1.6 µM) and [³H]poly(rA):poly(dT) as a substrate.

Single-stranded α-DNA oligonucleotides can also act as competitive inhibitors towards β-oligodeoxynucleotides being primers of purified M-MLV reverse transcriptase (34). 150 µM of α-[d(A$_{15}$)] is required to obtain total inhibition of the polymerase activity of the enzyme when poly(rU) (150 µM) and

$d(A_{10})$ (22 µM) are used as a template and a primer, respectively. Additionally, it has been shown that α-$[d(T_n)]$s (n>10) are potent inhibitors of HIV reverse transcriptase. With poly(rA) as a template and β-$[d(T_{12-18})]$ as a primer (7.5 µM), 50% inhibition was obtained with 1.2 µM of α-$[d(T_{16})]$. For α-dT_{10} the kinetics of HIV/RT inhibition were studied with respect to the natural primer β-dT_{10} and the Km values were found to be 2.7 µM for the natural substrate and 0.08 µM for the α analog (Ki) (36).

Stability in Biological Extracts and Cells.

The half-life of α-(16-mer) and β-(17-mer) single-stranded oligonucleotides was measured by HPLC after injection into Xenopus oocytes (37). It was 10 min for β-oligonucleotides and 480 min for the α-analogs. This result is consistent with previous data obtained with purified enzymes. Furthermore, it was shown in another study (38) that 5'-labeled α-oligonucleotides have a half-life around 24 hr in various sera as well as in rabbit reticulocyte lysates or in Hela cell postmitochondrial supernatant. It has already been shown that 5'-labeled β-oligonucleotides have a half-life which varies from more than 90 min to less than 15 min in various biological media (39). However, in these experiments, the real half-lives could be different from the measured ones because of 5'-phosphatase or 5'-exonuclease activities. Similar experiments were performed by thin layer chromatography on extracts of 3T3 mouse fibroblasts (40). α-$[d(A_6)]$ displays a half-life of 132 min whereas it is 54 min for β-$[d(A_6)]$. This last result indicates that some cell extracts can be rather active for the degradation of α-oligonucleotides, although they still display a higher lifetime than β analogs. Another very important point is that α-oligonucleotides can also confer resistance to hybridized nucleic acids in a β-conformation.

The resistance to nucleolytic activities of a duplex $r(A_{12})$:α-$[d(G_2T_{12}G_2)]$ was measured in rabbit reticulocyte and wheat germ extract (33). This duplex appears resistant to nuclease activities whereas the $r(A_{12})$: β-$[d(G_2T_{12}G_2)]$ is rapidly degraded in both extracts. After 4 hr at 25°C, there is no intact $r(A_{12})$ left in both extracts, when hybridized to the β-oligonucleotide, whereas $r(A_{12})$ is fully intact under the same conditions when hybridized to the α-oligonucleotide.

Inhibition of Gene Expression

α-Antisense oligonucleotides have been compared to β-anomeric ones in their capacity to inhibit the translation of several natural mRNAs in reticulocyte lysates (28,41), in wheat germ cell-free extracts (41) in microinjected Xenopus oocytes (41) and in HIV-infected MT4 cells (unpublished results). In most of these experiments, 9- to 20-mer α-oligonucleotides were chosen to hybridize to their mRNA target in the expected parallel orientation; a β-anomeric oligonucleotide of the same length hybridizing in the antiparallel orientation was used as a positive control and random α- or β-sequences as negative controls. As explained below, the eventual inhibition of translation by α-oligonucleotides depends on the mRNA targeted site, but in any case, their β-anomeric counterparts were efficient.

A 20-mer β-anomeric oligonucleotide complementary to the translation initiation region of the IL6 mRNA inhibited strongly its expression as a 26 kD protein in a reticulocyte cell-free lysate in the presence of minute amounts of RNase H. The inhibition was sequence-specific and was detectable at an oligonucleotide mRNA molar ratio as low as 1/10; no significant reduction of mRNA translation was observed in the absence of RNase H under these conditions. An α-anomeric oligonucleotide hybridizing to the same target in parallel orientation had no specific translation inhibitory activity even when added in a large excess to its mRNA target (molar ratio as high as 1/1000), regardless of whether or not exogeneous RNase H was added (28). Similarly, α-oligonucleotides did not inhibit the translation of vesicular stomatitis virus (VSV) N protein mRNA (28), β-globin mRNA (41) or Trypanosoma brucei mRNAs (C. Verspieren, unpublished results) in reticulocyte lysates under experimental conditions such that β-anomeric oligonucleotides are active as inhibitors.

Experiments performed on intact cells were disappointing in several systems in which a β-oligonucleotide or appropriate derivative was found effective. Leonetti et al. (42) linked 15-mer α- or β-oligonucleotides complementary to VSV N protein mRNA or VSV intergenic consensus region RNA to poly (L-lysine) (PLL) and examined their antiviral activity in intact cells. α-Oligonucleotide PLL conjugates were totally inactive while the β-analog provided a strong antiviral activity at a concentration of 1 µM in the culture medium as documented previously for this type of approach (43). Similarly, α-oligonucleotides did not inhibit the translation of β-globin mRNA when co-injected by micromanipulation into Xenopus oocytes (41) and did not exert any trypanocidal activity at doses up to 120 µM in in vitro cultures of these parasites (C. Verspieren, unpublished results).

In those experiments where the α-oligonucleotides were targeted against AUG or intergenic regions of mRNA, they were unable to arrest the translation. This could be consistent with the inability of RNase H to degrade the RNA strand in the RNA:α-DNA duplex. In contrast, an α-oligonucleotide (15-mer) complementary with a parallel orientation to the 5'-untranslated region of the β-globin mRNA downstream from the cap was recently found to inhibit the protein synthesis in the rabbit reticulocyte lysate (50% inhibition at 3µM oligo concentration). A similar result (50% of inhibition of translation) was found in wheat germ extract at 0.1 and 0.05µM for the α- and corresponding β-oligonucleotide respectively (44). These data suggest that α-oligonucleotides, which are unable to stop the peptide chain elongation, could inhibit the translation initiation.

Preliminary results obtained from HIV-infected MT4 cells treated with an α-oligonucleotide (18-mer) which is complementary to the primer binding site (PBS) of HIV-1 viral RNA have shown a delay in syncytia formation and in reverse transcriptase production at doses between 3 to 100 μM. These unpublished results suggest that the α-anti-PBS oligonucleotide has an antiviral effect at non-toxic doses. In contrast, two 18-mer α-oligonucleotides, the first complementary to a tat gene region and the second to a random sequence, have not shown any inhibitory effect. Other experiments are in progress to confirm these results. However, it is difficult to provide a clearcut explanation for these data since the exact mechanism(s) through which synthetic oligonucleotides exert their repressor activity on gene expression is still not clear in most cases.

CONCLUSION

The α-DNA analog is the first representative chimeric oligonucleotide where a chemical modification has been introduced into the sugar moiety. Such unusual oligomers are now commercially available (45).

It was shown by reversing the configuration of the C-1' atom in deoxynucleotides that α-oligodeoxynucleotides are easily synthesized by known methodologies. They are quite resistant to enzymatic degradation, and they bind strongly to complementary DNA or RNA but with a parallel polarity, thus forming a double helix.

Although α-DNA does not seem to be of use for hybridization arrest of translation when RNase H is involved, available preliminary data indicate that when RNase H activity is not required, α-oligonucleotides could function as effective inhibitors of translation. Furthermore, it has been shown that the corresponding α,β-DNA duplexes could be considered as inhibitors of enzymes which recognize double-stranded nucleic acids specifically or nonspecifically. This behavior could be related to the conformational similarity between a natural double-stranded nucleic acid and an α,β-DNA hybrid which, as was shown, belongs to the B family. We can presume also that the presence of the negative charge on the phosphate backbone, which makes those compounds isoelectronic with DNA, is important for such enzymatic interactions.

The field of applications for α-DNA remains to be fully explored, and these preliminary data on the first sugar modified chimeric nucleic acid open the way to other series which could be of importance for future developments in the field of molecular biology.

Acknowledgments: We express appreciation to all the main scientists who have contributed to development of the α-DNA field namely J.-R. Bertrand, M. Lavignon, C. Malvy, J. Paoletti and C. Paoletti (Institut Gustave Roussy, Villejuif), J.W. Lown (University of Alberta), J. Balzarini, E. DeClercq and R. Pauwels (Leuven University), C. Gagnor, B. Lebleu, M. Lemaitre and J.-P. Leonetti (Université de Montpellier). This research was supported by grants from Association pour la Recherche sur le Cancer and from PNRS and CNRS.

REFERENCES

1 Prusoff, W.H., Lin, T.S. and Zucker, M. (1986) Antiviral Res. 6, 311-328.

2 Miller, P.S., Agris, C.H., Blake, K.R., Murakami, A., Spitz, S.A., Reddy, M.P. and Ts'o, P.O.P (1983) Nucl. Acids Res. 11, 6225-6242.

3 Blake, K.R., Murakami, A., Spitz, S.A., Glave, S.A., Reddy, M.P., Ts'o, P.O.P. and Miller, P.S., (1985) Biochemistry 24, 6139-6145.

4 Smith, C.C., Aurelian, L., Reddy, M.P., Miller, P.S. and Ts'o, P.O.P. (1986) Proc. Nat. Acad. Sci. U.S.A. 83, 2787-2791.

5 Agris, C.H., Blake, K.R., Miller, P.S., Reddy, M.P. and Ts'o, P.O.P. (1986) Biochemistry, 25, 6268-6275.

6 Potter, B.V.L., Romaniuk, P.J. and Eckstein, F. (1983) J. Biol. Chem. 258, 1758-1760.

7 Anderson, D.J., Reischer, R.J., Taylor, A.J. and Wechter, W.J. (1984) Nucleosides and Nucleotides, 3, 499-512.

8 Sequin, U. (1983) Experientia 29, 1059-1062.

9 Sequin, U. (1974) Helv. Chim. Acta 57, 68-81.

10 Morvan, F., Rayner, B., Imbach, J.-L., Chang, D.-K. and Lown, J.W. (1986) Nucl. Acids Res., 14, 5019-5035.

11 Morvan, F., Rayner, B., Imbach, J.-L., Thenet, S., Bertrand, J.-R., Paoletti, J., Malvy, C. and Paoletti, C. (1987) Nucl. Acids Res. 15, 3421-3437.

12 Chassignol, M. and Thuong, N.T. (1987) C.R. Acad. Sci. Paris 305 (Serie II), 1527-1530.

13 Morvan, F., Rayner, B., Leonetti, J.-P. and Imbach, J.-L. (1988) Nucl. Acids Res 16, 833-847.

14 Yamaguchi, T. and Saneyoshi, M. (1984) Chem. Pharm. Bull. 32, 1441-1450.

15 Hubbard, A.J., Jones, A.S. and Walker, R.T. (1984) Nucl. Acids Res. 12, 6827-6837.

16 Ness, R.K. (1968) in Synthetic Procedures in Nucleic Acid Chemistry, Vol. 1 (Zorbach, W.W. and Tipson, R.S., eds.) pp. 183-187, John Wiley and Sons, New York, NY.

17 Robins, M.J. and Robins, R.K. (1969) J. Org. Chem. 34, 2160-2163.

18   Morvan F. (unpublished data).

19   Morvan, F., Rayner, B., Imbach, J.-L., Lee, M., Hartley, J.A., Chang, D.-K. and Lown, J.W. (1987) Nucl. Acids Res. 15, 7027-7044.

20   Morvan, F., Rayner, B., Imbach, J.-L., Chang, D.-K. and Lown, J.W. (1987) Nucl. Acids Res. 15, 4241-4255.

21   Durand, M., Maurizot, J.C., Thuong, N.T. and Hélène C. (1988) Nucl. Acids Res. 16, 5039-5053.

22   Paoletti, J., Bazile, D., Morvan, F., Imbach, J.-L. and Paoletti, C. (1989) Nucl. Acids Res. 17, 2693-2704.

23   Lancelot, G., Guesnet, J.-L., Roig, V. and Thuong, N.T. (1987) Nucl. Acids Res. 15, 7531-7547.

24   Praseuth, D., Chassignol, M., Takasugi, M., LeDoan, T., Thuong, N.T. and Hélène, C. (1987) J. Mol. Biol. 196, 939-942.

25   Sun, J.S., Asseline, U., Rozaud, D., Montenay-Garestier, T., Thuong, N.T. and Hélène, C. (1987) Nucl. Acids Res. 15, 6149-6158.

26   Praseuth, D., Perrouault, L., LeDoan, T., Chassignol, M., Thuong, N.T. and Hélène, C. (1988) Proc. Nat. Acad. Sci. U.S.A. 85, 1349-1353.

27   Thuong, N.T., Asseline, U., Roig, V., Takasugi, M. and Hélène, C. (1987) Proc. Nat. Acad. Sci. U.S.A. 84, 5129-5133.

28   Gagnor, C., Bertrand, J.-R., Thenet, S., Lemaitre, M., Morvan, F., Rayner, B., Malvey, C., Lebleu, B., Imbach, J.-L. and Paoletti, C. (1987) Nucl. Acids Res. 15, 10419-10436.

29   Sun, J.S., François, J.-C., Lavery, R., Saison-Behmoraras, T., Montenay-Garestier, T., Thuong, N.T. and Hélène, C. (1988) Biochemistry 27, 6039-6045.

30   Pattabiraman, N. (1986) Biopolymers, 25, 1603-1606.

31   Van de Sande, J.H., Ramsing, N.B., Germann, M.W., Elhorst, W., Kalisch, B.W., Kitzing, E.V., Pon, R.T., Clegg, R.C. and Jovin, T.M. (1988) Science 241, 551-557.

32   Gagnor, C., Rayner, B., Leonetti, J.P., Imbach, J.-L. and Lebleu, B. (1989) Nucl. Acids Res. 17, 5107-5134.

33   Sequin, U. and Tamm, Ch. (1972) Helv. Chim. Acta 55, 1196-1218.

34   Bloch, E., Lavignon, M., Bertrand, J.-R., Pognan, F., Morvan, F., Malvy, C., Rayner, B., Imbach, J.-L. and Paoletti, C. (1988) Gene 72, 349-360.

35   Lavignon, M., Bertrand, J.-R., Rayner, B., Imbach, J.-L., Malvy, C. and Paoletti, C. (1989) Biochem. Biophys. Res. Commun. (in press)

36   Pauwels, R., Debyzer, Z., Balzarini, J., Baba, M., Demyster, J., Rayner, B., Morvan, F., Imbach, J.-L. and De Clercq, E. (1988) Nucleosides and Nucleotides 8, 995-1000.

37   Cazenave, C., Chevrier, M., Thuong, N.T. and Hélène, C. (1987) Nucl. Acids Res. 15, 10507-10521.

38   Bacon, T.A., Morvan, F., Rayner, B., Imbach, J.-L. and
     Wickstrom, E. (1988) J. Biochem. Biophys. Methods 16, 311-
     317.
39   Wickstrom, E. (1986) J. Biochem. Biophys. Methods 13, 97-
     102.
40   Thenet, S., Morvan, F., Bertrand, J.-R., Gautier, C. and
     Malvy, C. (1988) Biochimie 70, 1729-1732.
41   Cazenave, C., Stein, C.A., Loreau, N., Thuong, N.T.,
     Neckers, L.M., Subasinghe, C., Hélène, C., Cohen, J.S. and
     Toulmé, J.-J. (1989) Nucl. Acids Res. 17, 4255-4273.
42   Leonetti, J.P., Rayner, B., Lemaitre, M., Gagnor, C.,
     Milhaud, P.G., Imbach, J.-L. and Lebleu, B. (1988) Gene 72,
     323-332.
43   Lemaitre, M., Bayard, B. and Lebleu, B. (1987) Proc. Nat.
     Acad. Sci. U.S.A. 84, 648-652.
44   Bertrand, J.-R. Imbach, J.-L., Paoletti, C. and Malvy, C.,
     Biochem. Biophys. Res. Commun (in press).
45   Appligene, route du Rhin, BP 72, 67402 Illkirch, Cedex,
     France.

THE UTILITY OF STREPTOMYCETES AS HOSTS FOR GENE CLONING

Paul K. Tomich and Yoshihiko Yagi

Chemical and Biological Screening
The Upjohn Company
Kalamazoo, MI 49001

INTRODUCTION

As in other biological areas, modern scientific techniques such as recombinant DNA methodology have helped to harvest the fruits of knowledge for those species belonging to the genus <u>Streptomyces</u>. This genus has long provided many useful products such as antibacterial and antineoplasia antibiotics (e.g., 1-4), immunomodulators (5) and a variety of enzymes such as proteases (6,7), restriction enzymes (8,9), amylases (10), xylanases (11), and agarases (12). In addition, <u>Streptomyces</u> ssp. deserve study in their own right because they have an interesting life cycle.

In its natural environment or on solid support media, <u>Streptomyces</u> spores germinate to produce hyphae. The hyphae branch out as substrate mycelia and attach to a solid support medium allowing the colony to grow. After several days of growth, aerial hyphae protrude upwards and undergo several morphological changes to produce and store newly generated spores (e.g., 13). (Generally, streptomycetes do not sporulate in liquid culture, ref. 14.) The shape and type of spore chains differ from species to species as well as the color of aerial and substrate mycelia. These variations are often associated with the growth conditions for a given species (14). This complex, colorful cycle provides another simple model for differentiation.

This chapter will provide an overview of the system available in <u>Streptomyces</u>, providing examples of useful vectors and hosts as well as updating some recently cloned genes. Several examples of biosynthetic pathways for potentially useful compounds will be described and some basic biological problems will be touched upon briefly.

Genetic Engineering, Vol. 12
Edited by J.K. Setlow
Plenum Press, New York, 1990

TRANSFORMATION OF DNA INTO STREPTOMYCETES

Transformation of streptomycetes has become routine since the original observation by Hopwood's group that PEG-mediated transformation of protoplasts was possible (15). The protocol has been improved for the original and still most frequently used hosts, Streptomyces lividans and S. coelicolor A3(2). The molecular biological techniques have been compiled in a manual that has become de rigueur (16).

In essence, streptomycetes are grown in an appropriate medium often in the presence of a low amount of glycine (0 to 2%). After harvesting, protoplasts are made by the addition of lysozyme (usually ca. 1 mg/ml final concentration) in an osmotically stabilized (normally sucrose) buffer. The solution is incubated at 37°C. Protoplast formation can be monitored under a phase contrast microscope. After the protoplasts are collected and washed, they can be transformed with DNA or frozen and stored for extended periods for future use. Transformation entails addition of DNA to protoplasts, followed by the addition of PEG in osmotically stable buffer, trituration, and plating at appropriate dilutions on regeneration plates.

Instead of isolating DNA and transforming protoplasts, protoplast fusion has had some success (e.g., 17,18). Electroporation also has been used to introduce DNA into streptomycete protoplasts as well (19). However, protoplasts still have to be made for electroporation and the only advantage for this procedure is a potentially increased transformation frequency. Regardless, the conditions for growth, protoplasting, regeneration, and if used, electroporation for each species must be determined. This can become a time-consuming process which explains why S. lividans remains the strain most frequently used for research purposes.

Recently, Mazodier et al. have developed conjugative vectors that transfer between E. coli and Streptomyces spp. (20). This vector contains the ColE1 and pIJ101 (24) origins of replication and the transfer origin for plasmid RK2. pPM801 contains the entire pIJ101 plasmid including functions that allow transfer between other streptomycetes whereas pPM803 lacks these. The ability to conjugate E. coli with streptomycetes requires transfer functions supplied in trans from an integrated derivative of RP4 in the E. coli chromosome. Transfer occurred on solid support media but not in liquid culture. Thus one can perform all the constructions in E. coli and transfer them back into Streptomyces. Addition of the lambda cos site would allow more facile generation of gene banks in E. coli. Furthermore, addition of the hft site from actinophage FP43 (see below) would allow for transduction of these constructions from S. lividans to other streptomycete strains rather than the more time-consuming plate conjugations. pIJ101 has recently been shown to transfer between streptomycetes in soil as well as on agar and can mobilize pIJ702 (see below) in the same conditions (21).

VECTORS

Plasmid Vectors

A number of streptomycete plasmids have been discovered. Insertion of suitable markers for selection and detection has generated useful cloning vectors. These markers include a streptomycete tyrosinase gene mel (melanin production, ref. 22) and antibiotic resistance genes (23) such as aph (aminoglycoside phosphotransferase, neomycin resistance), tsr (23S rRNA methylase, thiostrepton resistance) and vph (viomycin phosphotransferase, viomycin resistance). Although some investigators appear to prefer their own plasmids, the pIJ101 (24) derived high copy number plasmid vectors such as pIJ702 (22) and SCP2* derived low copy number plasmid vectors (25) have played a major role in cloning streptomycete DNAs.

I. High copy number vector. The high copy number plasmid pIJ702 (40 to 300 copies per chromosome, ref. 22) was constructed by inserting the mel gene, isolated from S. antibioticus, into a plasmid derived from pIJ101 (24). The inserted mel gene consists of two open reading frames, melC1 and melC2 (26). The latter gene codes for a copper-requiring apotyrosinase. The former encodes or regulates a copper-transfer protein which donates copper to the melC2 gene product to produce a catalytically active tyrosinase (27). Some of the organisms transformed with pIJ702 do not express or secrete tyrosinase (22,28-30) indicating that not all hosts can recognize the mel promoter (29,30).
In addition to mel, pIJ702 contains the thiostrepton resistance gene, tsr. Unique sites such as SphI, BglII and SstI in the mel gene and ClaI, EcoRV and PvuII in the tsr gene are available for cloning that inactivate each gene. Since pIJ702 became widely available, the list of suitable hosts (22,24) for the plasmid (and other pIJ101 related vectors) has expanded to include such related species as S. veolaceoniger (31), S. hygroscopicus (32), S. avermitilis (33), S griseofuscus (34), S. granaticolor (35), S. clavuligerus (36), Amycolatopsis orientalis (Nocardia orientalis) (37) and Micromonospora purpurea (28).
Because of its high copy number, gene dosage effects have been observed (38-42). Failure to observe such an effect probably indicates that the cloned DNA contains only a portion of the gene of interest as reported by Seno et al. (43), Murakami et al. (32) and Butler et al. (44). In these reported examples, successful complementation resulted from homologous recombination between the host chromosome and the cloned gene.

    Because pIJ702 exists as a high copy number plasmid, it is
not a suitable vector for cloning large molecular size fragments.
The largest fragment cloned into pIJ702 reported in the
literature is the 19 kb fragment of S. cacaoi which carries a
gene for β-lactamase (40). Iwasaki et al. reported cloning of a
18.1 kb DNA into pIJ702 from Streptomyces sp. no. 36a (39).
However, DNA fragments with a molecular size below 10 kb have
been cloned successfully and maintained stably.
    Deng et al. recently found a cis-acting site for "strong
incompatibility" or sti. It consists of about 200 bases on
pIJ101 and its derivatives (45). The sti locus, which appears to
be the initiation site for second DNA strand replication,
controls coexistence of the pIJ101 plasmids. Those plasmids with
sti in the same orientation (either the orientation expressing,
Sti+, or the nonexpressing reverse orientation, Sti-) or without
sti can coexist, while Sti+ and Sti- plasmids cannot. Thus,
pIJ702 and pIJ486 (described below), which are both Sti-, can
coexist. "Strong incompatibility" of pIJ101 and its derivatives
must be kept in mind when one wishes to introduce genes cloned
onto different pIJ101 derivatives into the same host.

    II. Low Copy Number Vectors. Compared to high copy number
plasmids, potential advantages of low copy number plasmids are:
1) an absence of possible gene dosage effects, and 2) an ability
to maintain stably large DNA fragments. The two plasmids SLP1
and SCP2* have been the source of many widely used low copy
number vectors.

(A) SLP1 Derivatives

    The SLP1 plasmids have a molecular size ranging from 10 to
14.5 kb. They were originally isolated from the recipient
strains of S. lividans when mated with the closely related
streptomycete species S. coelicolor, which served as the donor
(44). Southern blot hybridization revealed the presence of the
SLP1 sequences on the chromosome of S. coelicolor. It has been
proposed that during mating the SLP1 sequences on the chromosome
excise at different points to generate a series of autonomously
replicating SLP1 plasmids in S. lividans. Copy numbers for the
SLP1 plasmids are estimated to be ca. 3 to 5 (33,46).
    pIJ41 (15.8 kb) and pIJ61 (15.7 kb) were constructed from
SLP1.2 (14.5 kb) by Thompson et al. (47). pIJ41 carries two
resistance markers, aph and tsr. pIJ61 was derived from pIJ41 by
in vitro deletion of 0.09 kb of DNA, which eliminated one of the
two PstI sites in pIJ41 and left a unique PstI site in the aph
gene. This site as well as the BamHI site in aph are available
for insertional inactivation for this construct (47). The host
range for the SLP1 derivatives is rather limited (33,47).

(B)SCP2* Derivatives

SCP2* is a fertility factor found in S. coelicolor with a molecular size of 31.4 kb and exists in one to two copies per chromosome (48). A number of useful cloning vectors derived from SCP2* have been described in detail by Lydiate et al. (49). These vectors were assembled with different combinations of the genetic determinants for thiostrepton, viomycin and hygromycin resistance and tyrosinase. In some cases (such as pIJ940 and pIJ941), two markers are present, one for selection and the other for insertional inactivation. However, vectors with only one selectable marker have been used to clone successfully because of their stability and because of their ability to maintain inserts with large fragments of cloned DNA. For example, the entire biosynthetic pathway for the polyketide antibiotic actinorhodin, encoded on 35 kb of DNA, has been cloned onto an SCP2* derived plasmid (50).

SCP2* has a broad host range. Derivatives of SCP2* have been transformed stably into many species such as S. clavuligerus, S. glaucescens, S. peucetius and S. rimosus in addition to the standard hosts S. lividans and S. coelicolor (49). A derivative of pIJ943 was able to transform S. avermitilis (Yagi, unpublished). However, the pIJ922 derivative was unstably maintained after transformation into Streptomyces erythreus (51).

III. Promoter Probe Plasmids. Among the promoter probe plasmids for streptomycetes (52-57), pIJ486/7 (54,58) have been used extensively. Since they contain the pIJ101 replicon (24), pIJ486/7 exist as high copy number plasmids with a wide host range. A polylinker is attached to the 5' end of the promoterless kanamycin resistance gene(aph), derived from the transposon Tn5. Cloning of a DNA fragment with promoter sequences into the polylinker region activates the kanamycin resistance gene. Transcriptional read-through from plasmid promoters is prevented, or at least greatly decreased, by the presence of the terminator isolated from E. coli phage fd. The orientation of the polylinker in pIJ487 is opposite to that in pIJ486.

It has been reported that insertion of Streptomyces DNA larger than 2 kb results in kanamycin resistance (15 µg/ml in S. lividans) in more than 50% of the transformed clones (58). pIJ486/7 have been employed to detect active promoters on DNA fragments isolated from streptomycetes and their related organisms including S. coelicolor (59,60), S. glaucescens (61), Micromonospora echinospora (62), S. lividans (63), S. rimosus (64), S. albus G (65), S. alboniger (66) and the actinophage φC31 (P.K. Tomich, unpublished results). pIJ486/7 plasmids have also been employed as general purpose cloning vectors (67-70).

Another promoter probe derived from pIJ486, pRS1105, contains the luciferase genes luxA and luxB which originated from Vibrio harveyi (71). Besides indicating when a promoter is transcribed in the cell life cycle (e.g., temporal regulation based on the time of synthesis regulated, by different sigma factors), this promoter probe can also indicate where in the substrate mycelia, aerial hyphae, or encapsidating spore a gene product would reside (i.e., spatial expression) (71,72). In addition, after initial expression, the luciferase activity decreases gradually with time probably because the enzyme is somewhat unstable in streptomycetes. Thus it might also be used to study promoter shutoff (71).

Similarities between promoters for gene products involved in primary or secondary metabolism and differences between primary and secondary metabolite promoters have been covered in detail elsewhere (73-76). Obviously, promoters that can be regulated would have tremendous utility in expression of heterologous gene products in streptomycetes. One example is a promoter repressed by phosphate which can be activated by lowering external phosphate concentrations (76). The amount of phosphate needed depends on the streptomycete host used. Another example is a thiostrepton inducible promoter characterized from proteins induced after addition of thiostrepton (77). An interesting feature of the sequence encompassing this promoter is the potential formation of an imperfect stem-loop structure that could affect regulation of transcription, possibly via thiostrepton (63).

IV. **Plasmids Precede Phages?** Some plasmids such as pIJ408 (77), SLP1 (78), pSE101 (79), pMEA100 (80) and pSAM2 (81) can exist integrated in or autonomously from the streptomycete chromosome. Integration into the chromosome occurs at a specific chromosome attachment site (attB). These plasmids have attachment (attP) sites that are similar in size, ca. 50 bp (cited in 82). Besides SLP1, pSAM2 has been studied the most intensively (81,83,84). The host range of this plasmid is quite extensive. Interestingly, its attP site hybridized to a variety of Streptomyces species (81,83). An open reading frame upstream of the host attB site is followed by a conserved inverted repeat structure. This arrangement exists for a variety of strains. However, the sequence for the inverted repeat is not conserved although its structural organization and relative location are conserved (83).

The 58 bp attP site for pSAM2 also exhibits extensive homology not only with the attB site of its host S. ambofaciens but also with the other integrating streptomycete plasmids and with the lambdoid coliphages (83). Furthermore, the integration event is thought to be mediated by an integrase encoded by the plasmid and located near the attP site, as for other lambdoid phages (84). The possibility of phage evolution by assembly from modules existing in bacterial hosts has been proposed (85). The question of initial origin also occurs. Streptomyces, a related genus, or a more immediate ancestor may have been a reservoir of these phage modules just as Streptomyces may also have been an original source for the antibiotic resistance genes.

A practical application for these plasmids also exists. By placing the functions needed for plasmid integration onto a plasmid unable to replicate in streptomycetes, one has a way of introducing genes of interest into a variety of Streptomyces spp. in a stable, inheritable manner. This alleviates the need for constant selective pressure such as the use of antibiotics. The construction of such a plasmid has been accomplished (82).

Actinophage Vectors

Actinophages have been known for many years (86) and are still being actively pursued for specific hosts (87). Several bacteriophage vectors have found utility in Streptomyces spp. The most characterized actinophage, ØC31, has been used to clone random (88) and specific DNA fragments (89), to determine genetic organization of operons (43,90), to generate deletions (91) and to complement mutations of (92) or transfer mutations into (89) the chromosome. Promoterless reporter genes for ß-galactosidase (93) or viomycin (94) have been incorporated into this vector for use as a promoter probe. However, since the initial vectors created by Chater's group (e.g., 59), very few recent modifications have been made with this vector to expand its utility. (Thus, plasmid vectors have been used primarily in most recent cloning experiments, see above.) Phage vectors that could accept more than 10 kb of cloned DNA or an in vitro packaging system would have tremendous utility.

One recent ØC31 variant, KC591, had the minicircle transposon (see below) inserted into the ØC31 derivative KC515 (95). The salient features of KC515 are a phage deficient in its attachment site (attP), such that phage integration must occur by another mechanism, an active repressor gene for stable lysogeny and the cloned genes tsr and vph for selection. This construct, KC591, integrates via the transposon into the minicircle preferred site, even when strong selection for selenate or 2-deoxyglucose resistance was imposed. Thus this system cannot be used for insertional mutagenesis. Lysogens of this construct yield phage spontaneously in an imprecise excision event (95).

An in vivo packaging system utilizing the cos site for actinophage R4 on the plasmid pIJ702 is available (96). Actinophage R4 packages DNA at specific sites as does the coliphage lambda. Another in vivo packaging system exists for the actinophage FP43 originally isolated from Streptomyces griseofuscus (97). FP43 packages by a head-full mechanism (98) similar to the coliphage P1. These workers cloned the site needed for packaging (called hft, high frequency transduction)

onto pIJ702 and demonstrated that this vector could be transduced
into a variety of Streptomyces spp. as well as species from
related genera (98). They also demonstrated that this vector
could transduce into strains containing restriction endonucleases
at a much greater frequency than the corresponding pIJ702
parental plasmid (99). Since the vector is much smaller than the
actinophage, they postulate that multiple copies of the vector
are encapsulated into the phage head, allowing for the increased
recovery of intact plasmid after transduction. This fact
coupled with the wide host range, the ability to store gene banks
in relatively small volumes, greater facility of doing plaque
hybridizations and especially the greater ease of introducing DNA
into a recipient host, are advantages phage vectors have over
plasmid vectors.

SPECIFIC EXAMPLES OF CLONED STREPTOMYCES DNA

Since streptomycetes are unique organisms which produce a
plethora of therapeutic agents, especially antibiotics, it is
natural that considerable effort has been directed towards
cloning the genes for different antibiotic biosynthetic pathways.
Table 1 lists most of the cloned genes not found in previous
review articles (74,100,140). The complete set of genes for the
biosynthetic pathways of actinorhodin (50), tetracenomycin (38)
and oxytetracycline (101) were cloned as a cluster; that is, the
genes for these antibiotics exist contiguously on the
streptomycete chromosome. Closely linked genes have been
reported for the antibiotic pathways for puromycin (102),
streptomycin (103,104), erythromycin (105) and bialaphos
(32,110).

I. Glycerol Utilization Operon of S. coelicolor A3(2)

Glycerol kinase and glycerol phosphate dehydrogenase, which
are involved in glycerol utilization in S. coelicolor A3(2), are
coordinately induced and repressed by glycerol and glucose,
respectively (106). The 2.84 kb fragment of S. coelicolor A3(2)
cloned into pIJ702 (designated pIJ2116, ref. 43) restored the
normal phenotype to two known classes of mutants (gylA, glycerol
kinase, gylB, glycerol-3-phosphate dehydrogenase). Subsequently,
employing pIJ2116 as a probe and pSAE1 as a cosmid vector, a 40
kb fragment of S. coelicolor A3(2) was identified that contained
the complete gyl operon including the promoter (107). S1 mapping
with a large number of overlapping restriction fragments
confirmed the initial finding (43) that the gylA and gylB genes
are transcribed polycistronically to generate mRNA molecules of
4.3 kb (107). In addition, the gyl operon produced a 5.4 kb mRNA
which is a 3' extension product of the 4.3 kb mRNA. The gene
designated gylX, which specifies the 3' extension, is thought to

Table 1
Genes Cloned from _Streptomyces_ spp.

Gene/Gene Product	Source (Vector)	Host	Reference
Aminocyclitol acetyltransferase	S. rimosus (pIJ702)	S. lividans	(64)
Photolyase	S. griseus (pUC18)	E. coli	(141)
rRNA	S. ambofaciens (pIJ61)	S. lividans	(142,143)
Oxytetracycline resistance and biosynthetic genes	S. rimosus (pPFZ12)	S. rimosus	(44)
α-Amylase	S. griseus (pIJ699)	S. lividans	(145)
Alkaline phosphatase	S. griseus (pIJ699)	S. lividans	(144)
eryG(O-methyl-transferase?)	S. polyspora-erythraea (pIJ702,pRS1105)	S. lividans, S. polyspora-erythraea	(145)
aphE(Streptomycin-3"-phosphotransferase)	S. griseus (pIJ702)	S. lividans	(116)
tipA(thiostrepton-induced protein)	S. lividans (EMBL4, pIJ486)	E. coli S. lividans	(77)
afsR, a regulatory gene for pigment production and development	S. lividans (pJAS01)	S. lividans	(134)
Oxytetracycline biosynthetic genes	S. rimosus (pIJ916)	S. lividans	(101)
Isopenicillin N synthase	S. jumonjinensis S. lipmanii (D69)	E. coli	(112)
Deacetoxyceph-alosporin C synthase	S. clavuligerus (pKC462a)	E. coli	(114)

code for a facilitator protein for glycerol uptake. The third mRNA species of 0.9 kb, generated upstream of the start sites for both the 5.4 and 4.3 kb transcripts, was always found in glycerol-grown cultures and was repressed weakly by glucose. The 0.9 kb gene product, therefore, appears to be a positively acting gyl regulatory (gylR) protein (107,108).

The glycerol utilization operon contains 3 promoter regions: gylRp for the gene gylR and gylp1, p2 for the genes gylABX. The transcription initiation site of gylR is separated from that of gylp1 by approximately 1,000 bp. The open reading frame within gylR would code for a polypeptide of 254 amino acids, which contains sequences similar to the DNA binding region of the transcriptional activator AsnC (108). In this region, 97% of the codons contain a G or C in the third base position, reflecting a highly biased codon usage as previously found in other streptomycete sequences (109). The promoters of gylR and gylABX contain an operator-like element consisting of 17 bp palindromic sequences near the transcription start-sites. The gylABX promoter region also contains a different sequence of 16 bp with dyad symmetry which is situated 48 bp upstream from gylp1 (108). This latter sequence shows similarity to other cis-acting sites in several bacteria that bind regulatory proteins (108).

II.   Bialaphos Biosynthetic Gene Cluster of S. hygroscopicus

Bialaphos, produced by S. hygroscopicus, is a tripeptide antibiotic herbicide synthesized in 13 well-defined steps (32). It consists of 2 L-alanine molecules and phosphinothricin (an analog of L-glutamic acid). The pathway for the biosynthesis of bialaphos has been well studied. This work mainly involved examination of chemical intermediates produced and converted by non-producing mutants. Mutant strains are available for seven of the thirteen genes involved in bialaphos synthesis (32). Complementation studies were also performed. DNA fragments were cloned onto pIJ702 and transformants were selected for bialaphos production (32). One clone, carrying an approximately 3 kb insert, complemented the mutants for steps 1,3 and 4 while another with a 9.7 kb insert complemented mutants defective in steps 5,6,12 and 13. The latter clone also contained a bialaphos resistance gene (bar) which is involved in the tenth biosynthetic step. Subsequently, a cosmid clone was obtained which complemented all the known mutants for these seven bialaphos genes (bap). They were localized within a 16 kb DNA fragment.

Recent work (110) related to bap genes indicates that while they are located close to each other, they are not contiguous. The mutants defective in the alanylation step involved in bialaphos biosynthesis have been isolated, and five different genes have been identified that are involved in this process. The previously mapped eight genes (32,111) and four alanylation genes (110) are localized within a 35 kb long DNA segment; the fifth alanylation gene was not found in this region.

Regulation of the clustered bap genes has yet to be elucidated in detail; however, at least one apparent regulatory gene (designated brpA) has been identified (111). This gene is located downstream from the bap gene for biosynthetic step 13. BrpA encodes a molecule which initiates mRNA transcription for steps 5,10 and 13.

III. Genes from Streptomycetes Related to Other Genera

Interestingly, several of the genes listed in Table 1 have demonstrated homology to genes with identical or similar function between different streptomycete species and other genera as well as eukaryotic species. For example, isopenicillin N synthetase (IPNS) is involved in the biosynthesis of β-lactam antibiotics. IPNS exhibits amino acid homology of greater than 70% between three Streptomyces spp. and more than 60% to fungal IPNS enzymes (112,113). Southern hybridization showed that several non-β-lactam producing species also carried at least part of the IPNS gene (112). Furthermore, the enzyme deacetoxycephalosporin C synthase(DAOCS) involved in ring expansion in the synthesis of cephalosporin β-lactams has been cloned from S. clavuligerus. It has ca. 57% amino acid homology and 67% DNA homology with a fungal (Cephalosporium acremonium, ref. 114) enzyme having similar activity. The possibility exists that transfer of a (precursor) IPNS to a eukaryote occurred early in evolution (114). Credence for this possibility comes from recent experiments demonstrating transfer of genetic information between E. coli and Saccharomyces cerevisiae (115).

As mentioned previously, the thesis that antibiotic resistance genes derived from streptomycetes has been proposed. Heinzel et al. go one step further. They worked on the S. griseus streptomycin-3"-phosphotransferase (aphE, see Table 1). This gene shows extensive homology to the S. fradiae aphA gene, encoding an aminoglycoside 3'-phosphotransferase, both in the coding region and in the upstream and downstream regulatory sequences (116). They also compared the sequences of phosphotransferases to protein kinases and find four related domains. They propose that all these enzymes had a common ancestral gene that became disseminated over time.

EXPRESSION AND SECRETION OF FOREIGN GENES OF NONSTREPTOMYCETE ORIGIN IN STREPTOMYCES

Streptomycetes produce a wide variety of industrially important compounds which accumulate in the growth media (117). Therefore, they are attractive microorganisms as a possible host for expression of cloned genes and for secretion of gene

products. In order to develop a useful expression system, a
variety of promoters must be examined for efficient transcription
in streptomycetes. An early report by Bibb and Cohen indicated
that promoters of E. coli, S. marcescens and B. licheniformis
function in S. lividans expressing the E. coli chloramphenicol
acetyltransferase gene (52). Since then, other E. coli promoters
which could direct the synthesis of E. coli hygromycin
phosphotransferase (118), an E. coli β-lactamase (119) and human
interleukin-2 (120) were found to be functional in S. lividans.
Other nonstreptomycete promoters functional in S. lividans
include those identified in the Mycobacterium bovis BCG
chromosome (57) and the staphylokinase gene from Staphylococcus
aureus phage 42D (69).

Although not secreted, the genes for bovine growth hormone
(121) and human interferon-α2(hIFN-α2, 122) have been expressed
in S. lividans employing pIJ702 as a vector. The 600 bp PstI-
NcoI fragment directly upstream from the aph coding region
isolated from S. fradiae (123) was a source of the promoter in
both cases. Since a reasonably satisfactory production of
biologically active hIFN-α2 was obtained in S. lividans, use of
the ribosome binding site for the E. coli lipoprotein gene was
suggested as a possible means of increasing the translational
efficiency for other genes (122).

Secretion of nonstreptomycete gene products by S. lividans
has been reported for hIFN-α1 (69) and interleukin-1β (124). A
sak-(staphylokinase) expression/secretion unit was constructed.
It consists of the sak promoter and ribosome binding site and a
DNA sequence encoding 27 amino acid residues of the sak signal
peptide plus the first 4 amino acids of staphylokinase (69,125).
This assemblage was fused to a sequence for a mature hIFN-α1 gene
(1.1 kb). The entire construct was then cloned into the
polylinker region of pIJ487 at the EcoRI site and transformed
into S. lividans. The hIFN-α1 gene was transcribed from the sak
promoter, and more than 90% of the hIFN-α1 was found in the
culture media where it was stable for more than 60 hr (69).

To express and secrete interleukin-1β (124), the coding
sequence of interleukin-1β was fused to the Streptomyces β-
galactosidase gene consisting of the sequences for the promoter,
the signal peptide and the first eight amino acids of the mature
protein. Localization of the enzyme in the culture supernatant
was dependent on the β-galactosidase signal sequence as
determined by oligonucleotide directed mutagenesis.

TRANSPOSONS IN STREPTOMYCES

Since transposons have the ability to move from one location
to another, they are a powerful tool for studying the genetic
organization of both prokaryotic and eukaryotic cells (126). A
2.6 kb DNA sequence of S. coelicolor A3(2), shown by Lydiate et

al. (127), exists either as an autonomously replicating circular molecule (designated minicircle) or an integrated molecule in the chromosome. Two locations where integration occurred were discovered on the chromosome. These sites were also found to be the same among different strains of S. coelicolor indicating specificity of integration by this DNA sequence.

The transposable element IS110 was first identified as a 1.6 kb DNA which transposed from the chromosome of S. coelicolor (strain J1501) to the temperate phage øC31 (128). While the number and the location of IS110 vary among the strains of S. coelicolor, in strain J1501 the chromosome appears to contain three copies, one of which is located in a region where DNA rearrangements frequently occur. IS110 may have a site preference for its integration since transposition of IS110 on øC31 always took place in the same location. Southern hybridization revealed the presence of a DNA sequence homologous to IS110 in several other Streptomyces spp.

The transposon Tn4556 (6.8 kb) was discovered in S. fradiae (strain UC® 8592) which produces neomycin (129). Transposition was first noted to occur from the chromosome to a number of different locations on a 23.5 kb plasmid derived from the S. fradiae prophage pUC™ 13 (130). Southern hybridization demonstrated the presence of one copy of Tn4556 in the chromosome of the original host. The vph gene was inserted into Tn4556 giving rise to Tn4560. Tn4560 was used to show transposition from the plasmid into different locations of the chromosomes of S. lividans (129), S. lincolnensis (131) and S. avermitilis (Y. Yagi and L.A. Ruble, 7th Internat. Symp. Biol. of Actinomycetes, abstr. no. P5-22, 1988). In S. lividans, transposition of Tn4560 has been shown to occur from the chromosome to plasmids (131, Y. Yagi and L.A. Ruble, unpublished) as well as from plasmid to plasmid (131).

Sequence analysis of both ends of Tn4556 revealed the presence of 38 bp terminal inverted repeats with a single mismatched pair. These repeats share 70% sequence homology with the ends of Tn3 (132). Direct repeat sequences of 5 bp are generated as a result of Tn4556 transposition (132). The complete nucleotide sequence of Tn4556 (133) shows the presence of nine possible open reading frames of which four are predicted actually to encode proteins. Although more detailed studies remain to be performed, open reading frame 1 (2,676 bp, 892 amino acids) may code for a transposase with approximately 60% amino acid homology to that of Tn3 (131).

CONCLUDING REMARKS

Streptomyces have interested commercial concerns for many years as well as a limited number of academic institutions. The interest in these organisms has increased markedly over recent

years. With this in mind, this chapter attempted to introduce
the reader to the vast accomplishments and interesting problems
that remain to be solved for Streptomyces. Rather than give
excruciating detail on specific strains, a general overview of
how to transform these organisms was presented with adequate
references for those interested in pursuing work in this area.
The "tools" continually improve and these hosts may soon find
application in the production of heterologous proteins as well as
modification of secondary metabolites for which streptomycetes
are so well known.

The topics of regulation and regulatory elements (secretion
signals, promoters and sigma factor, transcription terminators,
etc.) for Streptomyces were not covered (see 73-75,109). Also,
the topic of streptomycete growth and differentiation, the
possible cross regulation between secondary metabolism (i.e.,
antibiotics) and differentiation through small molecules was also
deferred (e.g., see 134-136).

As in any system, caveats always exist. With streptomycetes
one caveat arises from the extensive rearrangements that occur in
this genus. Amplifications and extensive deletions are common
occurrences, especially in DNA not required for growth or primary
metabolism (see 137-139 for examples). In fact, deletions
exceeding 800 kb, ca. 15 to 18% of the entire chromosome, have
been reported for S. glaucescens (138). Nevertheless,
Streptomyces spp. should prove useful not only as hosts with
which to clone and to express genes of interest, but they may
also prove valuable in helping to bridge the evolutionary gap
between prokaryotes and eukaryotes.

REFERENCES

1 Waksman, S.A. (1959) Proc. Nat. Acad. Sci. U.S.A. 45, 1043-
   1046.
2 Pittenger, R.C. and Brigham, R.B. (1956) Antibiot.
   Chemother. 6, 642-647.
3 Porter, J.N., Hewitt, R.I., Hesseltine, C.W., Krupka, G.,
   Lowery, J.A., Wallace, W.S., Bohonos, N. and Williams, J.H.
   (1952) Antibiot. Chemother. 2, 409-410.
4 Ohkuma, H., Sakai, F., Nishiyama, Y., Ohbayashi, M.,
   Imanishi, H., Konishi, M., Miyaki, T., Koshiyama, H. and
   Kawaguchi, H. (1980) J. Antibiot. 33, 1087-1097.
5 Kino, T., Hatanaka, H., Hashimoto, M., Nishiyama, M., Goto,
   T., Okuhara, M., Kohsaka, M., Aoki, H. and Imanaka, H.
   (1987) J. Antibiot. 40, 1249-1255.
6 Henderson, G., Krygsman, P., Liu, C.J., Davey, C.C. and
   Malek, L.T. (1987) J. Bacteriol. 169, 3778-3784.
7 Laluce, C. and Molinari, R. (1977) Biotechnol. Bioeng. 19,
   1863-1884.

8   Fuchs, L.Y., Covarrubias, L. and Escalante, L. (1980) Gene 25, 39-46.

9   Goff, S.P. and Rambach, A. (1978) Gene 13, 347-352.

10 Long, C.M., Virolle, M.-J., Chang, S.-Y., Chang, S. and Bibb, M.J. (1987) Mol. Gen. Genet. 203, 79-88.

11 Mondou, F., Shareck, F., Morosoli, R. and Kluepfel, D. (1986) Gene 49, 323-329.

12 Bibb, M.J., Jones, G.H., Joseph, R., Buttner, M.J. and Ward, J.M. (1987) J. Gen. Microbiol. 133, 2089-2096.

13 Kutzner, H.J. (1981) in The Prokaryotes, Vol. 2 (Starr, M.P., Stolp, H., Truper, H.G., Balows, A. and Schlegel, H.G., eds.) pp. 2028-2090, Springer Verlag, NY.

14 Pridham, T.G. (1965) Appl. Microbiol. 13, 43-61.

15 Bibb, M.J., Ward, J.M. and Hopwood, D.A. (1978) Nature (London) 274, 398-400.

16 Hopwood, D.A., Bibb, M.J., Chater, K.F., Kieser, T., Bruton, C.J., Kieser, H.M., Lydiate, D.J., Smith, C.P., Ward, J.M. and Schrempf, H. (1985) Genetic Manipulation of Streptomyces: A laboratory manual, John Innes Foundation, Norwich, United Kingdom.

17 Prakash, R.K. and Cummings, B. (1988) Plant Mol. Biol. 10, 281-289.

18 Okamura, T., Nagata, S., Misono, H. and Nagasaki, S. (1988) Agric. Biol. Chem. 52, 1433-1438.

19 MacNeil, D.J. (1987) FEMS Microbiol. Lett. 42, 239-244.

20 Mazodier, P., Petter, R. and Thompson, C. (1989) J. Bacteriol. 171, 3583-3585.

21 Rafi, F. and Crawford, D.L. (1988) Appl. Environ. Microbiol. 54, 1334-1340.

22 Katz, E., Thompson, C.J. and Hopwood, D.A. (1983) J. Gen. Microbiol. 129, 2703-2714.

23 Thompson, C.J., Kieser, T., Ward, J.M. and Hopwood, D.A. (1982) J. Bacteriol. 151, 668-677.

24 Kieser, T., Hopwood, D.A., Wright, H.M. and Thompson, C.J. (1982) Mol. Gen. Genet. 185, 223-238.

25 Lydiate, D.J., Malpartida, F. and Hopwood, D.A. (1985) Gene 35, 223-235.

26 Berman, V., Filpula, D., Herber, W., Bibb, M. and Katz, E. (1985) Gene 37, 101-110.

27 Lee, Y.-H. W., Chen, B.-F., Wu, S.-Y., Leu, W.-M., Lin, J.-J., Chen, C.-W. and Lo, S.-J. (1988) Gene 65, 71-81.

28 Kelemen, G.H., Financsek, I. and Jarai, M. (1989) J. Antibiot. 42, 325-328.

29 Manome, T. and Hoshino, E. (1987) J. Antibiot. 40, 1440-1447.

30 Shinkawa, H., Sugiyama, M. and Nimi, O. (1988) in Biology of Actinomycetes '88 (Okami, Y., Beppu, T. and Ogawara, H., eds.), pp. 353-358, Japan Scientific Societies Press, Tokyo.

31 Marcel, T., Drocourt, D. and Tiraby, G. (1987) Mol. Gen. Genet. 208, 121-126.

32 Murakami, T., Anzai, H., Imai, S., Satoh, A., Nagaoka, K.
    and Thompson, C.J. (1986) Mol. Gen. Genet. 205, 42-50.
33 MacNeil, D.J. and Klapko, L.M. (1987) J. Ind. Microbiol. 2,
    209-218.
34 Epp, J.K., Burgett, S.G. and Schoner, B.G. (1987) Gene 53,
    73-83.
35 Petrícek, M., Smerckova, I. and Tichy, P. (1985) Folia
    Microbiol. 30, 474-478.
36 Garcia-Dominguez, M., Martin, J.F., Mahro, B., Demain, A.L.
    and Liras, P. (1987) Appl. Env. Microbiol. 53, 1376-1381.
37 Matsushima, P., McHenney, M.A. and Baltz, R. (1987) J.
    Bacteriol. 169, 2298-2300.
38 Motamedi, H. and Hutchinson, C.R. (1987) Proc. Nat. Acad.
    Sci. U.S.A. 84, 4445-4449.
39 Iwasaki, A., Kishida, H. and Okanishi, M. (1986) J.
    Antibiot. 39, 985-993.
40 Lenzini, M.V., Nojima, S., Dusart, J., Ogawara, H.,
    Dehottay, P., Frere, J.-M. and Ghuysen, J.-M. (1987) J. Gen.
    Microbiol. 133, 2915-2920.
41 Kendall, K. and Cullum, J. (1984) J. Bacteriol. 162, 406-
    412.
42 Dohottay, P., Dusart, J., Duez, C., Lenzini, M.V., Martial,
    J.A., Frere, J.-M., Ghuysen, J.-M. and Kieser, T. (1986)
    Gene 42, 31-36.
43 Seno, E.T., Bruton, C.J. and Chater, K.F. (1984) Mol. Gen.
    Genet. 193, 119-128.
44 Butler, M.J., Friend, E.J., Hunter, I.S., Kaczmarek, F.S.,
    Sugden, D.A. and Warren, M. (1989) Mol. Gen. Genet. 215,
    231-238.
45 Deng, Z., Kieser, T. and Hopwood, D.A. (1988) Mol. Gen.
    Genet. 214, 286-294.
46 Omer, C.A. and Cohen, S.N. (1984) Mol. Gen. Genet. 196, 429-
    438.
47 Thompson, C.J., Kieser, T., Ward, J.M. and Hopwood, D.A.
    (1982) Gene 20, 51-62.
48 Bibb, M.J. and Hopwood, D.A. (1981) J. Gen. Microbiol. 126,
    427-442.
49 Lydiate, D.J., Malpartida, F. and Hopwood, D.A. (1985) Gene
    35, 223-235.
50 Malpartida, F. and Hopwood, D.A. (1984) Nature 309, 462-464.
51 Yamamoto, H., Maurer, K.H. and Hutchinson, C.R. (1986) J.
    Antibiot. 39, 1304-1313.
52 Bibb, M.J. and Cohen, S.N. (1982) Mol. Gen. Genet. 187, 265-
    277.
53 Horinouchi, S. and Beppu, T. (1985) J. Bacteriol. 162, 406-
    412.
54 Ward, J.M., Janssen, G.R., Kieser, T., Bibb, M.J., Buttner,
    M.J. and Bibb, M.J. (1986) Mol. Gen. Genet. 203, 468-478.
55 Forsman, M. and Jaurin, B. (1987) Mol. Gen. Genet. 210, 23-
    32.

56 Feitelson, J. (1988) Gene 66, 159-162.
57 Kieser, T., Moss, M.T., Dale, J.W. and Hopwood, D.A. (1986) J. Bacteriol. 168, 72-80.
58 Hopwood, D.A., Bibb, M.J., Chater, K.F. and Kieser, T. (1987) in Methods in Enzymology (Wu, R., ed.) 153, pp. 116-166, Academic Press, San Diego, CA.
59 Hallam, S.E., Malpartida, F. and Hopwood, D.A. (1988) Gene 74, 305-320.
60 Buttner, M.J., Fearnley, I.M. and Bibb, M.J. (1987) Mol. Gen. Genet. 209, 101-109.
61 Vögtli, M. and Hütter, R. (1987) Mol. Gen. Genet. 208, 195-203.
62 Baum, E.Z., Love, S.F. and Rothstein, D.M. (1988) J. Bacteriol. 170, 71-77.
63 Murakami, T., Holt, T.G. and Thompson, C.J. (1989) J. Bacteriol. 171, 1459-1466.
64 López-Cabrera, M., Pérez-Gonzalez, J.A., Heinzel, P., Piepersberg, W. and Jímenez, A. (1989) J. Bacteriol. 171, 321-328.
65 Rodicio, M.R. and Chater, K.F. (1988) Mol. Gen. Genet. 213, 346-353.
66 Lacalle, R.A., Pulido, D., Vara, F., Zalacain, M. and Jiménez, A. (1989) Gene 79, 375-380.
67 Lydiate, D.J., Mendez, C., Kieser, H.M. and Hopwood, D.A. (1988) Mol. Gen. Genet. 211, 415-423.
68 VanPee, K.-H. (1988) J. Bacteriol. 170, 5890-5894.
69 Noack, D., Geuther, R., Tonew, M., Breitling, R. and Behnke, D. (1988) Gene 68, 53-62.
70 Thiara, A.S. and Cundliffe, E. (1988) EMBO J. 7, 2255-2259.
71 Schauer, A., Ranes, M., Santamaria, R., Guijarro, J., Lawlor, E., Mendez, C., Chater, K. and Losick, R. (1988) Science 240, 768-772.
72 Schauer, A. (1988) Trends in Biotech. 6, 23-27.
73 Tomich, P.K. (1988) Antimicrob. Agents Chemother. 32, 1472-1476.
74 Hutchinson, C.R. (1987) Appl. Biochem. Biotechnol. 16, 169-190.
75 Janssen, G.R., Bibb, M.J., Smith, C.P., Ward, J.M., Kieser, T. and Bibb, M. J. (1985) in Microbiology-1985, pp. 392-396, American Soc. Microbiol., Washington, D.C.
76 Rebollo, A., Gil, J.A., Liras, P., Austurias, J.A. and Martìn, J.F. (1989) Gene 79, 47-58.
77 Sosio, M., Madón, J. and Hütter, R. (1989) Mol. Gen. Genet. 218, 169-176.
78 Omer, C.A. and Cohen, S.N. (1986) J. Bacteriol. 166, 999-1006.
79 Brown, D.P., Chiang, S.-J.D., Tuan, J.S. and Katz, L. (1980) J. Bacteriol. 170, 2287-2295.
80 Moretti, P., Hintermann, G. and Hütter, R. (1985) Plasmid 14, 126-133.

81 Kuhstoss, S., Richardson, M.A. and Rao, R.N. (1989) J. Bacteriol. 171, 16-23.

82 Chater, K.F., Henderson, D.J., Bibb, M.J. and Hopwood, D.A. (1988) in 43rd Symposium of the Society of General Microbiology (Kingsman, A.M., Chater, K.F. and Kingsman, S.M., eds.) pp. 1-42, Cambridge Univ. Press, Cambridge.

83 Boccard, F., Smokvina, T., Pernodet, J.-L., Friedmann, A. and Guerineau, M. (1989) Plasmid 21, 59-70.

84 Boccard, F., Smokvina, T., Pernodet, J.-L., Friedmann, A. and Guerineau, M. (1989) EMBO J. 8, 973-980.

85 Campbell, A. and Botstein, D. (1983) in Lambda II (Hendrix, R.W., Roberts, J.W., Stahl, F.W. and Weisberg, R.A., eds.) pp. 365-380, Cold Spring Harbor Laboratory, New York.

86 Wieringa, K.T. and Wiebols, G.L.K. (1936) Tijdschr. Plantensiekten 42, 235-240.

87 Katz, L., Chiang, S.-J.D., Tuan, J.S. and Zablen, L.B. (1988) J. Gen. Microbiol. 134, 1765-1771.

88 Ikeda, H., Seno, E.T., Bruton, C.J. and Chater, K.F. (1984) Mol. Gen. Genet. 196, 501-507.

89 Chater, K.F. and Bruton, C.J. (1983) Gene 26, 67-78.

90 Chater, K.F. and Bruton, C.J. (1985) EMBO J. 4, 1893-1897.

91 Fisher, S.H., Bruton, C.J. and Chater, K.F. (1987) Mol. Gen. Genet. 206, 35-44.

92 Piret, J.M. and Chater, K.F. (1985) J. Bacteriol. 163, 965-972.

93 King, A.A. and Chater, K.F. (1986) J. Gen. Microbiol. 132, 1730-1752.

94 Rodicio, M.R., Bruton, C.J. and Chater, K.F. (1985) Gene 34, 283-292.

95 Lydiate, D.J., Ashby, A.M., Henderson, D.J., Kieser, H.M. and Hopwood, D.A. (1989) J. Gen. Microbiol. 135, 941-955.

96 Morino, T., Takagi, K., Nakamura, T., Takita, T., Saito, H. and Takahashi, H. (1986) Agric. Biol. Chem. 50, 2493-2497.

97 Cox, K.L. and Baltz, R.H. (1984) J. Bacteriol. 159, 499-504.

98 McHenney, M.A. and Baltz, R.H. (1988) J. Bacteriol. 170, 2276-2282.

99 Matsushima, P., McHenney, M.A. and Baltz, R.H. (1989) J. Bacteriol. 171, 3080-3084.

100 Ogawara, H. (1989) Actinomycetol. 3, 9-34.

101 Binnie, C., Warren, M. and Butler, M.J. (1989) J. Bacteriol. 171, 887-895.

102 Vara, J.A., Palido, D., Lacalle, R.A. and Jiménez, A. (1988) Gene 69, 135-140.

103 Distler, J., Braun, C., Ebert, A. and Piepersberg, W. (1987) Mol. Gen. Genet. 208, 204-210.

104 Ohnuki, T., Imanaka, T. and Aiba, S. (1985) J. Bacteriol. 164, 85-94.

105 Stauzak, R., Matsushima, P., Baltz, R.H. and Rao, R.N. (1986) Bio/Technol. 4, 229-232.

106 Seno, E.T. and Chater, K.F. (1983) J. Gen. Microbiol. 129, 1403-1413.

107 Smith, C.P. and Chater, K.F. (1988) Mol. Gen. Genet. 211, 129-137.

108 Smith, C.P. and Chater, K.F. (1988) J. Mol. Biol. 204, 569-580.

109 Hopwood, D.A., Bibb, M.J., Chater, K.F., Janssen, G.R., Malpartida, F.M. and Smith, C.P. (1986) in Regulation of Gene Expression-25 Years On: 39th Symp. Soc. Gen. Microbiol. (Booth, I. and Higgins, C.F., eds.), pp. 251-276, Cambridge University Press, Cambridge.

110 Hara, O., Anzai, H., Imai, S., Kumada Y., Murakami, T., Itoh, R., Takano, E., Satoh, A. and Nagaoka, K. (1988) J. Antibiot. 41, 538-547.

111 Anzai, H., Murakami, T., Imai, S., Satoh, A., Nagaoka, K. and Thompson, C.J. (1987) J. Bacteriol. 169, 3482-3488.

112 Shiffman, D., Mevarech, M., Jensen, S.E., Cohen, G. and Aharonowitz, Y. (1988) Mol. Gen. Genet. 214, 562-569.

113 Weigel, B.J., Burgett, S.G., Chen, V.J., Skatrud, P.L., Frolik, C.A., Queener, S.W. and Ingolia, T.D. (1988) J. Bacteriol. 170, 3817-3826.

114 Kovacevic, S., Weigel, B.J., Tobin, M.B, Ingolia, T.D. and Miller, J.R. (1989) J. Bacteriol. 171, 754-760.

115 Heinemann, J.A. and Sprague Jr., G.F. (1989) Nature 340, 205-209.

116 Heinzel, P.H., Werbitzky, O., Distler, J. and Piepersberg, W. (1988) Arch. Microbiol. 150, 184-192.

117 Williams, S.T., Goodfellow, M., Alderson, G., Wellington, E.M.H., Sneath, P.H.A. and Sackin, M.J. (1983) J. Gen. Microbiol. 129, 1743-1813.

118 Kuhstoss, S. and Rao, R.N. (1983) Gene 26, 295-299.

119 Jaurin, B. and Cohen, S.N. (1984) Gene 28, 83-91.

120 Muñoz, A., Perez-Aranda, A. and Barbero, J.L. (1985) Biochem. Biophys. Res. Commun. 133, 511-519.

121 Gray, G., Selzer, G., Buell, G., Shaw, P., Escanez, S., Hofer, S., Vogeli, P. and Thompson, C.J. (1984) Gene 32, 21-30.

122 Pulido, D., Vara, J.A. and Jiménez, A. (1986) Gene 45, 167-174.

123 Thompson, C.J. and Gray, G. (1983) Proc. Nat. Acad. Sci. U.S.A. 80, 5190-5194.

124 Lichenstein, H., Brawner, M.E., Miles, L.M., Meyers, C.A., Young, P., Simon, P.L. and Eckherdt, T. (1988) J. Bacteriol. 170, 3924-3929.

125 Behnke D. and Gerlach, D. (1987) Mol. Gen. Genet. 210, 528-534.

126 Shapiro, J.A. (1983) Mobile Genetic Elements, pp. 1-782, Academic Press, Inc., New York, NY.

127 Lydiate, D.J., Ikeda, H. and Hopwood, D.A. (1986) Mol. Gen. Genet. 203, 79-88.

128 Chater, K.F., Bruton, C.J. and Foster, S.G. (1985) Mol. Gen. Genet. 200, 235-239.
129 Chung, S.-T. (1987) J. Bacteriol. 169, 4436-4441.
130 Chung, S.-T. (1982) Gene 17, 239-246.
131 Chung, S.-T. and Crose, L.L. (1989) Microbiology-1989, Amer. Soc. Microbiol. (in press).
132 Olsen, E.R. and Chung, S.-T. (1988) J. Bacteriol. 170, 1955-1957.
133 Siemieniak, D.R., Slightom, J.L. and Chung, S.-T. Gene (in press).
134 Stein, D. and Cohen, S.N. (1989) J. Bacteriol. 171, 2258-2261.
135 Horinouchi, S., Malpartida, F., Hopwood, D.A. and Beppu, T. (1989) Mol. Gen Genet. 215, 355-357.
136 Babcock, M.J. and Kendrick, K.E. (1988) J. Bacteriol. 170, 2802-2808.
137 Häusler, A.,Birch, A., Krek, W., Piret, J. and Hütter, R. (1989) Mol. Gen. Genet. 217, 437-446.
138 Birch, A., Häusler, A., Vögtli, M., Krek, W. and Hütter, R. (1989) Mol. Gen. Genet. 217, 447-458.
139 Cullum, J., Flett, F. and Piendl, W. (1988) Microbiol. Sci. 5, 233-235.
140 Tomich, P.K. (1988) Antimicrob. Agents Chemother. 32, 1465-1471.
141 Kobayashi, T., Takao, M., Oikawa, A. and Yasui, A. (1989) Nucl. Acids Res. 17, 4731-4744.
142 Pernodet, J.-L., Boccard, F., Alegre, M., Blondelet-Rouault, M.-H. and Guérineau, M. (1988) EMBO J. 7, 277-282.
143 Pernodet, J.-L., Boccard, F., Alegre, M.-T., Gagnat, J. and Guérineau, M. (1989) Gene 79, 33-46.
144 Martín, J.F., Daza, A., Vigal, T., Alegre, T., Garcia, M., Liras, P. and Gil, J.A. (1989) Biochem. Soc. Transactions 17, 342-344.
145 Weber, J.M., Schoner, B. and Losick, R. (1989) Gene 75, 235-241.

FROM FOOTPRINT TO FUNCTION: AN APPROACH TO STUDY GENE EXPRESSION AND REGULATORY FACTORS IN TRANSGENIC PLANTS

Eric Lam

Center for Agricultural Molecular Biology
Waksman Institute of Molecular Biology
Box 759
Rutgers State University
Piscataway, NJ 08855

INTRODUCTION

Differential expression of genes plays an indispensable role in the developmental program of biological organisms. This is especially evident in the case of higher eukaryotes where each individual cell of an organism has a distinctive biochemical composition. In most cases, the phenotype and morphology of a particular cell are results of a regulated program of gene expression. Although different mechanisms of gene regulation are known to play important roles in development, such as differential splicing and mRNA stability, regulation at the level of transcription initiation has been the most intensely studied.

To study factors that are responsible for the expression pattern of a gene at the organismal level, the ability to generate transgenic organisms is a clear necessity. Plants offer a particularly versatile system for this approach. Significant advances in this field in recent years include the development of Agrobacterium tumefaciens-mediated transformation systems into routine application for various plant species (1), as well as the adaptation of the beta-glucuronidase (GUS) expression system for the histochemical detection of the expression pattern at the cell-specific level (2). With the ability to handle relatively large numbers of transformants at a time and the ease with which transformed lines can be stored in the form of seeds, the plant system offers some advantages over that of transgenic mice and Drosophila. The ability to express genes in a specific tissue and a particular stage of development will be invaluable for the future improvement of crop plants when applied to genes with agronomical interest (i.e., insect resistance, viral resistance, etc.).

The purpose of this chapter is to describe some of the recent work on the characterization of cis- and trans-acting elements in plants. In particular the major focus will be on results which relate to the architecture of plant promoters. Recent technological advances such as cloning of trans-acting factors by transposon tagging or expression library screening will also be reviewed. Lastly, future prospects in the study of plant gene expression with relation to development will be discussed.

AN INTEGRATED APPROACH TO STUDY GENE REGULATION

The strategy and scope of an experimental design for studying gene regulation in plants is summarized in Figure 1. An active promoter was chosen to carry out successive stages of analysis in order to identify and characterize the factors that are involved in its expression pattern. Well-studied plant promoters include the 35S promoter from cauliflower mosaic virus (CaMV) and two light-responsive promoters encoding pea ribulose 1,5-bisphosphate carboxylase (RbcS) and wheat chlorophyll a/b binding protein (Cab). Standard 5' deletion analysis defined the minimal region of these promoters which is necessary for expression in vivo (reviewed in ref. 3).

In order to more precisely localize the various sequence motifs that interact with cellular factors, methods have been established for plant cell extract preparations and DNase I footprinting with crude extracts (4). These techniques enabled the identification of sequence-specific DNA-binding factors which recognize elements within these promoters (5-8). In addition, site-specific mutations in some of these factor-binding sites and their functional characterization in transgenic tobacco shed light on the possible role of these factors (7,9). In addition to the study of the role of the factor-binding elements in the context of their cognate promoters, promoters which contain the tetramer of a specific factor-binding site fused upstream of truncated derivatives of the CaMV 35S promoter have been synthesized and their expression patterns in vivo characterized (8). These synthetic promoters, each with a distinct pattern of expression, thus provided tools for the future isolation of trans-acting mutants defective in the regulation of factors which bind to the multimerized elements. This can be accomplished by using the synthetic promoters to drive the expression of a selectable marker (i.e., the nptII gene for kanamycin resistance (10) or the tms2 gene for overproduction of auxin (11)). Transfer and expression of these selectable genes into plants will allow us to select for trans-acting mutants. Due to the low number of distinct factor binding-sites on these synthetic

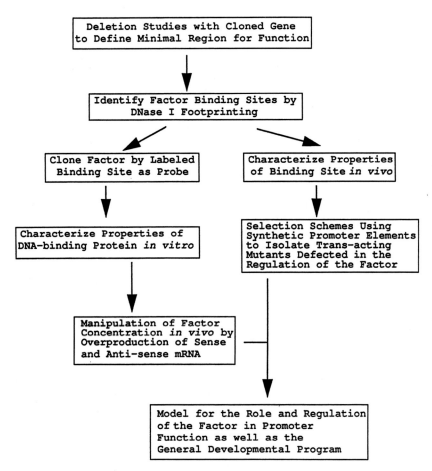

Figure 1. Schematic illustration of an approach for studying gene expression with transgenic plants.

promoters as compared to the natural promoters, the expected complexity of the mutants generated will also be lower and thus facilitate their analyses.

In addition to cis-element studies, the defined binding sites can also be used to isolate the trans-acting factors. This can be accomplished by affinity chromatography with immobilized multimers of the factor binding-site (12). The purified protein can then be used to generate antibodies or to obtain partial sequences of the protein. Subsequently, the antibody or a degenerate set of oligonucleotides can be used to isolate cDNA clones for the factor (12). This approach, however, is laborious and relatively time consuming. More recently, a new approach which involves direct screening of expression libraries with labeled factor-binding sites has been successful in cloning DNA-binding factors (13). This technique may be generally applicable since most factors characterized so far have a modular structure in which the DNA-binding domain is usually localized to a relatively small region of the protein sequence. Thus, DNA binding in vitro usually does not require the full-length cDNA clone. However, DNA-binding proteins which contain non-identical subunits, such as the mammalian CCAAT-box factor (14), will not be easily cloned by this method. This library screening method has been applied successfully to clone three different sequence-specific DNA-binding proteins from plants (15,16). The cloning of these factors provided the tools for the characterization of their expression pattern and their possible responses to environmental and physiological stimuli.

Once we have characterized the functional aspects of the factor-binding sites and the expression pattern of the factors which bind to these elements, we still need to demonstrate the role of the cloned factors in vivo. This may be tested by manipulation of their concentration by expression of the sense or anti-sense mRNA of the particular factor in transgenic plants. The ectopic expression of these genes can thus be used to test directly the role of these factors in the expression of the cis-elements as well as to provide clues to their functions in plant development. Similar approaches have been successful in the study of homeotic genes in Drosophila (17).

SEQUENCE-SPECIFIC DNA-BINDING FACTORS FROM PLANT EXTRACTS

In this section, I will describe some of the plant factors that we have characterized in the past few years. The techniques used in these studies such as extract preparation, DNase I footprinting and methylation interference have been described previously (4).

GT-1

GT-1 was first described by Green et al. (5). It binds to conserved motifs designated Box II and Box III which are found in the promoters of genes encoding ribulose 1,5-bisphosphate carboxylase (RbcS) from different plant species (5). The active form of GT-1, as assayed by in vitro binding, is similar in leaf extracts prepared from light-grown or dark-adapted plants (5,6). However, attempts to detect it in root nuclear extracts of tobacco had been unsuccessful (unpublished data). By site-specific mutation analysis of Box II, the binding site core has been defined as GGTTAA (6). In addition to the RbcS promoters, recently GT-1 has also been reported to bind to the promoter of the rice phytochrome gene (18). This is intriguing since the RbcS promoters are activated by light whereas the phytochrome promoter is repressed by light. Thus if GT-1 is involved in the light-responsiveness of these promoters, it must be capable of mediating both photophilic as well as photophobic expression. In either the RbcS or phytochrome promoters, the binding sites for GT-1 are usually found as pairs, and mutation in one site severely affects the binding of GT-1 to the other site (5,6). This indicates that GT-1 binding to these sites may be cooperative. It is not known at present if GT-1 is a heterogeneous set of proteins with similar sequence specificity or a single protein. Cloning of this factor and photo-crosslinking studies will be necessary to define more precisely the complexity of this DNA-binding activity.

ASF-1 AND HSBF

ASF-1 was detected first as a factor which binds to the -75 region of the 35S promoter of CaMV. It was found in root and leaf nuclear extracts of tobacco as well as pea (7). The binding site of this factor, as-1, was found to contain two tandem TGACG motifs with close homologies to the cAMP responsive element found in mammalian systems (19). Methylation interference assays and site-specific mutagenesis experiments demonstrated that these motifs are indeed involved in sequence-specific interaction with ASF-1 (7,20). In addition to the 35S promoter, ASF-1 also binds to functionally important regions of the nopaline synthase and octopine synthase promoters of the T-DNA from Agrobacterium tumefaciens (unpublished data). These binding sites for ASF-1 all appear to have a "dimeric" structure with two TGACG-like motifs, thus suggesting that this dimeric configuration for ASF-1 binding sites may be conserved. In contrast, we found that ASF-1 also binds to a conserved hexameric sequence, hex-1, found in the promoter of plant histone genes (16) which has only one TGACG motif. In addition, we found that hex-1 also binds a different factor, HSBF, which does not recognize the dimeric sites (15).

ASF-2

ASF-2 binds to the -100 region of the 35S promoter and its binding site as defined by DNase I footprinting is designated as-2 (8). It is easily detected in leaf nuclear extracts but not in root nuclear extracts from tobacco. Methylation interference assay and mutation analysis indicate that the guanines of two GATA motifs at the core of the binding site are critical for binding of ASF-2. By sequence comparison, it is found that a putative light regulatory element found to be conserved among light-responsive genes (21) also contains the GATA motif. In particular, a conserved GATA repeat has been found among Cab promoters from divergent plant species (22,23). We have then demonstrated that indeed ASF-2 binds to the GATA repeat sequence of the petunia Cab-22L promoter (8). Interestingly, we note that a GATA motif is also found in the binding site of an erythrocyte-specific factor (24). Thus, the DNA-binding domain of ASF-2 may in fact share homology with a mammalian factor.

CHARACTERIZATION OF IN VIVO PROPERTIES OF FACTOR-BINDING SITES

Box II

This is the best-characterized binding site for GT-1. It consists of the -151 to -138 sequence of the pea RbcS-3A promoter. Site-specific mutation of the guanines at the -147 and -146 positions has been shown to abolish activity of a -166 derivative of the promoter (9). The same mutation also severely attenuated GT-1 binding to Box II in vitro (5), thus directly correlating factor binding in vitro and activity of the promoter in vivo, and suggests GT-1 binding is required for light-responsive expression. However, this result does not show whether GT-1 is sufficient for light-responsive expression. Recently, we have addressed this question by examining the ability of a tetramer of Box II to confer expression on heterologous promoters. The general scheme of our approach is shown in Figure 2. The wild-type or the mutant Box II tetramers are fused upstream of the truncated derivatives (-46 to +8 or -90 to +8) of the 35S promoter from CaMV. These derivatives of the 35S promoter have previously been shown to be inactive in leaf of transgenic tobacco (25). The bacterial GUS gene is fused downstream as the reporter and the full 35S promoter-driven chloramphenicol acetyl-transferase gene serves as our reference. Transgenic tobacco plants with these constructs were examined for expression of the GUS gene. We found that a tetramer of Box II is unable to confer detectable expression when fused upstream of the -46 derivative of the 35S promoter. However, with the -90 derivative of the 35S promoter, which contains a single ASF-1 binding site, leaf expression was observed with the wild-type Box II tetramer. The mutant Box II tetramer did not confer significant leaf expression. Moreover, the expression of the synthetic promoter with the Box II tetramer is light dependent and is preferentially observed in chlorophyll-containing cells. Thus, these data demonstrate that GT-1 is likely to be directly involved in the light-activation process of gene expression. The apparent dependence of the activity of the Box II tetramer on an ASF-1 binding site suggests that Box II is a class B element. This class of cis-acting elements is active, even after multimerization, only in the presence of another element (26).

as-1 and hex-1

The 35S promoter (-343 to +8) of CaMV has been considered as a nominally constitutive promoter (3). We have sought to address

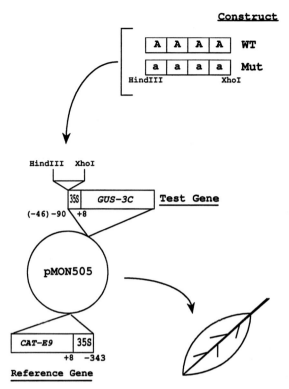

Figure 2. Characterization of tetramers of factor binding sites in transgenic plants. Tetramers of the wild-type and mutant binding site for a particular factor were synthesized with unique HindIII and XhoI restriction sites at the 5' and 3' ends, respectively. These tetramers were then inserted into the polylinker region of the test gene site in the pMON505 derivatives, X-GUS-46 and X-GUS-90. The test gene site promoter element is either the -46 to +8 (X-GUS-46) or the -90 to +8 (X-GUS-90) derivative of the 35S promoter. The GUS gene with the 3' end of pea RbcS-3C as polyadenylation signal is fused downstream from the promoter. The reference gene in these vectors consists of the chloramphenicol acetyl-transferase (CAT) gene fused downstream to the 35S promoter (-343 to +8). The 3' end of pea RbcS-E9 was placed 3' to the CAT gene. These vectors were then transferred into Agrobacterium tumefaciens and subsequently used to generate transgenic tobaccco.

the role of ASF-1 by site-specific mutation of its binding site. Mutation of 4 bases within the TGACG motifs abolishes ASF-1 binding to the 35S promoter (7). In vivo, these mutations severely attenuated root and stem expression of the 35S promoter. In contrast, the leaf expression of the 35S promoter is much less affected by the attenuation of ASF-1 binding. To demonstrate the positive role of ASF-1 in root expression, we have also inserted as-1 into the green tissue-specific RbcS-3A promoter and shown that the wild-type, but not the mutant, binding site confers high levels of root expression in transgenic tobacco (7). Since we have detected ASF-1 binding activity in leaf as well as root extracts, it is intriguing that the -90 derivative of the 35S promoter, which contains a single as-1 site, is expressed preferentially in root (27). One model is that the active form of ASF-1 may be limiting in leaf as compared to root tissues. Thus, a single binding site for ASF-1 will be sufficient to confer expression in root but not in leaf. Consistent with this hypothesis, we found that a tetramer of as-1 is able to confer leaf and root expression when fused upstream of either the -46 or -90 derivative of the 35S promoter (8). In addition, the expression of TGA1a, a putative cDNA clone of ASF-1, shows about 10-fold higher mRNA concentration in root when compared with leaf (15). Thus, by increasing the number of ASF-1 binding sites, we apparently overcame a limiting step that prevented leaf expression with a single as-1 site. This is analogous to the limitation of the range of hunchback expression by the gradient of its trans-regulator, bicoid, in Drosophila (28) and may be one common way that diversity in the pattern of expression can be generated from a single regulatory element. The ability of as-1 to confer expression in the absence of other upstream elements aside from the TATA box suggests that it is a type A enhancer element (26).

Although the hex-1 element binds ASF-1 in vivo, we found that it is functionally distinct from as-1. When a tetramer of hex-1 is fused to the -46 derivative of the 35S promoter, it is inactive in either root or leaf tissues (unpublished data). It is highly active in these tissues, however, when fused upstream of the -90 derivative of the 35S promoter, thus suggesting a functional dependence on the presence of ASF-1. The inability of the hex-1 tetramer to confer expression to the -46 derivative of the 35S promoter may be due either to prevention of ASF-1 binding by HSBF or to the fact that ASF-1 has different conformations depending on the sequence context of its binding site.

as-2

With the approach described in Figure 2, the functional properties of the ASF-2 binding site in transgenic tobacco have been examined (8). A tetramer of as-2 confers expression only when fused upstream of the -90 derivative of the 35S promoter. This activity is preferentially localized in the leaf and mutations at the GATA motifs abolish function of as-2 as well as ASF-2 binding in vitro, thus correlating factor binding to in vivo activity. Although ASF-2 also binds to light-responsive promoters, such as Cab, the activity of the as-2 tetramer construct is not light dependent and is found in all observed cell-types of the leaf. Thus, the as-2 tetramer apparently confers organ-specific but not cell-type-specific expression. It is also a class B element since it is unable to activate transcription in the absence of ASF-1. This functional dependence of ASF-2 on other factors binding to the promoter may enable it to participate in the leaf expression of genes with distinct cell-type specificity, such as the Cab genes, which are expressed preferentially in chlorophyll-containing cells.

CLONING AND CHARACTERIZATION OF PLANT TRANS-ACTING FACTORS

Transposon Tagging

At least two putative trans-acting factors have been cloned from maize by transposon tagging strategies. The cl gene, a regulator of anthocyanin biosynthesis, has been cloned with the use of the transposon Spm (29) and En (30). Sequence analysis of cl reveals homologies to the myb proto-oncogene at the DNA-binding domain and the presence of an acidic domain at the carboxy terminus (31). These observations suggest that cl is a transcription factor. Another plant factor that has been cloned by transposon tagging is the maize regulatory locus opaque-2. This is a positive, trans-acting locus which regulates the synthesis of zein seed storage-protein genes. Using the transposon Spm, Schmidt et al. (32) cloned the opaque-2 gene (O2) by screening for rare O2-mutables. Recent sequence analysis of cDNA for O2 revealed that it contains the leucine zipper motif found in transcription factors such as GCN4 and C/EBP. Although the cloning of these regulatory genes demonstrated the utility of the transposon tagging approach, this technique is limited by the need for a well-defined transposable element in the system of interest. In addition, for practical reasons, a visual phenotype is necessary for the screening procedure. For example, 530,000 seeds were screened to isolate the gene of O2 (32). This approach however has the important advantage that the pathway in which the factor is involved is known.

Direct Cloning of DNA-Binding Proteins

Several different sequence-specific DNA-binding proteins have now been cloned from plants. These are all isolated by the direct screening method of Singh et al. (13). Two clones that were isolated by this method were identified as ASF-1 and HSBF by their respective sequence specificity (15). Thus, one clone (TGA1a) binds to both as-1 as well as hex-1 while another clone (TGA1b) binds hex-1 specifically. Using the hex-1 sequence as probe, Tabata et al. (16) also reported the cloning of the wheat HBP-1 factor. At present, however, we do not know whether or not

HBP-1 binds as-1. The predicted amino acid sequences of these
clones all show significant homology at their putative DNA-
binding domain to that of the cAMP-responsive element binding
protein (CREB) as well as to GCN4 and the proto oncogene JUN
(Figure 3). These trans-acting factors all contain a leucine
zipper motif, known to be involved in dimerization, and a highly
basic region that is likely to be important for DNA recognition
(33). In addition, these proteins all appear to recognize
variants of the TGACG motif. For comparison, the primary
structure of O2 is also shown in Figure 3. Based on amino acid
sequence homologies, it is likely that O2 will also bind to
target sequences containing a TGACG-like motif. In any case, the
structure of this type of DNA-binding domain appears to be
conserved among such divergent species as yeast, humans and
plants.

Although the direct screening method for cloning sequence-
specific DNA-binding proteins has been rather successful, it is
still a rather laborious and time-consuming technique. A recent
selection method (34) with the use of promoter interference in
bacteria may increase the rate by which these proteins can be
cloned in the future.

FUTURE PROSPECTS

Promoter Architecture and Factor Networks

From our studies with defined factor-binding sites, we have
evidence for the following general characteristics of plant
promoters:

1). There are different classes of cis-elements as
discussed previously by Fromental et al. (26). Thus, class A
elements (e.g., as-1) are functional in the absence of other
elements aside from the basic transcription factors. In
contrast, class B elements (e.g., as-2 and Box II) are functional
only in the presence of another distinct element.

2). Oligomerization of factor-binding sites usually can
lead to higher activity and can expand the range of expression.
This is also observed in mammalian cells (26) and Drosophila
(28).

3). In nature, functional cooperation among factors binding
to a promoter is likely to play a major role in the generation of
diverse patterns of expression from the combinatorial action of a
limited number of trans-acting factors. Thus, even in the
absence of cooperative binding, factors on the same promoter can
synergistically activate transcription through the combination of
their activation domains (35).

4). Common factors are used by promoters with different
patterns of expression. This is exemplified by our study with
ASF-1 and ASF-2 which are factors that bind to a nominally

```
GCN4     ESSDPAALKRARNTEAARRSRARKLQRMKQLEDKVEELLSKNYHLENEVARL      TGACTCA
CREB     EAARKREVRLMKNREAARECRKKKEYVKCLENRVAVLENQNKTLIEELKAL       TGACGTCA
TGA1a    KPVEKVLRRLAQNREAARKSRLRKKAYVQQLENSKLKLIQLEQELERARKQG      TGACG
TGA1b    DEDEKKRARLVRNRESAQLSRQRKKHYVEELEDKVRIMHSTIQDLNAKVAYI      TGACG
HBP-1    ERELKKQKRKLSNRESARRSRLRKQAECEELGQRAEALKSENSSLRIELDRI      TGACG
O2       PTEERVRKRKESNRESARRSRMRKAAHLKELEDQVAQLKAENSCLLRRIAAL      ?
JUN      QERIKAERKRMRNRIAASKSRKRKLERIARLEEKVKTLKAQNSELASTANML      TGACTCA
```

Figure 3.   Sequence comparison of leucine zipper-containing
proteins.   The amino acid sequence deduced from cDNA clones of
GCN4, CREB, TGA1a, TGA1b, HBP-1, opaque-2 (O2), and JUN are
aligned at their first leucine of their leucine-repeat.   The
binding site core sequence for each clone is shown on the right
and the leucines of the zipper region are underlined.   Shown in
boxes are amino acid residues which are conserved among all seven
sequences in addition to the leucines.   Conservative changes,
arginine (R) to lysine (K), are also included.

constitutive promoter.   ASF-1 is found to be a positive factor
involved in root expression (7,27), whereas ASF-2 is demonstrated
to confer leaf expression preferentially (8).   Both factors are
also found to bind to conserved motifs in promoters with distinct
patterns of expression.   It should be pointed out that if common
factors are used by different promoters and if factors can bind
to promoters in a cooperative manner, it is possible that
alteration of a critical factor (X) binding to one class of
promoters can affect the expression of another class of promoter
by competing away the limiting factor (Y).   This is schematically
shown in Figure 4.   Thus, one should consider the transcription
apparatus as a network of factors in which the expression of
different classes of genes may be linked by common trans-acting
factors.   This may provide an economical way for the cell to
simultaneously turn on one set of genes and turn off another set.

In Vivo Engineering

An approach that may provide new insights into promoter
architecture and factor regulation is the synthesis of new cis-
elements by combination of different enhansons (36).   This
approach entails juxtaposition of different factor-binding sites
to generate new functional units.   Another approach is the
generation of new heterodimers of transcription factors by
splicing the protein-protein interaction domain of one class of
factor onto the DNA-binding domain of another (33).   This
approach may generate new trans-acting factors with
characteristics that may shed new light on the regulation of the
cognate factors.

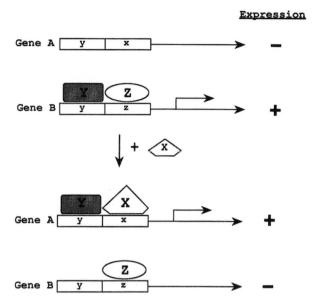

Figure 4. Network model for gene expression. Two classes of promoters (A and B) are considered in this model. Both promoters depend on the presence of a limiting factor, Y, for positive expression. In addition, the promoter for Gene A depends on the factor X and the promoter for Gene B depends on factor Z. X is a regulated factor whereas Z is a constitutive factor in this simplified presentation. In the absence of X, Y binds preferentially to the promoter of Gene B due to interaction with Z, thus Gene B is active and Gene A is not. In the presence of X, Y is effectively competed away from gene B due to its interaction with X. This results in activation of Gene A and repression of Gene B. It should be noted that the interaction between X, Y and Z needs not be direct. This interaction can be mediated by other proteins binding to the respective promoters.

Using the synthetic oligomers for a specific factor-binding site, one can also start to screen for regulatory mutants for a specific class of DNA-binding protein by fusing these promoters to selection genes. This approach will be especially applicable to the small weed Arabidopsis due to its small genome and the availability of a reproducible transformation system (37). Suppressors for these mutants can also be selected and their characterization will be invaluable for the elucidation of the complex regulation of these factors.

Acknowledgments: I would like to thank Dr. Nam-Hai Chua (Rockefeller University, NY) for his advice and guidance during my stay in his laboratory and many members of the Chua laboratory for stimulating discussions over the past years, especially Drs. Philip Benfey, Philip Gilmartin and Fumiaki Katagiri. I also thank the NSF, NIH and Monsanto Co. for their support during my postdoctoral years and Dr. Johan Memelink for critical reading of the manuscript.

REFERENCES

1   Fraley, R.T., Rogers, S.G., Horsch, R.B., Eichholtz, J.S., Flick, C.L., Finch, W.L., Hoffmann, N.L. and Sanders, P.R. (1985) Bio/Technology 3, 629-635.
2   Jefferson, R.A., Kavanagh, T.A. and Bevan, M.W. (1987) EMBO J. 6, 3901-3907.
3   Benfey, P.N. and Chua, N.H. (1989) Science 244, 174-181.
4   Green, P.J., Kay, S.A., Lam, E. and Chua, N.H. (1989) in Plant Molecular Biology Manual, Suppl. II (Gelvin, S. and Schilperoort, R.A., eds.), pp. 303-310, Klumer Academic Publishers, Dordrech.
5   Green, P.J., Kay, S.A. and Chua, N.H. (1987) EMBO J. 6, 2543-2549.
6   Green, P.J., Yong, M.H., Cuozzo, M., Kano-Murakami, Y., Silverstein, P., and Chua, N.H. (1988) EMBO J. 7, 4035-4044.
7   Lam, E., Benfey, P.N., Gilmartin, P.M., Fang, R.X. and Chua, N.H. (1989) Proc. Nat. Acad. Sci. U.S.A. (in press).
8   Lam, E. and Chua, N.H. (1989) (unpublished data).
9   Kuhlemeier, C., Cuozzo, M., Green, P.J., Goyvaerts, E., Ward, K. and Chua, N.H. (1988) Proc. Nat. Acad. Sci. U.S.A. 85, 4662-4666.
10 Horth, M., Negrutiu, I., Burny, A., Van Montagu, M. and Herrera-Estrella, L. (1987) EMBO J. 6, 2525-2530.
11 Klee, H.J., Horsch, R.B., Hinchee, M..A., Hein, M.B. and Hoffmann, N.L. (1987) Genes and Development 1, 86-96.
12 Kadonaga, J., Carner, K.C., Masiarz, F.R and Tjian, R. (1987) Cell 51, 1079-1090.
13 Singh, H., LeBewitz, J.H., Baldwin, A.S. and Sharp, P.A. (1988) Cell 52, 415-423.
14 Chodosh, L.A., Baldwin, A.S., Carthew, R.W. and Sharp, P.A. (1988) Cell 53, 11-24.
15 Katagiri, F., Lam, E. and Chua, N.H. (1989) Nature 340, 727-730.
16 Tabata, T., Takase, H., Takayama, S., Mikami, K., Nakatsuka, A., Kawata, T., Nakayama, T. and Iwabuchi, M. (1989) Science 245, 965-967.
17 Schneuwly, S., Klemenz, R. and Gehring, W.J. (1987) Nature 325, 816-818.
18 Kay, S.A., Keith, B., Shinozaki, K., Chye, M.L. and Chua, N.H. (1989) The Plant Cell 1, 351-360.
19 Hoeffler, J.P., Meyer, T.E., Yun, Y.D., Jameson, L.J. and Habener, J.F. (1988) Science 242, 1430-1433.

20 Lam, E., Benfey, P.N. and Chua, N.H. (1989) in Plant Gene Transfer, UCLA Symposia on Molecular and Cellular Biology (Lamb, C. and Beachy, R., eds.), Alan R. Liss, Inc., New York, NY (in press).

21 Grob, U. and Stuber, K. (1987) Nucl. Acids Res. 15, 9957-9973.

22 Castresana, C., Staneloni, R., Malik, V.S. and Cashmore, A.R. (1987) Plant Mol. Biol. 10, 117-126.

23 Gidoni, D., Brosio, P., Bond-Nutter, D., Bedbrook, J. and Dunsmuir, P. (1989) Mol. Gen Genet. 215, 337-344.

24 Tsai, S.F., Martin, D., Zon, L., D'Andrea, A., Wong, G. and Orkin, S. (1989) Nature 339, 446-451.

25 Fang, R.X., Nagy, F., Sivasubramaniam, S. and Chua, N.H. (1989) The Plant Cell 1, 141-150.

26 Fromental, C., Kanno, M., Nomiyama, H. and Chambon, P. (1988) Cell 54, 943-953.

27 Benfey, P.N., Ren, L. and Chua, N.H. (1989) EMBO J. 8, 2195-2202.

28 Struhl, G., Struhl, K. and Macdonald, P.M. (1989) Cell 57, 1259-1273.

29 Cone, K.C., Burr, F.A. and Burr, B. (1986) Proc. Nat. Acad. Sci. U.S.A. 83, 9631-9635.

30 Paz-Ares, J., Wienand, U., Peterson, P. and Saedler, H. (1986) EMBO J. 5, 829-833.

31 Paz-Ares, J., Ghosal, D., Wienand, U., Peterson, P. and Saedler, H. (1987) EMBO J. 6, 3553-3558.

32 Schmidt, R.J., Burr, F.A. and Burr, B (1987) Science 238, 960-963.

33 Sellers, J.W. and Struhl, K. (1989) Nature 341, 74-76.

34 Elledge, S.J., Sugiono, P., Guarente, L. and Davis, R.W. (1989) Proc. Nat. Acad. Sci. U.S.A. 86, 3689-3693.

35 Ptashne, M. (1988) Nature 355, 683-689.

36 Ondek, B., Gloss, L. and Herr, W. (1988) Nature 333, 40-45.

37 Valvekens,D., Van Montagu, M. and Van Lijsebettens, M. (1988) Proc. Nat. Acad. Sci. U.S.A. 85, 5536-5540.

PURIFICATION OF RECOMBINANT PROTEINS WITH METAL CHELATE ADSORBENT

Erich Hochuli

F. Hoffmann-La Roche AG
Central Research Units
CH-4002 Basel, Switzerland

INTRODUCTION

In 1975 Porath and co-workers introduced immobilized metal ion affinity chromatography (IMAC) for the purification of peptides and proteins (1). The principle of this technique is the coordination between the electron donor groups on a protein (peptide) surface and immobilized transition metal ions. The tridentate chelator, iminodiacetic acid, is coupled via a spacer arm to a solid support and used for the immobilization of metal ions such as Ni(II), Cu(II) or Zn(II). Porath postulated that the histidine, cysteine and tryptophan residues in proteins (peptides) are most likely to form stable coordination bonds with metal chelates at neutral pH. Present experience (2) indicates that histidine residues on protein surfaces are the predominant electron donor groups.

In recent years very attractive genetic approaches have been developed to facilitate purification of recombinant proteins. The concept is based on the preparation of hybrid proteins by fusing the coding sequence of the protein of interest with the coding sequence of a protein with high affinity for an immobilized ligand, together with the sequence of a specific cleavage site. The expressed fusion proteins are then purified by taking advantage of the specific binding of the affinity protein (affinity tag) to the adsorbent. After purification of the hybrid protein, the affinity tag can be split off at the designed cleavage site. This general method for purification of recombinant proteins has been demonstrated with the use of a number of different affinity protein-ligand systems (3-8). Alternative purification strategies involve the synthesis of fusion proteins containing, for example, poly-arginine at their carboxy terminus

which allows purification by cation-exchange chromatography (9).
Gene fusion systems using immobilized metal ion affinity chromatography to facilitate the purification of recombinant proteins
have been described recently. Smith and co-workers have purified
a fusion protein comprising a His-Trp chelating peptide fused to
proinsulin with the use of immobilized iminodiacetic acid charged
with Ni(II) ions (10). We have developed a general purification
method for recombinant proteins based on the selective interaction between a poly-His peptide fused to the protein of interest
and an immobilized nitrilotriacetic acid derivative charged with
Ni(II) ions (11,12). The effects of placing poly-His tags of
different lengths at the carboxy as well as at the amino terminus
have been investigated. Here we present a protocol for the
preparation of the Ni(II)-nitrilotriacetic acid adsorbent (NTA-
Ni(II) adsorbent) and its use for the purification of poly-His
fusion proteins.

PREPARATION OF THE NTA-Ni(II) ADSORBENT

The NTA-adsorbent is a quadridentate chelator which occupies
four positions in the metal coordination sphere of, e.g. Ni(II).
The remaining two ligand positions in the octahedral coordination
sphere are available for selective interactions with adjacent
histidines of a fusion protein (Figure 1).

The NTA-adsorbent is prepared from N^ϵ-protected lysine and
bromoacetic acid. After the protecting group is removed, the
lysine derivative is immobilized on epoxy-activated Sepharose CL-
6B. Bromoacetic acid, N^ϵ-benzyloxycarbonyl-L-lysine and

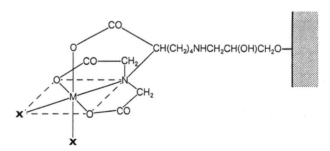

Figure 1. Schematic representation of the nitrilotriacetic acid
adsorbent (NTA-M adsorbent). The quadridentate chelator occupies
four positions in the metal coordination sphere of a metal ion
with a coordination number of six, e.g., M = Ni(II). The remaining two ligand positions X in the octahedral coordination sphere
are available for selective interactions with adjacent histidines
of a fusion protein.

epibromohydrine are obtained from Fluka (Buchs, Switzerland). Sepharose CL-6B is purchased from Pharmacia. All other chemicals are of reagent grade. A 41.7 g amount of bromoacetic acid is dissolved in 150 ml of 2 M sodium hydroxide solution and cooled to 0°C. A solution of 42 g of N^ϵ-benzyloxycarbonyl-L-lysine in 225 ml of 2 M sodium hydroxide solution is slowly added dropwise at 0°C with stirring. After 2 hr the cooling is stopped and the mixture is stirred overnight. It is then held at 50°C for 2 hr and 450 ml of 1 M hydrochloric acid are subsequently added. After the mixture has been cooled, the crystals that formed are filtered off. The product is dissolved in 1 M sodium hydroxide solution and again precipitated with the same amount of 1 M hydrochloric acid and filtered off. White crystals (about 40 g) of N-(5-benzyloxycarbonylamino-1-carboxypentyl)iminodiacetic acid are obtained, m.p. 172-174°C (dec.), $[\alpha]_D = +9.9°$ (c = 1; 0.1 M sodium hydroxide). Calculated for $C_{18}H_{24}N_2O_8$ (mol. wt. 396.4): C 54.54, H 6.10, N 7.07%; observed: C 54.50, H 6.22, N 7.05%. 10 g of the lysine derivative is then dissolved in 100 ml 0,5 M sodium hydroxide solution and, after the addition of 500 mg 5% Pd/C, hydrogenated at room temperature and normal pressure. The catalyst is filtered off and the filtrate acidified to pH 2 with 5 M hydrochloric acid. Then 100 ml ethanol are added and the product crystallized at 0°C. White crystals (about 6 g) are formed m.p. 216-217°C. Calculated for $C_{10}H_{18}N_2O_6$ (mol. wt. 262.26): C 45.80, H 6.92, N 10.68%; observed: C 46.01, H 6.72, N 10.63%.

100 ml of Sepharose CL-6B are washed twice on a glass suction filter with about 500 ml of water. The gel is then transferred to a 500 ml round bottom flask; the volume is made up to 200 ml with water and treated at 30°C for 4 hr with 16 ml of 4 M sodium hydroxide solution and 8.22 ml of epibromohydrine. The activated Sepharose is subsequently filtered off, washed to neutrality with water, returned to the reaction vessel and the volume made up to 200 ml with water. Then 5.7 g of the lysine derivative from the previous step and 10.6 g of sodium carbonate dissolved in 50 ml of water are added and the mixture is stirred slowly at 60°C overnight. The resulting NTA-absorbent is filtered off and washed with water.

For charging with metal ions, the resin is packed in a chromatography column and the following solutions are subsequently pumped through it: 100 ml of aqueous 1% (w/w) $NiSO_4 \cdot 6H_2O$, 200 ml of water, 200 ml of 0.2 M acetic acid (containing 0.2 M sodium chloride and 0.1% Tween 20) as well as 200 ml of water.

The ligand density (content of the lysine derivative) of the chelating adsorbent is determined by nitrogen elemental analysis and the amount of chelated Ni(II) ions is measured by atomic adsorption spectroscopy (11). The results of a representative preparation are given in Table 1.

Table 1
Ligand and Metal Content of the NTA-Ni(II) Adsorbent

	Microequivalent/ml adsorbent
NTA-ligand	7.1
Ni(II)	9.5

The ratio of the ligand and metal ion concentration indicates that a 1:1 metal-ligand complex is formed in the adsorbent.

CLONING AND EXPRESSION OF THE POLY-HIS FUSION PROTEINS

The principle for cloning and expression of the poly-His fusion proteins is exemplified by mouse dihydrofolate reductase with six histidines at the carboxy terminus (DHFR-(His)$_6$). For the construction of the plasmid pDHFR-(His)$_6$ (12) the BglII/HindIII fragment of plasmid pDS78/RBSII (13,14) was replaced by the synthetic adaptor encoding six histidines (Figure 2). The oligonucleotide-forming adaptor was synthesized chemically on controlled pore glass. The plasmid pDHFR-(His)$_6$ was transformed into E. coli M 15 cells containing plasmid pDMI.1. The E. coli transformant was grown at 37°C in LB medium containing 100 µg/ml ampicillin and 25 µg/ml kanamycin. At OD 600 nm ~0.7 isopropyl-β-D-thiogalactopyranoside (IPTG) was added to a final concentration of 2 mM and the cells incubated for an additional 5 hr, before being harvested by centrifugation. The synthesis of the fusion protein DHFR-(His)$_2$ was carried out analogously to the synthesis of DHFR-(His)$_6$, with the synthetic adaptor encoding two histidines.

In principle these plasmids can be used for the expression of any fusion protein with a poly-His tag. Thus the gene coding for DHFR can be replaced by the gene encoding a new protein of interest yielding a plasmid for the expression of a new protein with the affinity tag.

PURIFICATION OF THE POLY-HIS FUSION PROTEINS

Adsorption of the fusion proteins to the metal chelate adsorbent has to be carried out at a pH at which the imidazole residues of the histidines are not protonated. The apparent dissociation constant pK_2 (imidazole) of histidine is 5.97. Thus at pH 7 more than 91% of the imidazole residues are unprotonated.

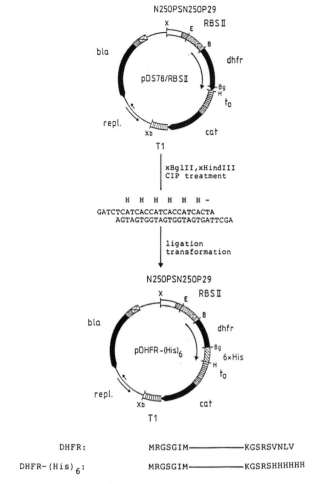

Figure 2. Construction of the plasmid used for the expression of DHFR-(His)$_6$. The upper part of the figure shows plasmid pDS78/ RBSII and the construction of plasmid pDHFR-(His)$_6$. All plasmids contain the regulatable promotor/operator element N250 PS N250 P29 and the synthetic ribosomal binding site RBSII (13,14). Replacement of BglII/HindIII fragment by the synthetic adaptor results in plasmid pDHFR-(His)$_6$. The genes for β-lactamase, chloramphenicol acetyltransferase and mouse dihydrofolate reductase are denoted bla, cat and dhfr, respectively. Calf intestinal alkaline phosphatase is abbreviated with CIP. B, Bg, E, H, X and Xb denote cleavage sites for restriction enzymes BamHI, BglII, EcoRI, HindIII, XhoI and XbaI. The amino acid sequences of DHFR (encoded by plasmid pDS78/RBSII) and DHFR-(His)$_6$ (plasmid pDHFR-(His)$_6$) are shown in the lower part of the figure.

Therefore the fusion proteins should be loaded on the metal chelate adsorbent at neutral or slightly alkaline pH. The metal chelate adsorbent is packed into a chromatography column and equilibrated with a phosphate buffer of pH 7 to 8 prior to the loading with protein. The equilibration buffer, which itself forms no chelate with the metal ion, can contain a detergent, e.g., of the polyoxyethylene-sorbitan-fatty acid ester type (Tween). In addition the equilibration buffer can contain high concentrations of dissociating agents as 8 M urea or 6 M guanidine hydrochloride. In order to quench electrostatic interactions between protein and metal chelate adsorbent the buffers usually contain sodium chloride. The elution can be carried out at a constant pH-value by ligand exchange, or with linear or step pH gradients. In the first protocol the imidazole residues of the histidines in the protein are displaced by imidazole dissolved in the elution buffer. An imidazole concentration up to 100 mM may be used. In the second protocol the imidazole residues of the histidines are protonated by lowering pH and thus will reverse the capability to bind to the immobilized metal ion. Buffers, preferably acetate, up to a concentration of 0.2 M and a pH of 3 to 5, are used for this procedure. Recombinant proteins expressed in bacteria often occur in an aggregated and denatured form in the cytoplasm as granules or inclusion bodies. In this form the proteins are protected from proteolysis, but the inclusion bodies can only be dissolved with detergents or highly concentrated guanidine hydrochloride or urea solutions. The recovery of active proteins from such dissociating solutions is possible when they can be refolded from their unfolded conformation. The refolding of a denatured protein in a crude extract may be much more complicated than the refolding of a purified protein. However, most of the conventional chromatographic methods cannot be used with solutions of 6 M guanidine hydrochloride or 8 M urea. Therefore the method of choice may be chromatography in 6 M guanidine hydrochloride on the NTA-Ni(II) adsorbent.

Chromatography of a (His)$_6$ Fusion Protein in 6 M Guanidine Hydrochloride

Materials   37 ml NTA-Ni(II) adsorbent packed into a chromatography column (2.6 x 10 cm).

Extraction buffer: 0.1 M potassium phosphate/ 0.15 M NaCl, pH 8.5, containing 6 M guanidine/HCl.

Wash buffer: extraction buffer with the pH adjusted to 5.7.

Elution buffer: 0.1 M sodium acetate/0.15 M
                       NaCl, pH 4.8, containing 6 M
                       guanidine/HCl.

3 g E. coli cells, containing a (His)$_6$ fusion protein
recovered from the broth medium by centrifugation (4,000 x g, 10
min, 4°C), are stirred in 30 ml extraction buffer for 1 hr at
room temperature with a magnetic stirrer.   After centrifugation
(10,000 x g, 30 min, 4°C), the supernatant is directly loaded
with a peristaltic pump (flow rate 70 ml/hr) onto the NTA-Ni(II)
column equilibrated with 150 ml extraction buffer.   Column
effluent is monitored with a UV monitor (280 nm) and strip chart
recorder.   The loaded column is washed with 200 ml extraction
buffer, then wash buffer until UV-absorbance is back to zero
again.   Then the column is eluted with elution buffer and
fractions collected with a fraction collector.   The fractions are
analyzed for protein content, for purity by SDS-PAGE (15) and for
activity.   The solvent of the purified protein may be exchanged
by dialysis or by gel filtration.   Proteins prepared with this
protocol are suitable for direct immunization to generate
antibodies.   As a representative example the chromatography of
DHFR-(His)$_6$ in 6 M guanidine hydrochloride is shown in Figure 3.
    The efficiency of the different poly-His affinity tags is
dependent on the solvent system used throughout the chromatography (12).   The (His)$_6$-tag is perfectly suited to chromatography
in 6 M guanidine hydrochloride.   But in physiological buffer
systems, such as phosphate buffer without dissociating agents,
the (His)$_6$-fusion proteins often bind too strongly to the NTA-
Ni(II) absorbent and elution of the product may be difficult.   On
the other hand the (His)$_2$ tag does not bind in 6 M guanidine
hydrochloride but works perfectly in physiological buffers.
Therefore the (His)$_2$ tag at the carboxy terminus may be the
chelating peptide of choice for the purification of recombinant
proteins produced in a soluble form.   These proteins are not
protected from proteolysis, and use of protease inhibitors in the
extraction buffer may be necessary.

Chromatography of a (His)$_2$ Fusion Protein in a Physiological Buffer

Materials   260 ml NTA-Ni(II) adsorbent packed into a
            chromatographic column (5 x 16 cm).

            Extraction buffer: 0.1 M sodium phosphate/0.1 M
                                  NaCl, pH 8, containing 0.1%
                                  Tween-20 and 10 µM PMSF.

            Elution buffer: 0.1 M sodium acetate/0.1 M
                                  NaCl, pH 5.0, containing 0.1%
                                  Tween-20.

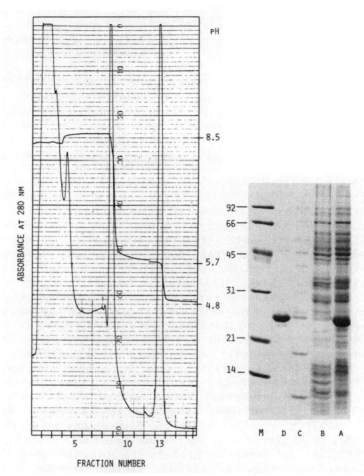

Figure 3.   Chromatography of DHFR-(His)$_6$ in 6 M guanidine
hydrochloride.   26 ml crude extract obtained from 3 g biomass
were purified on a 37 ml NTA-Ni(II) column (2.6 x 10 cm).   The
column was developed with a pH step gradient.   Collected fractions (35 ml) were analyzed by SDS-PAGE.   M: standard molecular
weight markers.   A: E. coli extract containing DHFR-(His)$_6$, B:
fractions 1-5, C: fractions 8-11, D: fraction 13.

30 g E. coli cells, containing a soluble fusion protein with
two His at the carboxy terminus, are recovered from the broth
medium by centrifugation (4,000 x g, 10 min, 4°C), and broken
open by sonication in 180 ml extraction buffer.   A sonicator W-
375 from Ultrasonic Inc, New York, with a 0.5 inch probe or a
similar apparatus is used for 15 min at 0°C.   After removal of
the cell debris by centrifugation (10,000 x g, 30 min, 4°C), the

Table 2
Purification of DHFR-(His)$_2$ on the NTA-Ni(II) column

	Activity (U)	Recovery (%)	Sp. Act. (U/mg)
Crude extract	7306	100	2.4
Column flowthrough	704	9	0.3
Eluted DHFR-(His)$_2$	6439	88	9.0*

The DHFR activity was assayed by reduction of dihydrofolate with NADPH (12). One unit (U) converts 1 μmole of dihydrofolate to tetrahydrofolate per minute.
*DHFR purified on a Methotrexate-Sepharose affinity column had a specific activity of 10.3 U/mg. The DHFR-(His)$_2$ had a purity of 90%.

supernatant is directly loaded with a peristaltic pump (flow rate 180 ml/hr) onto the NTA-Ni(II) column equilibrated with 1 liter extraction buffer. Column effluent is monitored with a UV monitor (280 nm) and strip chart recorder. The loaded column is washed with extraction buffer until UV-absorbance is back to zero again. The column is eluted with a 1 liter pH gradient from 8 to 5 (500 ml extraction buffer to 500 ml elution buffer) with the use of a gradient mixer. Fractions are collected with a fraction collector and analyzed for protein content, purity (by SDS-PAGE) and activity. The purification of DHFR-(His)$_2$ with this protocol is summarized in Table 2.

Cleavage of the His-tag with Carboxypeptidase A (CPA)

CPA rapidly releases amino acids with an aromatic or large aliphatic side chain from the carboxy termini of polypeptide chains. This enzyme can therefore be used to cleave off the His-tag at the carboxy terminus. The purified hybrid protein is dialyzed against 0.05 M Tris/HCl, pH 8. The protein, at a concentration of about 2 mg/ml, is then incubated at room temperature with bovine pancreatic CPA (1 mg CPA for 50 mg protein). The reaction is followed by measurement of released histide. When the theoretical amount of amino acids has been removed the protein can be precipitated, such as with $(NH_4)_2SO_4$.

When purified DHFR-(His)$_2$ was treated with CPA, two equivalents of His and one equivalent of Ser were released (12). This result is in good agreement with the expected release of amino acids according to the carboxyterminal amino acid sequence

                          ····Ser-Arg-Ser-His-His

of the fusion protein, since Arg is expected not to be digested
with CPA. However, this method has limitations. The reaction
with CPA only stops completely at the position of a basic amino
acid. Therefore the procedure is only suited for the production
of an authentic protein with a basic amino acid at the carboxy
terminus. Other enzymatic and chemical methods to cleave off
affinity tags have been described (3). All these methods have
limitations. Present experience indicates that there is no
general cleavage method available. The optimal procedure must be
determined on a case-by-case basis, after analysis of the amino
acid sequence of the protein of interest.

                 REGENERATION OF THE NTA-Ni(II) COLUMN

<u>Materials</u> Rinse buffer: 0.2 M acetic acid/0.2 M NaCl
                                        containing 0.1% Tween-20.

            EDTA solution: 0.1 M disodium tetraacetate.

            Ni(II) solution: 1% (w/w) $NiSO_4 \cdot 6H_2O$.

     After every chromatography cycle the NTA-Ni(II) column is
washed with two column volumes (CV) of rinse buffer followed by
two CV of water. Then the column is equilibrated with the
appropriate extraction buffer. After about five chromatography
cycles it is convenient to strip the nickel ions with 1 CV of
EDTA solution. Then the column is washed subsequently with the
following solutions: four CV of water, one CV of Ni(II) solution, two CV of water, two CV of rinse buffer and two CV of
water. Then the column is equilibrated with the appropriate
extraction buffer.

                          CONCLUDING REMARKS

     We have shown that fusion proteins comprising a protein of
interest and a poly-His peptide can be purified very easily on a
NTA-Ni(II) adsorbent. Placement of 2 to 6 histidines at the
carboxy as well as at the amino terminus has been investigated.
The efficiency of different poly-His affinity tags depends on the
buffer solutions used throughout the chromatography. The $(His)_6$
tag at the carboxy or amino terminus of the protein of interest,
in combination with a buffer containing 6 M guanidine hydrochloride, is appropriate for the purification of proteins which are
formed in inclusion bodies. For proteins which are produced in a
soluble form the $(His)_2$ tag at the carboxy terminus is recommended together with a physiological buffer. Orienting

experiments have shown that the purification protocol is also compatible with reducing agents such as mercaptoethanol at concentrations up to 10 mM. Such buffers are usually prepared freshly every day.

To split off the affinity tag from the protein of interest, a specific chemical or enzymatic cleavage site may be introduced at the junction (3). These methods are successful in several instances, but often have limitations by poor cleavage yields or by unwanted cleavage that occurs within the desired protein sequence. If the protein of interest already comprises the amino acid sequence of the cleavage site, the product will be degraded. We exploited carboxypeptidase A for the removal of the $(His)_2$ tag at the carboxy terminus. This enzyme can be used for the production of authentic proteins which comprise an amino acid with basic side chain at the carboxy terminus. From present experience it can be concluded that the optimal procedure for removal of the affinity tag must be determined on a case-by-case basis, after analysis of the amino acid sequence of the protein of interest.

Several affinity protein-ligand systems have been developed for the purification of recombinant proteins. However the relatively large affinity proteins described might be a general disadvantage. Small affinity peptides can be more favorable. The observation that the $(His)_2$ tag at the carboxy terminus has only a small effect on the enzymatic activity of DHFR (see Table 2) indicates that the poly-His peptides have minimal effect on the structure of the protein of interest.

For many applications the removal of the affinity tag might not be necessary. Poly-His fusion proteins are well suited for direct immunization to generate antibodies against the protein of interest.

Recombinant proteins expressed in bacteria often form inclusion bodies. A refolding step is necessary for the purification of active protein from inclusion bodies. Proteins formed in inclusion bodies can easily be dissolved in 6 M guanidine hydrochloride and then purified on the NTA-Ni(II) column. In addition it is possible to do refolding experiments in situ in the column. In the adsorbed form the protein molecules are separated in space and therefore the problem of protein aggregation during the refolding process is minimized.

In conclusion we believe that the method described is an attractive addition to the range of modern purification procedures for heterologously-produced proteins.

REFERENCES

1   Porath, J., Carlsson, J., Olsson, I. and Belfrage, G. (1975) Nature (London) 258, 598-599.

2   Hemdan, S.E., Zhao, Y., Sulkowski, E. and Porath, J. (1989)
     Proc. Nat. Acad. Sci. U.S.A. 86, 1811-1815.

3   Moks, T., Abrahmsen, L., Oesterlöf, B., Josephson, S.,
     Oestling, M., Enfors, S.O., Persson, I., Nilsson, B. and
     Uhlen, M. (1987) Bio/Technology 5, 379-382.

4   Nygren, P.-A., Eliasson, M., Palmcrantz, E., Abrahmsen, L.
     and Uhlen, M. (1988) J. Mol. Recognition 1, 6-74.

5   Ullman, A. (1984) Gene 29, 27-31.

6   di Guan, C., Li, P., Riggs, P.D. and Inouye, H. (1988) Gene
     67, 21-30.

7   Smith, D.B. and Johnson, K.S. (1988) Gene 67, 31-40.

8   Germinio, I. and Bastia, D. (1984) Proc. Nat. Acad. Sci.
     U.S.A. 81, 4692-4696.

9   Sassenfeld, H.M. and Brewer, S.J. (1984) Bio/Technology 2,
     76-80.

10   Smith, M.C., Furman, T.C., Ingolia, T.D. and Pidgeon, C.
     (1988) J. Biol. Chem. 263, 7211-7215.

11   Hochuli, E., Döbeli, H. and Schacher, A. (1987) J.
     Chromatogr. 411, 177-184.

12   Hochuli, E., Bannwarth, W., Döbeli, H., Gentz, R. and
     Stüber, D. (1988) Bio/Technology 6, 1321-1325.

13   Bujard, H., Gentz, R., Lanzer, M., Stüber, D., Müller, M.,
     Ibrahimi, I., Haeuptle, M.-T. and Dobberstein, B. (1987)
     Methods in Enzymology 155, 416-433.

14   Lanzer, M. and Bujard, H. (1988) Proc. Nat. Acad. Sci.
     U.S.A. 85, 8973-8977.

15   Laemmli, U.K. (1970) Nature (London) 227, 680-685.

DETERMINANTS OF TRANSLATION EFFICIENCY OF SPECIFIC

mRNAs IN MAMMALIAN CELLS

David S. Peabody

Departments of Cell Biology and Biochemistry
University of New Mexico School of Medicine
Albuquerque, NM 87131

INTRODUCTION

The means by which the ribosomes of mammalian cells recognize translation initiation sites and the structural features of an mRNA that define its overall translation efficiency are only partially understood. Even so, an effort to engineer the efficient translation of a specific sequence in mammalian cells can benefit from existing information about mechanisms of translation initiation. The purpose of this article is to review briefly our present knowledge of the factors that influence translation efficiency and to suggest some means by which this knowledge can be exploited to influence the translation efficiency of a specific mRNA.

It is convenient to think of the translation efficiency of a given mRNA as being determined by factors that may be grouped into three categories: (1) features of mRNA structure that influence the ability of the 40S ribosome to gain access to the initiation codon; (2) factors that define the inherent efficiency of the initiation site itself; that is, having gained access to the initiation site, what is the probability that the ribosome will recognize it? (3) alterations in the translational machinery (e.g., chemical modification of initiation factors) that influence the rate of protein synthesis. Each of these aspects will

be discussed after a brief overview of the translation initiation
process.

BASIC MECHANISMS OF INITIATION

For a time it seemed that the signals that specify ribosome
binding sites in eukaryotic mRNAs might be simple. In 1978 Kozak
proposed the scanning hypothesis, which, in its original form,
held that an early event in initiation was the interaction of the
40S ribosomal subunit with the 5' end of the mRNA, followed by
migration in a 5' to 3' direction until an AUG triplet was
encountered (1). At the time of its proposal the scanning
hypothesis, and its corollary, the first AUG rule, were consistent with nearly all the mRNA sequences then known. Moreover,
biochemical experiments indicated the requirement for a 5' end
and suggested the capacity of 40S subunits to migrate in the
required fashion on mRNA (2-4). The widespread application of
recombinant DNA techniques, however, soon resulted in an explosion in the number of known mRNA sequences and uncovered a number
of exceptions to the rule that ribosomes always initiate at the
AUG nearest the 5' end (5). Poliovirus provided an early and
dramatic example of internal initiation (6). In addition to such
natural examples, the use of recombinant DNA techniques for the
construction of novel genes, and their introduction into mammalian cells via viral and plasmid vectors, also demonstrated that
in some cases translation could efficiently initiate at internal
sites (for examples see references 7-9). The following variations of the scanning hypothesis were proposed to account for the
exceptional cases.

Translational reinitiation. When an upstream initiator AUG
is followed by an in-frame termination codon located in the
vicinity of an internal AUG, ribosomes are apparently able to
reinitiate translation at the internal site (10-13). This model
requires no modification of the scanning model beyond the
provision for reinitiation, a phenomenon long known in bacteria.

Leaky scanning. The sequence context of an AUG can modulate
its ability to serve as an initiation codon (14-17). AUGs in
poor contexts will tend to be invisible to the 40S subunit, thus
permitting initiation at a downstream site. Some non-AUG
triplets are able also to serve as initiation codons (18-25).
However, since initiation at such sites is never 100% efficient,
non-AUG initiators are always leaky.

Direct internal ribosome binding. Recent results from the
study of picornaviruses may require a third mechanism by which
ribosomes could bind directly to internal positions on the mRNA
and then scan to the next available initiation site (26-31).
Picornaviral RNA may contain elements that obviate the normal
requirement for a 5' end.

FACTORS THAT INFLUENCE THE ABILITY OF THE RIBOSOME TO GAIN ACCESS TO THE INITIATION SITE

The 5' Cap and the Role of RNA Secondary Structure

A wealth of evidence now indicates that stable secondary structure in the 5' untranslated region (UTR) is generally inhibitory of translation initiation (32-38). The extent of inhibition is apparently a function of the stability of the folded structure and its location in the mRNA. To understand how secondary structure can influence translation efficiency, it is necessary to review certain aspects of the biochemistry of the initiation process.

One of the peculiarities of eukaryotic mRNA structure is the 5' terminal cap structure, $me^7G(5')ppp(5')N$. Its presence increases the translation efficiency of most mRNAs many-fold. However, the extent of cap-dependence varies widely amongst individual mRNAs. Although many mRNAs show a pronounced effect of the cap, some other messengers are translated in a virtually cap-independent fashion. For example, although the alfalfa mosaic virus RNA 4 normally contains a cap, its removal does not substantially alter translation efficiency (39,40). Moreover, some mRNAs (e.g., poliovirus) are naturally uncapped (41,42).

A number of experiments have now established that the degree of cap-dependence is correlated with the extent to which the 5' UTR of the mRNA is involved in stable secondary structure. Messengers whose efficient translation does not depend on the presence of a cap tend to have a 5' UTR with substantial single-stranded character. This behavior is a consequence of the participation of specific initiation factors in the ribosome-binding process. The messenger is prepared for ribosome binding by the action of three initiation factors, eIF4F, eIF4A and eIF4B (reviewed in references 43 and 44). eIF4F (also called cap-binding complex) is composed of three subunits. One is a cap-binding protein of molecular weight about 24,000 that confers on eIF4F its ability to interact with the mRNA at its capped 5' terminus. A second subunit is called p220 because of its approximate size of 220,000 daltons. In poliovirus-infected cells eIF4F is inactivated by the proteolytic cleavage of p220 (45). eIF4A exists in two forms. One has been called $eIF4A_c$ (43) and is complexed with the cap-binding protein and p220 where it is regarded as the third eIF4F subunit. However, eIF4A is also present in a free, uncomplexed form called $eIF4A_f$. eIF4A is an ATP-dependent single-stranded RNA-binding protein with the ability to melt RNA secondary structure. In a typical mRNA with a moderate amount of secondary structure, eIF4F interacts with the 5' cap of the mRNA via the 24 kD cap-binding protein and then $eIF4A_c$ melts the 5' secondary structure in an ATP-dependent reaction (46-48). An additional factor, eIF4B, stimulates eIF4A and eIF4F, and may play a role in locating the initiation site

(47,49,50). Thus prepared, the mRNA is able to bind the 40S subunit to form a preinitiation complex. In this model, discussed in a recent review by Sonenberg (43), the scanning function previously attributed to the 40S ribosome may in fact be a property of eIF4F. Given a 5' entry site for eIF4F and the apparent processivity of its RNA unwinding function, ribosome binding sites are revealed in a directional fashion. The 40S subunit may then bind directly to sites thus exposed on the RNA. In mRNAs with significant secondary structure, the unveiling of ribosome binding sites by RNA unwinding is cap-dependent, because in such cases the binding of eIF4A does not occur directly, but is mediated by cap-binding protein in the eIF4F complex. The more stable the secondary structure, the more difficulty eIF4F has in unwinding it. Thus, mRNAs with stable secondary structure tend to be less efficiently translated, and in a relatively cap-dependent manner. On the other hand, mRNAs devoid of stable 5' UTR secondary structure tend to be more efficiently translated and in a relatively cap-independent fashion, since free eIF4A is able to bind the RNA directly without the necessity of an initial binding event at the 5' end, and without expending effort in melting secondary structure.

Since an absence of secondary structure in the 5' UTR is able to confer cap-independent translation, it has been proposed that such an unstructured region upstream from an internal initiator AUG may permit ribosomes to initiate directly there, ignoring secondary structure and potential initiators in upstream positions (51). It is proposed that this mechanism is utilized in the translation of picornaviruses like poliovirus (26,43), for example, whose initiation codon is preceded by numerous AUG triplets in an 800 nucleotide 5' UTR. According to this model eIF4A is able to bind this internal single-stranded region without the assistance of other components of the cap-binding complex, eIF4F, thus creating ribosome-binding sites without regard to upstream sequences. In fact, one aspect of the mechanism by which poliovirus shuts off host cell protein synthesis is inactivation of the p220 subunit of eIF4F (45). Thus, in infected cells the translation of mRNAs that don't require intact eIF4F is favored, and messengers that lack stable 5' UTR secondary structure tend to be less susceptible to poliovirus-induced shutoff (52,53).

Internal initiation mediated by a picornaviral sequence has potential applications in biotechnology. Often it is desirable to construct permanent cell lines that synthesize the product of a heterologous gene. This is usually accomplished by co-transfection of the cloned gene in the presence of a plasmid containing a selectable marker (e.g., G418 resistance) followed by screening for independent stable transformants that have integrated both the selected and unselected sequences into the genome. However, recent experiments show that sequences from the genomes of poliovirus and encephalomyocarditis virus are able to confer translatability to the downstream sequence in a dicistronic mRNA. Therefore, any coding sequence that produces a product of interest could be linked to a downstream selectable marker in a dicistronic transcription unit with the internal initiation sequence between them. Selection for the expression of the downstream, internally initiated product would virtually ensure the expression of the upstream unselected sequence. Moreover, amplification of the selectable gene would lead to increased copy number of the linked, unselected sequence, and result in its overexpression.

The exact step of the initiation process that is inhibited by RNA secondary structure is probably a function of the position of such structure within the 5' UTR. One study indicates that although the interaction of the cap-binding component of eIF4F is not inhibited by 5' proximal secondary structure, a subsequent rearrangement that leads to eIF4A and eIF4B interaction with mRNA is drastically reduced when secondary structure extends all the way to the cap (54). In these experiments hybridization of synthetic deoxyribonucleotides to mRNA was used to mimic the effects of mRNA secondary structure. Hybridization to the first 15 nucleotides of an mRNA does not appreciably inhibit initial cap-binding, but does seem to inhibit the subsequent interaction of other eIF4F components involved in melting secondary structure. On the other hand, hybridization to sites downstream of nucleotide 15 does not markedly affect either the initial cap-binding step or the subsequent interaction with eIF4A and eIF4B. Messengers with single-strandedness at the extreme 5' end, on the other hand, are apparently more suitable substrates for the eIF4A and eIF4B interaction. The single-stranded character in the 5' proximal 15 nucleotides is probably sufficient to provide the required foothold. Analysis of the leader sequence of adenovirus late mRNAs indicates that the 5' terminal 25 nucleotides is single-stranded (55). This is thought to account for the efficient translation and relative resistance to poliovirus-induced shutoff of translation of these mRNAs (53).

Secondary structure in a 5' UTR apparently presents a barrier to the RNA unwinding function of eIF4F; the greater the degree of secondary structure, the greater the difficulty of melting it to reveal sites for ribosome binding. This would be reflected in a reduced rate access of 40S subunits to the initiation site. But how stable is too stable? In the Human Immunodeficiency Virus (HIV), Parkin et al. observed that a structure in the 5' UTR with a stability of -53 kcal/mol dramatically inhibits translation (34). Similarly, Kozak found that under normal conditions an artificially introduced hairpin of stability of -30 kcal/mol in preproinsulin mRNA does not represent a substantial barrier to translation, but a similar hairpin of greater stability (-50 kcal/mol) reduces translation efficiency by about 90% (33). This suggests that a hairpin must be quite stable before an inhibitory effect is obvious. However, it should be pointed out that under conditions of hypertonic stress an inhibitory effect of the less stable hairpin is also observed. This may indicate a reduced fitness for competition with other mRNAs.

All this suggests that stable secondary structure, which probably means any uninterrupted structure with a stability in the vicinity of -30 to -50 kcal/mol or better, should be avoided if possible in the 5' UTR. This, of course, includes interactions of the 5' UTR with downstream portions of the mRNA. Such elements can be removed either by deletion or by the introduction of point mutations. One assumes that occasional interruptions of the folding pattern (i.e., bulges or loops) are sufficient to accomplish the desired end. A structure so interrupted is likely to be more easily melted than a perfect hairpin even if the two structures have similar overall stability.

Sometimes it may be possible to confer the translational activity of a 5' UTR to a heterologous sequence. For example, the presence of the adenovirus tripartite leader sequence seems to elevate the translational activity of mRNAs to which it is artificially linked (56). The tripartite leader apparently folds as a relatively independent entity (55). On the other hand, the translational properties of some mRNAs could be determined by interactions between the 5' UTR and sequences 3' of the initiation codon. Thus, in linking a given leader to a heterologous coding sequence one may destroy long-range interactions, or create new ones.

The Effects of Leader Length

Increasing the length of the 5' UTR might be expected to decrease the translation efficiency of an mRNA. It is possible, for example, that requiring the translation apparatus to scan a great distance might increase the chance of dissociation from the mRNA before the initiation site is encountered. At present it is not possible to give a simple answer to this question. The problem is that if one increases leader length, the amount of secondary structure may also be increased and translation inhibited by a mechanism that has nothing to do with length per se. On the other hand, increasing the length of the 5' UTR with sequences that are unable to form secondary structure may confer on the mRNA the ability to initiate translation by the cap-independent mechanisms that we have already discussed. An example of this phenomenon was recently observed in the author's laboratory. The length of the 5' UTR of chloramphenicol acetyltransferase mRNA was increased by two methods. In the first, the 70 nucleotide 5' UTR from the SV40 early mRNA was fused to the CAT sequence. The leader was then further elongated by duplicating and then quadruplicating this 70 nucleotide sequence. Thus CAT derivatives with 70, 140, and 280 additional 5' nucleotides were created. In these experiments the in vitro translation efficiency of the CAT mRNA decreased up to five-fold as the leader length increased. Computer predictions of secondary structure indicate a dramatic increase in stable folding (from -21 to -56 to -126 kcal/mol) as the 70 nucleotide leader was multiplied. On the other hand, when 380 nucleotides were added to the 5' end of CAT by the addition of 76 repeats of the sequence GAAGA, a sequence incapable of the formation of any conventional secondary structure, the translation efficiency actually increased slightly. The opposing effects of lengthening the 5' UTR are probably the result of the different capacities of the added sequences to form secondary structure.

Leader length may play a more critical role when the 5' UTR becomes very short. This is a point that has never been systematically investigated, but in naturally-occurring mRNAs the distance from cap to initiation site is seldom less than 20 nucleotides (5). It is possible that when a leader is fewer than 10 nucleotides long the initiation site may become less visible to the translation machinery (57-60), lowering the efficiency of initiation at that site and creating the potential for internal initiation events.

Upstream AUGs can Serve as a Barrier to Initiation

As already mentioned, the vast majority of mammalian mRNAs initiate translation at the AUG triplet nearest the 5' end. Moreover, the presence of upstream AUGs generally reduces initiation at internal sites substantially. The efficiency with which an upstream AUG is a barrier to internal initiation is apparently a reflection of its own efficiency as an initiator. Thus, an upstream AUG that resides in a favorable sequence context for initiation may efficiently inhibit translation of a downstream reading frame (13). However, the extent of inhibition is also a function of the position at which the upstream reading frame is terminated relative to the internal initiation site. If the upstream reading frame terminates upstream, or a short distance downstream of an internal AUG, the inhibitory influence of the upstream AUG may be partially or even completely relieved (11-13). Such internal initiations are thought generally to occur by a translation reinitiation mechanism. In most instances, though, it has been difficult to rule out categorically the alternative possibility that active translation through a potential initiation site occludes ribosomes that are gaining access to the site by leaky scanning, or perhaps even by direct internal binding.

In any event, it is clear that to engineer the efficient translation of a sequence cloned in a mammalian gene transducing vector, it will usually pay to remove, by deletion or by point mutation, any AUG triplets upstream of the desired initiation codon. In some instances this will not be possible. For example, one may be shotgun cloning random DNA fragments upstream of an indicator gene (e.g., lacZ or CAT) looking for promoter activity. Frequently fragments with the desired transcriptional activity could be overlooked because they produce transcripts containing AUGs upstream of the reporter sequence, thus inhibiting its translation. In such instances the vector should be designed to contain translational terminators in all three reading frames between the promoter cloning site and the reporter sequence, in hopes of at least partially relieving the inhibitory effect. Synthetic oligonucleotides containing such terminator sequences are readily available from commercial sources or may be custom made to specific requirements.

As an example of the dramatic effects that appropriately positioned terminators of an upstream reading frame can have on the expression of an internally-encoded product, consider the expression of DHFR from the SV40 late region (12). When the DHFR sequence was cloned downstream of the VP2 and VP3 initiation codons in the vector called SVGT7, the efficient translation of DHFR depended on the termination of the VP2/3 reading frame upstream of the DHFR initiation codon. Under optimal conditions the DHFR protein could reach steady state levels that represented a few percent of total cellular protein.

FACTORS THAT DETERMINE THE INHERENT EFFICIENCY OF THE TRANSLATION INITIATION SITE

Utilization of Triplets other than AUG

An obviously important aspect of the translation initiation site is the initiation codon itself. For some time, AUG was considered to be the only initiation codon in eukaryotic cells (61). This was in contrast to bacteria where utilization of non-AUG triplets for translation initiation was well documented. Although in nearly every mammalian mRNA examined so far the initiator triplet is AUG, there are a few examples of initiation at non-AUG triplets (18-25). ACG and CUG are the most efficient of the non-AUG triplets so far tested and represent all the known natural examples of non-AUG initiators (18,21-23,25). In every case initiation at a non-AUG triplet is less efficient than at an AUG in an equivalent sequence context. Since initiation at such sites is leaky, non-AUG initiators provide a mechanism for the production of two proteins from a single RNA.

The Importance of the Sequence Context of the Initiation Codon

Inspection of the sequences of hundreds of genes supports the idea that sequences immediately surrounding an AUG triplet can influence its ability to serve as an initiation codon. The most recent statistical analyses of nucleotide sequences in the vicinity of translational start sites have resulted in the idea

that GCCGCC(A/G)CCAUGG is the optimal sequence for translation
initiation (5).   In her mutational analyses of a synthetic
translation initiation site fused to preproinsulin, Kozak has
experimentally verified the importance of much of this sequence
for efficient translation (14-16).   By far the most important
element is the purine residue at -3.   The G at +4 is also of some
importance, but the contribution of the other elements of this
consensus sequence is observed only when the important -3
position is perturbed. The importance of the -3 purine is also
indicated by the existence of a type of thalassemia in which
globin synthesis is reduced by the mutation of the A residue
normally present at -3 to a C (62).   Although these results
clearly indicate that this consensus sequence is important in
some cases, recent results in other systems raise doubts that
this is the full story.   For example the S1 and S4 mRNAs of
reovirus are translated with very different efficiencies, with S4
producing 5- to 20-fold more protein per mRNA than S1.   The S4
initiation site conforms to the consensus and contains the
crucial purine (G) at -3.   Meanwhile, the S1 sequence contains C
at that position.   To determine whether this accounts for the
difference in translation efficiencies of the two mRNAs,
Munemitsu and Samuel (63) improved the fit of the S1 sequence to
the consensus sequence by changing the -3 C to G.   At the same
time they attempted to worsen the S4 initiation sequence by
changing the -3 G to C. The synthesis of the specific products of
these genes was quantitated after their transfection into
mammalian cells.   Surprisingly, none of the mutations had any
effect on the translation efficiency of the S1 and S4 mRNAs.
Similar results have been observed for translation of mRNA for
mouse dihydrofolate reductase.   When its translation initiation
sequence, which naturally conforms closely to the consensus
sequence, is altered by placing pyrimidine residues at both the
-3 and +4 positions, synthesis of DHFR is reduced by no more than
50% in vivo (G. Pickett and D.S. Peabody, unpublished results).
Moreover, the same mutations have no effect at all on AUG selection or overall translation efficiency when DHFR is translated in
vitro.   However, if the translation initiation site is weakened
by converting the AUG to ACG, initiation becomes exquisitely
sensitive to context mutations (20).   Such observations raise the
possibility that additional elements may modulate the efficiencies of initiation codon utilization in some mRNAs.

Effects of Secondary Structure on AUG Selection

In principle, secondary structure might influence the
fidelity of initiation codon recognition. Little experimental
information is available on this point, but at least two
mechanisms can be envisioned which could result in such an
effect.   If an AUG resides within an element of stable secondary
structure, perhaps the initiation machinery will tend to bypass

it. This, of course, depends on whether the progress of the
initiation apparatus toward the initiation site requires the
complete melting of all secondary structure in its path, like
pulling a thread through the eye of a needle. However, at least
one experiment argues against the idea that secondary structure
can cause the ribosome to skip an initiator AUG (33). A strong
initiator AUG represents just as effective a barrier to downstream initiations when sequestered within secondary structure as
when it is not. Alternatively, secondary structure could affect
AUG selection by altering the rate of scanning. If the scanning
apparatus is stalled over an AUG, an increased kinetic opportunity for its recognition as an initiation site would be
created. On the other hand, in regions where scanning proceeds
rapidly, AUGs could tend to be bypassed. It is presently unclear
whether such mechanisms exist, but mRNA secondary structure has
been claimed to affect AUG selection in at least one case (64).

TRANSLATIONAL ENHANCEMENT OF TRANSFECTED SEQUENCES BY COTRANSFECTION WITH THE ADENOVIRUS VA$_I$ GENE

As a response to adenovirus infection host cell protein
synthesis is shut off. This is the result of a complex series of
events that include the inhibition of transcription and the
reduced export of non-viral mRNAs from the nucleus to the
cytoplasm. There are also translational alterations so that
adenoviral mRNAs are disproportionately represented on polysomes.
Adenovirus encodes two small RNAs from tandem transcription
units, one of which (VA$_I$) seems to be responsible for ability of
the virus to escape the translation inhibition (reviewed in ref.
65). The VA RNAs are produced in large amounts at late times of
adenovirus infection as products of transcription by RNA
polymerase III. Deletions that inactivate the VA genes result in
about a ten-fold depression of translation compared to that
observed in cells infected with the wild-type virus (66,67). This
shutoff of protein synthesis is the result of activation of
Double-Stranded RNA Activated Inhibitor (DAI), or Pl/eIF-2
kinase, an enzyme which, in its activated form, phosphorylates a
subunit of the translation initiation factor eIF2 (68,69). The
extent of protein synthesis inhibition is determined by the
extent to which the balance of the phosphorylated and non-
phosphorylated forms of eIF2 is tipped in favor of phosphorylation. To understand the mechanism of this inhibition, it is
necessary to describe briefly the role of eIF2 and its interactions with other components of the initiation process.
In an early event during initiation, eIF2 forms a ternary
complex with GTP and methionyl-tRNA$_i$, a complex which then
associates with the 40S ribosome into a preinitiation complex
(reviewed in ref. 70). Subsequent reactions result in mRNA
binding and association of the 60S subunit to form an 80S

initiation complex. During this process the eIF2-bound GTP is hydrolyzed to GDP, with the eventual liberation of an eIF2-GDP complex. In order to recycle eIF2 for use in additional cycles of initiation the GDP must be exchanged for GTP, a reaction accomplished by GEF, or Guanine-Nucleotide Exchange Factor (also known as eIF2B). It is this exchange reaction that is inhibited by the phosphorylation of eIF2. The phosphorylated form of eIF2 associates tightly with eIF2B, preventing the GTP-GDP exchange, and inhibiting translation initiation at the level of ternary complex formation. However, in the presence of the VA_I RNA, activation of the kinase is inhibited, eIF2 phosphorylation is prevented, and GTP-GDP exchange occurs normally.

The kinase is apparently also activated by transfection into animal cells of genes cloned on certain plasmid vectors, so that translation of a transfected sequence is often enhanced by the cotransfection of VA_I. The VA_I mediated enhancement of translation of transfected sequences was first demonstrated with cloned adenovirus genes. The presence of the cotransfected VA_I gene resulted in a 10- to 20-fold increase in expression of the products of the adenovirus E3 and E2A genes cloned on a plasmid (71). Large translational enhancements were also reported for non-adenovirus sequences linked to the adenovirus tripartite leader or to the 5' UTR of the SV40 early mRNA (56). Subsequently, it was shown that translation of transfected genes that contain no adenovirus sequences at all can also be stimulated by VA_I cotransfection (72), or by cotransfection of a mutant gene for a non-phosphorylatable eIF2 (73). In the absence of VA sequences similar enhancements are achieved by including 2-aminopurine, an inhibitor of the P1/eIF2 kinase, in the culture medium (74). Surprisingly, these enhancements of protein synthesis seem to be restricted to plasmid-derived mRNAs, since the synthesis of most cellular proteins seems unaffected (73).

Not all plasmid vectors activate the P1/eIF2 kinase. For example, Kaufman et al. (73) showed that sequences cloned in a vector containing a pBR322 backbone required cotransfection of VA_I for efficient translation of plasmid-derived sequences, but a similar vector based on pUC18 had no such requirement.

CONCLUSIONS

Because of our partial knowledge of the determinants of mRNA translational activity, the deliberate modulation of specific translational efficiency is still a tricky business. At this point it is difficult to guarantee that any single approach will be successful in any given case. However, by paying attention to a few aspects of mRNA structure, the investigator can increase the chances of success. In some instances the level of synthesis of the product of a transfected gene can be impressive. For example, Kaufman and colleagues created a vector system that contains a variety of features designed to optimize expression (73,75,76). These include genetic signals for the efficient replication of the vector and for the high level synthesis and efficient processing of mRNA. Also included are sequences designed to maximize the translation efficiency of cloned sequences. The vector contains the adenovirus VA genes so that they are introduced into every cell that receives the plasmid. It also produces mRNAs which contain the adenovirus tripartite leader. When the DHFR sequence is present in this vector, the enzyme has been produced at levels around 28% and eIF2 at about 15% of total cellular protein (73,74).

REFERENCES

1   Kozak, M. (1978) Cell 15, 1109-1123.
2   Kozak, M. (1979) Nature 280, 82-85.
3   Kozak, M. (1978) J. Biol. Chem. 253, 6568-6577.
4   Kozak, M. (1979) J. Biol. Chem. 254, 4731-4738.
5   Kozak, M. (1987) Nucl. Acids Res. 15, 8125-8148.
6   Kitamura, N., Semler, B.L., Rothberg, P.G., Larsen, G.R., Adler, C.J., Dorner, A.J., Emini, E.A., Hanecak, R., Lee, J.J., van der Werf, S., Anderson, C.W. and Wimmer, E. (1981) Nature 291, 547-553.
7   Mulligan, R.C. and Berg, P. (1981) Mol. Cell. Biol. 1, 449-459.
8   Subramani, S., Mulligan, R.C. and Berg, P. (1981) Mol. Cell. Biol. 1, 854-864.
9   Southern, P.J. and Berg, P. (1982) J. Mol. Appl. Genet. 1, 327-341.
10 Liu, D., Simonsen, C.C. and Levinsion, A.D. (1984) Nature 309, 82-85.
11 Peabody, D.S. and Berg, P. (1986) Mol. Cell. Biol. 6, 2695-2703.
12 Peabody, D.S., Subramani, S. and Berg, P. (1986) Mol. Cell. Biol. 6, 2704-2711.
13 Kozak, M. (1984) Nucl. Acids Res. 12, 3873-3893.
14 Kozak, M. (1984) Nature 308, 241-246.
15 Kozak, M. (1986) Cell 44, 283-292.
16 Kozak, M. (1987) J. Mol. Biol. 196, 947-950.
17 Lutcke, H.A., Chow, K.C., Mickel, F.S., Moss, K.A., Kern, H.F. and Scheele, G.A. (1987) EMBO J. 6, 43-48.
18 Becerra, S.P., Rose, J.A., Hardy, M., Baroudy, B.M. and Anderson, C.W. (1985) Proc. Nat. Acad. Sci. U.S.A. 82, 7919-7923.
19 Anderson, C.W. and Buzash-Pollert, E. (1985) Mol. Cell. Biol. 5, 3621-3624.
20 Peabody, D.S. (1987) J. Biol. Chem. 262, 11847-11851.
21 Hann, S.R., King, M.W., Bentley, D.L., Anderson, C.W. and Eisenman, R.N. (1987) Cell 52, 185-195.

22   Curran, J. and Kolakofsky, D. (1988) EMBO J. 7, 245-251.
23   Gupta, K.C. and Patwardhan, S. (1988) J. Biol. Chem. 263,
     8553-8556.
24   Peabody, D.S. (1989) J. Biol. Chem. 264, 5031-5035.
25   Prats, H., Kaghad, M., Prats, A.C., Klagsbrun, M., Lélias,
     J.M., Liauzun, P., Chalon, P., Tauber, J.P., Amalric, F.,
     Smith, J.A. and Caput, D. (1989) Proc. Nat. Acad. Sci.
     U.S.A. 86, 1836-1840.
26   Pelletier J. and Sonenberg N. (1988) Nature 334, 320-325.
27   Jang, S.K., Krausslich, H.-G., Nicklin, M.J.H., Duke, G.M.,
     Palmenberg, A.C. and Wimmer, E. (1988) J. Virol. 62, 2636-
     2643.
28   Bienkowska-Szewczyk, K. and Ehrenfeld, E. (1988) J. Virol.
     62, 3068-3072.
29   Pelletier, J. and Sonenberg, N. (1989) J. Virol 63, 441-444.
30   Jang, S.K., Davies, M.V., Kaufman, R.J. and Wimmer, E.
     (1989) J. Virol. 63, 1651-1660.
31   Herman, R.C. (1989) TIBS 14, 219-222.
32   Pelletier, J. and Sonenberg, N. (1985) Cell 40, 515-526.
33   Kozak, M. (1986) Proc. Nat. Acad. Sci. U.S.A. 83, 2850-2854.
34   Parkin, N.T., Cohen, E.A., Darveau, A., Rosen, C.,
     Haseltine, W. and Sonenberg, N. (1988) EMBO J. 7, 2831-2837.
35   Pelletier, J. and Sonenberg, N. (1987) Biochem. Cell. Biol.
     65, 576-581.
36   Payvar. R. and Schimke, R.T. (1979) J. Biol. Chem. 254,
     7637-7642.
37   Morgan, M.A. and Shatkin, A.J. (1980) Biochemistry 19, 5960-
     5966.
38   Kozak, M. (1980) Cell 19, 79-90.
39   Gehrke, L., Auron, P.E., Quigley, G.J., Rich, A. and
     Sonenberg, N. (1983) Biochemistry 22, 5157-5164.
40   Sonenberg, N., Guertin, D., Cleveland, D. and Trachsel, H.
     (1981) Cell 27, 563-572.
41   Hewlett, M.J., Rose, J.K. and Baltimore, D. (1976) Proc.
     Nat. Acad. Sci. U.S.A. 73, 327-330.
42   Nomoro, A., Lee, Y.F. and Wimmer, E. (1976) Proc. Nat. Acad.
     Sci. U.S.A. 73, 375-380.
43   Sonenberg, N. (1988) Prog. Nucl. Acids Res. Mol. Biol. 35,
     173-207.
44   Sonenberg, N. (1987) Adv. Virus Res. 33, 175-204.
45   Etchison, D., Milburn, S.C., Edery, I., Sonenberg, N. and
     Hershey, J.W.B. (1982) J. Biol. Chem. 257, 14806-14810.
46   Ray, B.K., Lawson, T.G., Kramer, J.C., Cladaras, M.H.,
     Grifo, J.A., Abramson, R.D., Merrick, W.C. and Thach, R.E.
     (1985) 260, 7651-7658.
47   Grifo, J.A., Abramson, R.D., Satler, C.A. and Merrick, W.C.
     (1984) J. Biol. Chem. 259, 8648-8654.
48   Abramson, R.D., Dever, T.E., Lawson, T.G., Ray, B.K., Thach,
     R.E. and Merrick, W.C. (1987) J. Biol. Chem. 262, 3826-3832.

49   Gross, D.J., Woodley, C.L. and Wahba, A.J. (1987) Biochem-
      istry 26, 1551-1556.
50   Butler, J.S. and Clark, Jr., J.M. (1984) Biochemistry 23,
      809-815.
51   Abramson, R.D., Dever, T.E. and Merrick, W.C. (1988) J.
      Biol. Chem. 263, 6016-6019.
52   Sonenberg, N., Guertin, D. and Lee, K.A.W. (1982) Mol. Cell.
      Biol. 2, 1633-1638.
53   Dolph, P.J., Racaneillo, V., Villamarin, A., Palladino, F.
      and Schneider, R. (1988) J. Virol. 62, 2059-2066.
54   Lawson, T.G., Ray, B.K., Dodds, J.T., Grifo, J.A., Abramson,
      R.D., Merrick, W.C., Betsch, D.F., Weith, H.L. and Thach,
      R.E. (1986) J. Biol. Chem. 261, 13979-13989.
55   Zhang, Y., Dolph, P.J. and Schneider, R.J. (1989) J. Biol.
      Chem. 264, 10679-10684.
56   Kaufman, R.J. (1985) Proc. Nat. Acad. Sci. U.S.A. 82, 689-
      693.
57   Sedman, S.A., Good, P.J. and Mertz, J.E. (1989) J. Virol.
      63, 3884-3893.
58   Strubin, M., Long, E.O. and Mach, B. (1986) Cell 47, 619-
      625.
59   Peterson, C.A. and Piatigorsky, J. (1986) Gene 45, 139-147.
60   Kelley, D.E., Coleclough, C. and Perry, R.P. (1982) Cell 29,
      681-689.
61   Sherman, F., McKnight, G. and Stewart, J.W. (1980) Biochim.
      Biophys. Acta 609, 343-346.
62   Morlé, F., Lopez, B., Henni, T. and Godet, J. (1985) EMBO J.
      4, 1245-1250.
63   Munemitsu, S.M. and Samuel, C.E. (1988) Virology 163, 643-
      646.
64   van Diujn, L.P., Holsappel, S., Kasperaitis, M., Bunschoten,
      H., Konings, D. and Voorma, H.O. (1988) Eur. J. Biochem.
      172, 59-66.
65   Schneider, R.J. and Shenk, T. (1987) in Translational
      Regulation of Gene Expression (Ilan, J., ed.) pp. 431-445,
      Plenum Press, New York, NY.
66   Thimmappaya, B., Weinberger, C., Schneider, R.J. and Shenk,
      T. (1982) Cell 31, 543-551.
67   Schneider, R.J., Weinberger, C. and Shenk, T. (1984) Cell
      37, 291-298.
68   Schneider, R.J., Safer, B., Munemitsu, S.M., Samuel, C.E.
      and Shenk, T. (1985) Proc. Nat. Acad. Sci. U.S.A. 82, 4321-
      4325.
69   O'Malley, R.P., Mariano, T.M., Siekierka, J. and Matthews,
      M.B. (1986) Cell 44, 391-400.
70   Safer, B. (1983) Cell 33, 7-8.
71   Svensson, C. and Akusjarvi, G. (1984) Mol. Cell. Biol. 4,
      736-742.
72   Svensson, C. and Akusjarvi, G. (1985) EMBO J. 4, 957-964.

73   Kaufman, R.J., Davies, M.V., Pathak, V.K. and Hershey, J.W.B. (1989) Mol. Cell. Biol. 9, 946-958.
74   Kaufman, R.J. and Murtha, P. (1987) Mol. Cell. Biol. 7, 1568-1571.
75   Kaufman, R.J., Murtha, P. and Davies, M.V. (1987) EMBO J. 6, 187-194.
76   Wang, G.G., Witek, J.S., Temple, P.A., Wilkens, K.M., Leary, A.C., Luxemberg, D.P., Jones, S.S., Brown, E.L., Kay, R.M., Orr, E.C., Shoemaker, C., Golde, C.W., Kaufman, R.J., Hewick, R.M., Wang, E.A. and Clark, S.C. (1985) Science 228, 810-815.

THE POLYMERASE CHAIN REACTION

Norman Arnheim

Molecular Biology
University of Southern California
Los Angeles, CA 90089-1340

INTRODUCTION

The development of a new technology can lead to novel approaches to answering current scientific questions. A new technology can also often lead to the framing of scientific questions that previously were considered intractable and not amenable to experimental investigation. The recombinant DNA revolution, Southern blotting and DNA sequencing are examples of technologies that have had such an impact on the biological sciences. Recently, a new molecular biological technique, the polymerase chain reaction (PCR) was developed. This method of in vitro gene amplification can in many circumstances simplify the standard procedures for cloning, analyzing and modifying nucleic acids. However, besides simplifying many recombinant DNA methodologies, PCR has opened the door to the investigation of a number of biological questions that heretofore were considered unanswerable. In this review, the basic principles of PCR along with modifications of the basic scheme will be discussed. The application of these methods to a number of standard genetic engineering procedures will be reviewed. Important factors that need to be taken into consideration in designing experiments using PCR are also examined.

THE BASIC METHOD

The polymerase chain reaction involves the enzymatic amplification of DNA in vitro (1,2) and was originally developed at the Cetus Corporation. This method is capable of increasing the amount of a target DNA sequence in a sample by synthesizing

Genetic Engineering, Vol. 12
Edited by J.K. Setlow
Plenum Press, New York, 1990

many copies of the DNA segment. PCR is carried out in discrete
cycles and each cycle of amplification can, if 100% efficient,
double the amount of target DNA. The target is exponentially
amplified such that after n cycles there is 2^n times as much
target as was present initially.

The principle of the PCR method is shown in Figure 1. A
target DNA sequence to be amplified is chosen first (Figure 1-I).
It is not necessary to know the nucleotide sequence of the target
DNA but two small stretches of DNA of known sequence that flank
this target segment must be found and used to design two
oligonucleotide primers. Each synthetically made single-stranded
primer is usually 20 nucleotides long. The sequence of each
primer is chosen so that one of them has base pair
complementarity with one of the flanking sequences while the
other is complementary to the other flanking sequence but on the
opposite strand. Thus, after denaturation of the double-stranded
target DNA into single strands, the primers will be able to
hybridize to their complementary sequences flanking the target
gene (Figure 1-II). The primers are oriented so that when they
form a duplex with the flanking sequences, their 3' hydroxyl ends
face the target sequence. Following primer annealing, the
addition of a DNA polymerase will result in an extension of the
primers through the target sequence thereby making copies of the
target (Figure 1-III). This series of steps, DNA denaturation,
primer hybridization and DNA polymerase extension, represent one
PCR cycle and each of the three steps is carried out at an
appropriate temperature. If the extension product of each primer
is long enough, it will include the sequences complementary to
the other primer. Thus, the extension products themselves can
act as templates in future cycles (Figure 1-IV). It is this fact
that accounts for the exponential increase in PCR product with
cycle number. After the first few cycles the major product is a
DNA fragment which is exactly equal in length to the sum of the
lengths of the two primers and the intervening target DNA. If
the amount of target exactly doubles with each cycle, as few as
20 cycles will generate about a million times more target
sequence than is present initially.

In a mammalian genome containing in excess of 10^9
nucleotides, a 100 base pair target would represent only 10^{-7} of
the available DNA. The annealing of the primers to other
sequences besides the target in such complex templates, even if
it occurred rarely, could result in amplification of non-target
sequences as well as the target itself, thereby reducing the
purity of the target in the final product. The extent to which
this imperfect annealing and extension can occur depends upon the
temperature at which the primer annealing and polymerase
extension steps are carried out, since the specificity of primer
annealing is greater at higher temperatures. The earliest PCR
experiments (1,2) utilized the Klenow fragment of E. coli DNA
polymerase I at a temperature of 37°C and consequently often

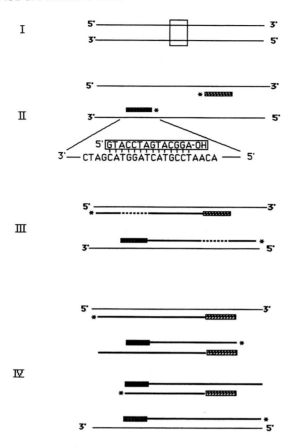

Figure 1. Fundamental principle of PCR. I: DNA double helix
with boxed target segment. II: Hybridization of PCR primers to
opposite strands of the region flanking the target. The 3' end
of each primer is indicated by *. The base-base interactions of
one primer (boxed) with the region flanking the target is shown
below. III: Result of extending each primer with a DNA
polymerase. The dotted region of each extension product
indicates the portion that is complementary to the other primer.
IV: PCR products after the material in III was subjected to
another round of amplification.

produced incompletely pure target product as judged by gel
electrophoresis.   The recent isolation of a heat resistant DNA
polymerase from Thermus aquaticus (Taq) allows primer annealing
and extension to be carried out at an elevated temperature (3).
This favors the annealing of the primers to the regions flanking
the specific target, and reduces mismatched annealing to
sequences elsewhere in the genome.   This added selectivity now
allows the experimenter to produce large amounts of virtually

pure target DNA for further characterization. In addition, the thermostability of the Taq polymerase allows it, unlike the Klenow enzyme, to escape inactivation during the DNA denaturation step. The fact that Taq polymerase need not be added at each cycle led to the development of automated PCR machines which, in turn, have been a significant factor in the rapid application of this technology by the scientific community. Many different makes of automated thermocyclers are on the market and a number of designs for constructing "home made" machines have been published (4-6).

In practice, carrying out PCR is straightforward. The reagents are commonly used in molecular biology and, in addition to the Taq polymerase, even the oligonucleotide primers can be purchased commercially. Amplification by PCR is also extremely rapid. Twenty-five cycles can be carried out in just over one hour. Beginning with 1 µg of human DNA, which contains about 300,000 copies of each unique sequence, 25 cycles of PCR can generate up to several µg of a specific product several hundred base pairs in length. While purified DNA is used in many applications it is not required. Crude cell lysates provide excellent templates (7). The DNA need not even be intact relative to the requirements of other standard molecular biological procedures as long as some molecules exist which contain sequences complementary to both primers. Under some circumstances even this is not absolutely required. If no molecules are left completely intact but a series of overlapping subfragments exist which, together, completely span the target region, PCR is capable of reconstructing the original fragment by "jumping" (8).

PCR is capable of producing large amounts (picomoles) of a specific DNA sequence using a nucleic acid sample of high complexity as a template. Even samples containing only the DNA present in a single haploid cell, that is, a single molecule of unique target sequence, can be amplified to detectable levels (9). The exquisite sensitivity of PCR is one of its most powerful attributes yet this can also be its Achilles heel. In carrying out PCR experiments one needs to be acutely aware of the possibility of contamination. The usual suspected source of contamination is PCR product derived from earlier experiments. If one considers that a single PCR reaction tube can contain several picomoles of product, or on the order of 10^{12} molecules, it is easy to imagine how even one millionth of that product (a million molecules) could reek havoc if it were to come in contact with a 1 microgram sample of genomic DNA which contains on the order of 300,000 copies of a unique target sequence. It is necessary therefore that special precautions be taken when carrying out PCR experiments so as to reduce the formation of aerosols and other possible contamination routes. Physically separating the areas in the laboratory where PCR reactions are set up from those where the PCR products are analyzed goes a long

way towards reducing contamination. Having a set of pipettes
which are dedicated to setting up PCR reactions is also
indicated. A number of other precautions are discussed by Kwok
and Higuchi (10).

THE APPLICATION OF PCR TO BASIC METHODS OF GENETIC ENGINEERING

DNA Cloning

    PCR can rapidly produce large amounts of a specific DNA
fragment starting with a complex genomic DNA template. It takes
considerably longer to accomplish the same task by traditional
cloning which uses in vivo replication of the target sequence
integrated into a cloning vector in a host organism. It should
be kept in mind however, that whereas in vivo cloning results in
isolation of large amounts of one of the original molecules
present in a sample, PCR produces large amounts of product
derived from all of the copies of the target present in the
original sample. Individual molecules present in the PCR
product, however, can be cloned by traditional methods and with
significantly less effort. To simplify cloning, the PCR primers
can be constructed so that in addition to the DNA sequence
homologous to the region flanking the target, additional
sequences containing a restriction endonuclease cleavage site can
be appended to the primers' 5' end (Figure 2; 11). Cutting the
PCR product with the appropriate restriction enzyme would allow
it to be easily ligated to an appropriate cloning vector. Blunt
end cloning of PCR product is also possible. However, the Taq
DNA polymerase, as well as other DNA polymerases, adds a non-
template-encoded nucleotide to the 3' end of duplex DNA rather
efficiently (12). Unless the added base is removed by an enzyme
with 3' to 5' exonuclease activity, the efficiency of blunt end
ligation can be reduced drastically.

Inverse PCR

    The basic method of PCR utilizes flanking DNA primers whose
3' ends face the target. This of course necessitates that DNA
sequence information be available on both sides of the sequence
to be amplified. The method of "inverse PCR" eliminates this
requirement (13-15). Thus it is possible to amplify a target
sequence under conditions where DNA sequence information is known
on only one side of the target. Although counter-intuitive, this
approach is topologically sound. Consider two primers which lie
within a region of known sequence. The unknown sequences on
either side of this known region comprise the target region
(Figure 3A). The primers are designed so that their 3' ends face
away from each other. Carrying out PCR will not be expected to
amplify the target sequence since the extension products of

PCR CLONING

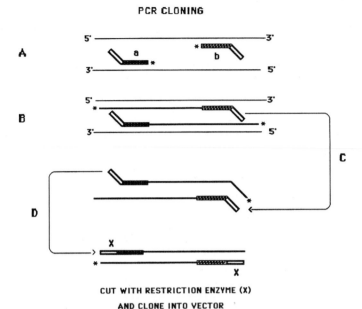

CUT WITH RESTRICTION ENZYME (X)
AND CLONE INTO VECTOR

Figure 2.   Use of modified primers to aid in cloning PCR
products.   A: Primers (a and b) used for PCR are modified by
addition, to the 5' end, of a sequence containing a restriction
endonuclease recognition site.   The 3' ends of the primers are
indicated with *.   B: Extension products of A.   C: The result
of using primer a to extend one of the products of extension in
B.   Note that the appended portion of primer b is copied at this
step.   D: The extension, by primer b, of one of the PCR products
from the previous round of amplification.   Note that the
resulting double-stranded structure contains a restriction enzyme
site at each end.

either primer will not contain sequences complementary to the
other primer.   However, if the linear DNA molecule containing the
target and primer sequences is first circularized by restriction
enzyme digestion followed by ligation, the target can be
amplified by PCR.   Figure 3B shows how circularization of the DNA
fragment results in the two primers now having their 3' ends
pointing towards each other across the target region.   The linear
product of the first round of PCR (Figure 3C) can be amplified in
the usual way.
     This clever modification of the basic PCR technique allows
amplification of unknown DNA sequences to one or the other or
both sides of a known DNA segment (depending upon which
restriction enzyme sites are chosen) with a result analogous to

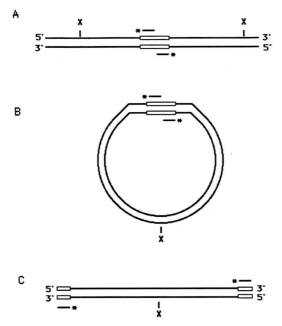

Figure 3. Inverse PCR. A: Original target of inverse PCR with
the restriction enzyme sites (X) indicated. Note that the
orientation of the PCR primers is such that the 3' ends (*) are
directed away from the region of known sequence (boxed). B: PCR
template after restriction enzyme digestion and self-ligation.
C: Linear product of the first PCR cycle of the circular
template. Note that now the 3' ends of the primers face each
other as in conventional PCR.

chromosome walking by traditional cloning methods. Of course
limitations on the size of the PCR product, about 10 kb (16),
also limit the amount of DNA that can be traversed by each step.
Inverse PCR has been used successfully to study the bacterial
integration sites of insertion sequences (14). Discussion of
additional applications can be found in Ochman et al. (17) and
the section, The Purposeful Modification of DNA by PCR.

Whole Genome PCR

The exquisite specificity of PCR allows the amplification of
only one sequence among a mixture of sequences. There are some
circumstances however where it is advantageous to amplify all of
the sequences present in a mixture with PCR, for example, if it
were desired to obtain large amounts of all of the sequences
present in a very small amount of starting material. In one
application of this approach (18) a DNA library was constructed

starting from specific segments of human chromosomes dissected
from whole chromosomes on a microscope slide. Since the amount
of starting DNA is so little, the construction of such a DNA
library is usually very inefficient. However the DNA purified
from the microdissected segments after restriction enzyme
digestion was ligated to a plasmid and plasmid sequences flanking
the insert were used for priming PCR. Amplification resulted in
levels of chromosomal DNA being produced that permitted efficient
library construction. Whole genome PCR has also been applied to
the analysis of protein-DNA interactions and is discussed below
in the section Analysis of DNA-Protein Interactions.

Analysis of RNA Populations by cDNA Amplification

Because RNA sequences can be copied by reverse transcriptase
(RT) into DNA, the polymerase chain reaction can be used to
amplify sequences present in RNA populations. A number of
different approaches have been used to amplify cDNAs and they
differ with respect to the nature of the sequence information
that is available for the construction of PCR primers. One
straightforward approach aims at the detection of an RNA species
whose sequence is known. In this case the reverse transcription
can be carried out with the use of random hexamers or a sequence
found in the body of the message itself as primer. Once the
reverse transcript is made, PCR can be initiated with a pair of
primers specific to the original message. Early applications of
this procedure included the detection of leukemia-specific mRNA
sequences present in myeloid and lymphoid cells in blood (19),
the cloning of class II HLA derived mRNAs (20), the analysis of
RNA processing (21) and quantitation of mRNA expression (22).
Additional applications of PCR to RNA analysis have recently been
reviewed (23,39).

Kawasaki and Wang (23) have wisely pointed out that,
wherever possible, primers used for amplifying cDNAs of mRNAs
should not be positioned in the same exon. This is because the
size of the PCR product produced from even the slightest amount
of contaminating DNA would be the same as the product expected
from the cDNA target. If a non-processed RNA is the PCR target,
DNAse I digestion needs to be carried out to destroy possible
contaminating DNA sequences and this needs to be confirmed with
appropriate controls including samples amplified without a prior
reverse transcription step.

PCR of RT copies of specific RNAs can also be carried out
even if the available sequence data for constructing primers are
limited. Thus, these modifications permit amplification even if
the available sequence data provide information for the design of
only one specific PCR primer. The basis for these modifications
is shown in Figure 4. This approach has been termed "single-
sided PCR" (24) or "RACE" (25). One method (Figure 4, A and B)
makes use of the fact that most mRNAs contain a polyA tail at the

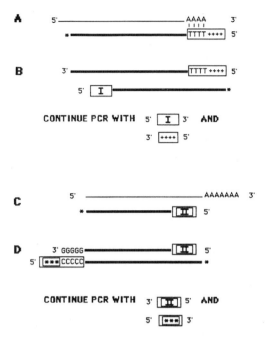

Figure 4.   Anchored PCR of cDNAs.   A and B: The use of 3'
anchors.   A: An RNA molecule with a polyA tail at the 3' end is
reverse transcribed with a oligo(dT) primer containing unique
sequences [++++] at its 5' end.   The 3' end of the extended
molecule is indicated by *.   B: Second strand synthesis is
carried out with a primer [I] which is specific for the mRNA.
PCR is continued with primer [I] and a primer identical to that
used for reverse transcription but lacking the oligo(dT)
sequences [++++].   C and D: PCR with a 5' anchor.   C: Reverse
transcription with a mRNA specific primer [II].   D: Following
addition of a polyG tail to the 3' end of the RT product, second
strand synthesis is initiated with a primer containing a polyC
run at the 3' end and a unique sequence at the 5' end [****CCCC].
PCR is continued with primer [II] and a primer containing only
the unique portion of the primer [****] used for second strand
synthesis.

3' end.   The RT step is carried out with a complex primer or
"anchor" that contains an oligo(dT) region at the 3' end and any
unique sequence at its 5' end.   Amplification of the single-
stranded DNA extension product is then carried out with two new
primers.   One primer comes from a sequence contained in the mRNA
itself, 5' of the polyA tail, and is completely responsible for
the specificity of the PCR reaction.   The other primer is
complementary to the sequence that was present at the 5' end of

the complex primer used for the RT step and which is present on
every RT extension product. Thus, this form of single-sided PCR
makes use of one specific and one nonspecific primer. This
approach would be useful if specific sequence information is
known only at the 5' end of the mRNA.

One can also amplify mRNA if specific sequence information
for primer construction is only available at the 3' end of the
message (Figure 4, B and C). This method makes use of the fact
that the 5' end of the RT product can be modified by addition of
nucleotides with terminal transferase or TdT. Reverse
transcription is initiated with a primer specific to the 3' end
of the target message. The TdT modified RT product containing,
for example, a string of dGs at the 5' end (24) can be copied
with an "anchor" primer containing, in addition to a unique
sequence, a polyC region at its 3' end. PCR is carried out with
a primer containing the unique portion of the "anchor" and the
original message-specific primer from the 3' end.

Single-sided PCR involves the use of one message-specific
primer and one primer complementary to all the RT products. As a
result only the desired RNA is amplified. It may also be
advantageous to amplify all of the messenger RNAs present in an
RNA preparation if, for example, one wanted to make a cDNA
library from a very small amount of starting material. In this
case a primer specific for the polyA tract and one specific for
the TdT addition to the 5' end of the RT product could be used
for amplification (26).

PCR Cloning of cDNA with Degenerate Primers

PCR amplification has also been used to clone specific cDNAs
in the complete absence of any available nucleic acid sequence
information. In these experiments, protein sequence data were
used to design the primers for PCR in a manner analogous to the
design of mixed oligonucleotides for screening cDNA libraries by
hybridization. In this case (27) however, the oligonucleotide
mixture is used for priming the PCR reaction. Two mixtures, one
for each flanking sequence, are required. Each mixture is
expected to contain at least one primer perfectly complementary
to the message as well as a number of primers differing only by
one or a few bases. The concentration of the primers that would
be expected to produce specific PCR product of course depends
upon the degeneracy of the primer mixture. Individual primers
containing over 200,000 different sequences have been used
successfully (28). This approach is especially useful in the
search for members of gene families with similar physiological
functions. Based upon protein sequence data new members of
several protein families have been discovered (28a-31).

Degenerate primers have also been used as an approach to
detecting uncharacterized viruses related to viruses that are
already known. Mack and Sninsky (32) used a Hepadnavirus model

system to demonstrate the utility of this method.   Using this
procedure, Shih et al. (33) have detected sequences related to
retroviral reverse transcriptases in human DNA.

Misincorporation of Bases by Taq Polymerase

The Taq polymerase has been cloned and characterized (34).
It lacks 3' to 5' exonuclease "proofreading" activity (35,36) and
therefore can misincorporate bases while making PCR product.
Estimates of the misincorporation rate come from two sources.
The first is the analysis of the DNA sequence of PCR products.
Individual PCR products were cloned and the number of base
sequence differences determined.   Approximately 0.25% of the
bases sequenced resulted from a misincorporation error using the
Taq polymerase after 30 cycles of PCR (3,37).   From this estimate
of the frequency of observed errors, a misincorporation rate per
PCR cycle can be calculated in analogy to the calculation of
bacterial mutation rates per generation based upon the frequency
of observed mutant colonies.   The estimated rate of Taq
misincorporation from the sequence data is about 1 per 10,000
nucleotides synthesized per PCR cycle (3).   A totally independent
estimate of the mutation rate agrees with this calculation.
Tindall and Kunkel (35) used a standard M13 lacZ alpha
complementation assay following in vitro DNA synthesis and
calculated a rate of 1 base substitution for every 9000 bases
polymerized.   They also were able to show that frameshift errors
occur at a frequency of 1 per 41,000.   Both kinds of studies
showed a significant excess of transition misincorporations.

The misincorporation of bases by Taq polymerase has
implications for analyzing PCR product.   The actual fraction of
amplified molecules that contain a mutation is related to the
number of bases in each target molecule, the rate of
misincorporation per base per cycle and the number of cycles.   A
mathematically precise theory for estimating the frequency of
molecules without any misincorporations has recently been put
forward (38).   Thus, for a 200 base pair target, after 30 cycles,
approximately 25% of the PCR products will contain a
misincorporated base.   Of course most of the molecules would not
contain exactly the same mutation.

The analysis of PCR product can be carried out by a variety
of molecular methods including gel electrophoresis followed by
ethidium bromide staining, DNA hybridization, denaturating
gradient gel electrophoresis, RNAse A mismatch analysis and DNA
sequencing (see 39,39a,40).   The consequences of Taq
misincorporations need to be considered in interpreting the
results of each method.   While misincorporations appear to have
relatively little effect on analysis by most of these methods,
some approaches to determining the exact nucleotide sequence of
PCR product are sensitive to this error and are discussed below
in the section Sequencing the PCR Product.

Recently, additional estimates of the Taq misincorporation rate have been made using different experimental conditions. As discussed by Gelfand and White (41) the two new estimates (42,43) came from experiments that used nucleotide and magnesium concentrations more typical of current PCR protocols (200 μM of each dNTP and 1.5 mM magnesium). The average misincorporation rate was about 40 times lower than previous estimates. If these observations are substantiated by additional studies there need be considerably less concern about the misincorporation by Taq for most applications of PCR.

There may be some applications where it is very important to have as low a misincorporation rate as possible. Under these circumstances other DNA polymerases might be used. In fact Keohavong et al. (54) showed that T4 DNA polymerase was suitable for PCR. Of course this enzyme, like the Klenow fragment of E. coli DNA polymerase, is heat sensitive and would have to be added fresh every cycle. However, compared to Taq and Klenow, the T4 enzyme has a misincorporation rate about 1000 times lower (see 55).

Sequencing the PCR Product

A number of different strategies were originally devised for directly sequencing PCR product (44-48). The double-stranded product can be sequenced after gel purification or, if the PCR product is pure enough, immediately after removing unincorporated dNTPs and primers by standard methods. Direct sequencing of the amplified product is preferred since, despite misincorporations, one will obtain the "average" sequence which is the same as that of the starting template. Standard sequencing protocols are satisfactory although some advantages have been proposed for using the Taq polymerase itself in the sequencing reaction (49). A chemical degradation method for directly determining the sequence of amplified DNA fragments which have incorporated phosphorothioate nucleotides during PCR has also been demonstrated (50).

Recently, a modification of the PCR protocol has been introduced to facilitate direct sequencing (51). This procedure, called "asymmetric PCR" is designed to generate a large amount of single-stranded rather than double-stranded product. In this protocol unequal amounts of the two primers are used during the PCR. During the early PCR cycles double-stranded product is produced. One primer is eventually used up and exponential amplification of the target ends. The remaining primer continues to be extended each cycle so that an excess of one of the strands of the PCR product is generated. The amount of single-stranded product increases linearly with cycle number.

In diploid organisms where allelic differences might exist, the PCR product of a single gene could be heterogeneous with respect to DNA sequence. Direct sequencing of double-stranded

PCR product from heterozygotes has been shown to be able to
identify those positions at which a polymorphism exists (48).
For some loci however, alleles differ by multiple substitutions.
It is impossible to know therefore which of the alternative bases
at each polymorphic position are associated with which alleles by
sequencing pooled PCR product. A number of different strategies
to overcome this problem have been discussed by Gyllensten (52).
One alternative is to clone the PCR product and sequence a number
of individual clones so that a "consensus" sequence for each
allele can be determined. On the other hand, denaturating
gradient gel electrophoresis (53) can be used to separate the
allelic products prior to sequencing. In the case of very highly
polymorphic loci such as some in the mammalian major
histocompatibility complex it is possible either to 1) use
sequencing primers that will extend only from one of the two
alleles or 2) use PCR primers that will only amplify one of the
two alleles.

The Purposeful Modification of DNA by PCR

PCR is a primer-directed DNA amplification process. Usually
the sequences of the two primers are made to be perfectly
complementary to the molecules being amplified. However, as
discussed above in the section PCR Cloning of cDNA with
Degenerate Primers, there is, within limits, no absolute
requirement for perfect complementarity. As a consequence, with
the use of a primer with a base pair substitution, an insertion,
or a deletion relative to the target sequence, large amounts of
mutant PCR product can be generated. One of the first
applications of this mutagenesis approach was to produce a
variety of DNA fragments to study the effects of base-pair
substitutions on the binding of E. coli RNA polymerase to the
phage T7 Al promotor (56). A mutation in any position in any DNA
fragment can thus be introduced by a primer containing the
altered sequence. Using an end-labeled primer or labeled
deoxynucleoside triphosphates during PCR also eliminates an
independent labeling step in this type of experiment (56). This
approach would be expected to be especially useful for studies on
DNA-protein interactions (see also section on Analysis of DNA-
Protein Interactions).

The ability to produce mutant-containing DNA fragments by
PCR can also simplify the introduction of mutations into plasmids
for in vivo functional studies. The ease with which this can be
carried out depends upon the availability of convenient
restriction enzyme sites. Assume the region of the plasmid to be
mutated was flanked closely by two restriction enzyme sites
unique to the plasmid and these recognition sequences were
contained within the primers along with the desired mutation.
Restriction enzyme digestion of both the plasmid and the PCR
product by these enzymes and cloning would allow substitution of

the non-mutated segment by the desired fragment (56,57).
Unfortunately, the chance that one unique restriction enzyme site
will be found close enough to the site to be mutated so that both
sequences can be contained within a single PCR primer is not very
great. One way to solve this problem is to make use of the idea
that PCR extension products themselves can act as primers under
appropriate conditions (58). This principle is shown in Figure
5-I. With the use of this strategy, mutations have been placed
well inside long PCR products (56,59) which have convenient
restriction enzyme sites that can be used for cloning.

Recently a new approach (60) for introducing mutations into
a plasmid has been developed which is considerably simpler than
the procedure described above in that it does not require
recloning of a PCR-produced fragment into a vector or
appropriately placed restriction enzyme sites. The method
depends on the concept of inverse PCR as described in the section
Inverse PCR. The mutagenesis protocol is shown in Figure 6 as
applied to the introduction of a mutation in the lacZ alpha
coding segment in the plasmid pBSIISK+. As required for inverse
PCR, the two primers face away from each other. One primer
contains the desired nucleotide substitution. Also, in this
application, the 5' ends of the two primers abut against each
other so that after the first round of PCR all of the nucleotides
within the original plasmid target have been copied into a linear
product. Continued amplification is followed by 3'-5'
exonuclease trimming of the ends required by base addition to the
3' ends of blunt end DNA by the Taq DNA polymerase as discussed
above in the section DNA Cloning. After blunt-end ligation and
transformation, colonies containing the mutated plasmid are
produced. This procedure is limited only to the extent that
amplification of long DNA stretches (greater that 4 or 5 kb) is
not very efficient. A similar approach for introducing deletions
into plasmids was noted by Mole, Iggo and Lane (61).

Finally, PCR can be used to recombine unrelated DNA
sequences (Figure 5-II). One primer for each target must have
appended to its 5' end a sequence that is complementary to the
other target. When the PCR products of the two targets are mixed
the overlapping homologies allow PCR products to be produced
which contain both original sequences. This procedure has been
applied to the construction of a coding sequence for a mosaic
fusion protein (62).

Allele-Specific PCR

As mentioned above in sections PCR Cloning of cDNA with
Degenerate Primers and The Purposeful Modification of DNA by PCR,
under most circumstances, single base-pair mismatches between
primer and template allow production of the desired PCR product.
However there are a number of circumstances where it would be
advantageous to be able to amplify only one of two sequences

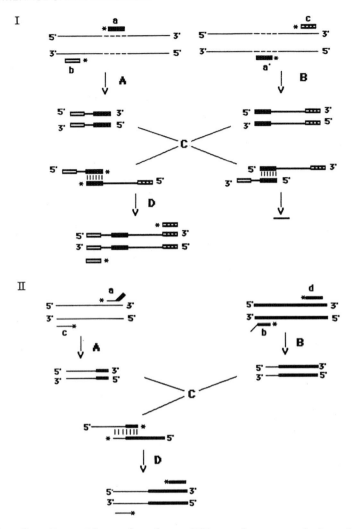

Figure 5. I: Formation of a long PCR product resulting from PCR mediated recombination between two short PCR fragments. A and B: Two separate PCR reactions are carried out on the same target DNA. Primer a, used in reaction A, is complementary to the same region of the target (dashed lines) as in primer a' which is used in reaction B. The PCR products formed in reactions A and B are mixed, denatured, and allowed to reanneal, forming in addition to the two original homoduplexes, two heteroduplexes (C) due to their complementary regions (vertical dashes). One of these can be extended by DNA polymerase (D). To amplify the full length molecule, PCR is carried out with primers b and c. II: Fusion of two unrelated DNA sequences. PCR is carried out on two unrelated sequences (thick and thin lines in (A) and (B) respectively). Primers a and b contain sequences from both targets, specifically

(Figure 5. cont.) the regions of both targets where the joint is to be created.   Primers a and b contain sequences specific for its target at the 3' end and sequences specific for the other target at the 5' end.   They are designed to be completely complementary to each other but only partially complementary to their respective targets.   Mixing the PCR products of the independent reactions (A and B) followed by denaturation and annealing (C) results in the formation of a hybrid molecule with the complementary regions shown as vertical dashes.   This can be extended and amplified as discussed in (I) above.

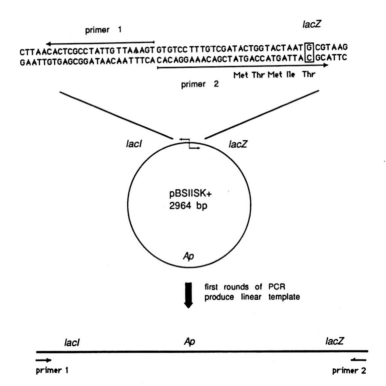

Figure 6.   Site-specific mutagenesis of a whole plasmid with the use of inverse PCR.   The sequence of the two primers and the plasmid's complementary region is shown.   The boxed nucleotides are those that will be deleted by incorporation of primer 2 (which lacks the C) into the PCR product resulting in a frameshift mutation.   Inverse PCR followed by 3' end trimming, 5' phosphorylation, ligation and transformation will produce colonies that lack the LacZ-alpha peptide complementing activity. For details see Hemsley et al. (60).

which differ by as little as a single nucleotide substitution. Two basic concepts can be used too attain this goal. One makes use of the fact that a base-pair mismatch can reduce the thermal stability of the primer-template complex. The other relies on the relative inefficiency of a DNA polymerase lacking a 3' to 5' exonuclease activity to extend a primer if there is a mismatch at the 3' end. A number of different strategies employing these concepts have achieved allele-specific amplification (63-66). In most applications, the DNA sample is divided into two aliquots. Each aliquot is amplified with one common primer and one of the two alternative allele-specific primers. The allelic nature of the sample can be deduced by which of the aliquots produce a PCR product. The work of Ehlen and Dubeau (66) demonstrated the importance of lowering the total dNTP concentration by almost a factor of 100 to enhance allele-specificity. This modification, which follows what is known about the fundamental enzymology of mismatch extension (67), allows very significant discrimination such that extension of the mismatched primer is virtually eliminated. Allele-specific priming has also been applied to determining the allelic nature of DNA sequences in single cells (Cui, Li and Arnheim, unpublished data). Recent studies on the effects of mismatches at and near the 3' end of PCR primers provides important information for the design of allele-specific primers (Kwok et al., unpublished data).

Analysis of DNA-Protein Interactions

PCR has been used to select, among all of the sequences in the genome, those that can bind to a specific protein (68). In this application total human genomic DNA is sheared and, after the ends are trimmed, ligated to a linker/primer. The ligated DNA is then digested with a restriction enzyme whose recognition sequence is contained within the linker/primer. The linker/primer is designed in such a way that only genomic DNA containing a linker/primer at each end can be amplified by PCR. The DNA preparation was combined with the eukaryotic RNA polymerase III transcription factor TFIIIA and the protein-DNA complexes formed were precipitated with antibody to TFIIIA. The precipitated DNA sequences were then amplified with primers complementary to the original linker/primer and another round of antibody precipitation followed by PCR was carried out. The amplified DNA sequences were then cloned and clones selected for nucleotide sequencing by hybridization with a radioactive probe composed of the selected and amplified DNA originally used for the cloning. All of the seven clones selected for analysis contained a sequence with at least 89% identity to the known TFIIIA consensus sequence. This approach to discovering which genomic DNA segments bind to a specific protein could contribute significantly to our understanding of how specific transcription factors can coordinate the expression of unlinked genes.

Another approach to studying protein-DNA interactions with PCR makes use of the gel retardation procedure (69,70) to separate DNA fragments that bind to protein from those that do not bind. DNA from the retarded DNA-protein complex is used as a source for amplification (Prentki et al., unpublished data). In this application, the binding of a mixture of synthetic double-stranded oligonucleotides varying in sequence at several positions at a known binding site for the bacterial protein IHF was studied. The oligonucleotide mixture was designed to have common sequences at each end so that they could all be amplified by a common set of primers. The sequences that were retarded were found, after PCR, to be enriched for those which had a specific affinity for IHF. Significant information about the sequence determinants of the interaction should be obtainable by this method. An advantage of this gel retardation approach is that it does not necessarily require that the protein to be studied has been purified. In addition, it "selects", among a large number of possible sequences, those that show significant binding activity.

Genomic Footprinting

A very clever approach to genomic footprinting with PCR has recently been devised (71). In this particular application isolated nuclei were treated with DMS followed by piperidine. Instead of directly carrying out the standard genomic, sequencing protocol (72) Mueller and Wold denatured the DNA, annealed it to a gene-specific primer and extended it using sequenase. The extension products would be expected to terminate at positions of piperidine cleavage. The double-stranded extension products were ligated to a double-stranded linker/primer which was made up of one 25 base pair strand annealed to an 11 base pair complementary piece to form one blunt end. Only the 3' end of the 25 base long strand can form a phosphodiester bond with the extended fragments. The 11 base pair strand cannot be ligated since its 5' end is not phosphorylated. After ligation and denaturation, PCR was initiated with a second gene-specific primer and the 25 base pair linker fragment. This would be expected to produce PCR products that span the region from the gene-specific primer to a point 25 base pairs away from the piperidine cleavage site. Finally, a third radioactively labeled primer was used to extend these PCR fragments and the products were run on a sequencing gel. Because of the amplification step very small amounts of material can be used for genomic footprinting.

CONCLUSIONS

The ease of applying PCR, its speed, sensitivity and versatility make it ideally suited for application to many problems in biology. Two recent reviews and a book indicate the extent to which PCR has had an impact on fields other than molecular biology including human genetics, forensic science, evolutionary biology and ecology and population biology (39,40,73). In addition to allowing the application of molecular biological procedures to areas that had not exploited these techniques previously other applications of PCR have produced information that would previously have been thought impossible to obtain. Among these applications include studies on mRNA levels during the earliest stages of mammalian embryogenesis (74), the analysis of genetic recombination by studying the DNA sequences present in individual sperm (9,75) and studies on mRNA levels of various growth factors in a macrophage population actually isolated from a wound undergoing healing (76). PCR has also made it possible rapidly to carry out retrospective medical studies on DNA in pathological samples stored for decades as formalin-fixed paraffin-embedded tissues without losing the histological information contained therein (77-79). DNA from tissues of extinct animals has also been studied by PCR (8). Only time will tell what new areas of biology may be influenced by the further development and application of this technique.

REFERENCES

1   Saiki, R.K., Scharf, S., Faloona, F., Mullis, K.B., Glenn, T., Horn, G.T., Erlich, H.A. and Arnheim, N. (1985) Science 230, 1350-1354.
2   Mullis, K.B. and Faloona, F.A. (1987) Methods Enzymol. 155, 335-350.
3   Saiki, R.K., Gelfand, D.H., Stoffel, S., Scharf, S.J., Higuchi, R., Horn, G.T., Mullis, K.B. and Erlich, H.A. (1988) Science 239, 487-491.
4   Rollo, F., Amici, A. and Salvi, R. (1988) Nucl. Acids Res. 16, 3105-3106.
5   Foulkes, N.S., Pandolfi de Rinaldis, P.P., Macdonnell, J., Cross, N.C.P. and Luzzatto, L. (1988) Nucl. Acids Res. 16, 5687-5688.
6   Wittwer, C.T., Fillmore, G.C. and Hillyard, D.R. (1989) Nucl. Acids Res. 17, 4353-4357.
7   Saiki, R.K., Bugawan, T.L., Horn, G.T., Mullis, K.B. and Erlich, H.A. (1986) Nature 324, 163-166.
8   Paabo, W., Higuchi, R.G. and Wilson, A.C. (1989) J. Biol. Chem. 264, 9709-9712.
9   Li, H., Gyllensten, U.B., Cui, X., Saiki, R.K., Erlich, H.A. and Arnheim, N. (1988) Nature 335, 414-417.
10 Kwok, S. and Higuchi, R. (1989) Nature 339, 237-238.
11 Scharf, S.J., Horn, G.T. and Erlich, H.A. (1986) Science 233, 1076-1078.
12 Clark, J.M. (1988) Nucl. Acids Res. 16, 9677-9686.

13   Triglia, T., Peterson, M.G. and Kemp, D.J. (1988) Nucl.
     Acids Res. 16, 8186.
14   Ochman, H., Gerber, A.S. and Hartl, D.L. (1988) Genetics
     120, 621-623.
15   Silver, J. and Keerikatte, V. (1989) J. Cell. Biochem.
     Suppl. 13E. Abstract WH239.
16   Jeffreys, A.J., Wilson, V., Neumann, R. and Keyte, J. (1988)
     Nucl. Acids Res. 16, 10953-10971.
17   Ochman, H., Ajioka, J.W., Garza, D. and Hartl, D.L. (1989)
     in PCR Technology: Principles and Applications for DNA
     Amplification (Erlich, H., ed.) pp. 105-111, Stockton Press,
     New York, NY.
18   Ludecke, H.-J., Senger, G., Claussen, U. and Horsthemke, B.
     (1989) Nature 338, 348-350.
19   Kawasaki, E.S., Clark, S.S., Coyne, M.Y., Smith, S.D.,
     Champlin, R., Witte, O.N. and McCormick, F.P. (1988) Proc.
     Nat. Acad. Sci. U.S.A. 85, 5698-5702.
20   Todd, J.A., Bell, J.I. and McDevitt, H.O. (1987) Nature 329,
     599-604.
21   Powell, L.M., Wallis, S.C., Pease, R.J., Edwards, Y.H.,
     Knott, T.J. and Scott, J. (1987) Cell 50, 831-840.
22   Chelly, J., Kaplan, J.C., Maire, P., Gautron, S. and Kahn,
     A. (1988) Nature 333, 858-860.
23   Kawasaki, E.S. and Wang, A.M. (1989) in PCR Technology:
     Principles and Applications for DNA Amplification (Erlich,
     H., ed.) pp. 89-97, Stockton Press, New York, NY.
24   Loh, E.Y., Elliott, J.F., Cwirla, S., Lanier, L.L. and
     Davis, M.M. (1989) Science 243, 217-220.
25   Frohman, M.A., Dush, M.K. and Martin, G.R. (1988) Proc. Nat.
     Acad. Sci. U.S.A. 85, 8998-9002.
26   Belyavsky, A., Vinogradova, T. and Rajewsky, K. (1989) Nucl.
     Acids Res. 17, 2919-2932.
27   Lee, C.C., Wu, X., Gibbs, R.A., Cook, R.G., Muzny, D.M. and
     Caskey, C.T. (1988) Science 239, 1288-1291.
28   Gould, S.J., Subramani, S. and Scheffler, I.E. (1989) Proc.
     Nat. Acad. Sci. U.S.A. 86, 1934-1938.
28a Libert, F., Parmentier, M., Lefort, A., Dinsart, C., Van
     Sande, J., Maenhaut, C., Simons, M.-J., Dumont, J.E. and
     Vassart, G. (1989) Science 244, 569-572.
29   Gautam, N., Baetscher, M., Aebersold, R. and Simon, M.I.
     (1989) Science 244, 971-974.
30   Wilks, A.F. (1989) Proc. Nat. Acad. Sci. U.S.A. 86, 1603-
     1607.
31   Kamb, A., Weir, M., Rudy, B., Varmus, H. and Kenyon, C.
     (1989) Proc. Nat. Acad. Sci. U.S.A. 86, 4372-4376.
32   Mack, D.H. and Sninsky, J.J. (1988) Proc. Nat. Acad. Sci.
     U.S.A. 85, 6977-6981.
33   Shih, A., Misra, R. and Rush, M.G. (1989) J. Virol. 63, 64-
     75.

34   Lawyer, F.C., Stoffel, S., Saiki, R.K., Myambo, K.,
      Drummond, R. and Gelfand, D.H. (1989) J. Biol. Chem. 264,
      6427-6437.
35   Tindall, K.R. and Kunkel, T.A. (1988) Biochemistry 27, 6008-
      6013.
36   Gelfand, D.H. (1989) in PCR Technology: Principles and
      Applications for DNA Amplification (Erlich, H., ed.) pp. 17-
      22, Stockton Press, New York, NY.
37   Dunning, A.M., Talmund, P. and Humphries, S.E. (1988) Nucl.
      Acids Res. 16, 10393.
38   Krawczak, M., Reiss, J., Schmidtke, J. and Rosler, U. (1989)
      Nucl. Acids Res. 17, 2197-2201.
39   White, T.J., Arnheim, N. and Erlich, H.A. (1989) Trends
      Ganet. 5, 185-189.
39a   Myers, R. and Maniatis, T. (1986) Cold Spring Harbor Symp.
      Quant. Biol. 51, 275-283.
40   Erlich, H. (ed.) (1989) PCR Technology: Principles and
      Applications for DNA Amplification. Stockton Press, New
      York, NY.
41   Gelfand, D.H. and White, T.J. (1989) in PCR Protocols, A
      Guide to Methods and Applications (Innis, M.A., Gelfand,
      D.H., Sninsky, J.J. and White, T.J., eds.) pp. 129-141,
      Academic Press, New York, NY.
42   Goodenow, M., Huet, T., Saurin, W., Kwok, S., Sninsky, J.
      and Wain-Hobson, S. (1989) J. AIDS 2, 344-352.
43   Fucharoen. S., Fucharoen, G., Fucharoen, P. and Fukumaki, Y.
      (1989) J. Biol. Chem. 264, 7780-7783.
44   Wrischnik, L.A., Higuchi, R.G., Stoneking, M., Erlich, H.A.,
      Arnheim, N. and Wilson, A.C. (1987) Nucl. Acids Res. 15,
      529-542.
45   Stoflet, E.S., Koeberl, D.D., Sarkar, G. and Sommer, S.S.
      (1988) Science 239, 491-494.
46   Engelke, D.R., Hoener, P.A. and Collins F.S. (1988) Proc.
      Nat. Acad. Sci. U.S.A. 85, 544-548.
47   McMahon, G., Davis, E. and Wogan, G.N. (1987) Proc. Nat.
      Acad. Sci. U.S.A. 84, 4974-4978.
48   Wong, C., Dowling, C.E., Saiki, R.K., Higuchi, R.G., Erlich,
      H.A. and Kazazian, H.H. (1987) Nature 330, 384-386.
49   Innis, M.A., Myambo, K.B., Gelfand, D.H. and Brow, M.A.D.
      (1988) Proc. Nat. Acad. Sci. U.S.A. 85, 4936-9440.
50   Nakamaye, K.L., Gish, G. Echstein, F. and Vosberg, H.-P.
      (1988) Nucl. Acids Res. 16, 9947-9959.
51   Gyllensten, U.B. and Erlich, H.A. (1988) Proc. Nat. Acad.
      Sci. U.S.A. 85, 7652-7656.
52   Gyllensten, U.B. (1989) in PCR Technology: Principles and
      Applications for DNA Amplification (Erlich, H., ed.) pp. 45-
      60, Stockton Press, New York, NY.
53   Fischer, S.G. and Lerman, L.S. (1983) Proc. Nat. Acad. Sci.
      U.S.A. 80, 1579-1583.

54   Keohavong, P., Kat, A.G., Cariello, N.F. and Thilly, W.G. (1988) DNA 7, 63-70.

55   Loeb, L.F. and Kunkel, T.A. (1982) Annu. Rev. Biochem. 51, 429-457.

56   Higuchi, R., Krummel, B. and Saiki, R.K. (1988) Nucl. Acids Res. 16, 7351-7367.

57   Vallette, F., Mege, E., Reiss, A. and Adesnik, M. (1989) Nucl. Acids. Res. 17, 723-733.

58   Mullis, K.B., Faloona, F.A., Scharf, S.J., Saiki, R.K., Horn, G.T. and Erlich, H.A. (1986) Cold Spring Harbor Symp. Quant. Biol. 51, 263-273.

59   Ho, S.N., Hunt, H.D., Horton, R.M., Pullen, J.K. and Pease, L.R. (1989) Gene 77, 51-59.

60   Hemsley, A., Arnheim, N., Toney, M.D., Cortopassi, G. and Galas, D.J. (1989) Nucl. Acids Res. 17, 6545-6551.

61   Mole, S.E., Iggo, R.D. and Lane, D.P. (1989) Nucl. Acids Res. 17, 3319.

62   Horton, R.M., Hunt, H.D., Ho, S.N., Pullen, J.K. and Pease, L.R. (1989) Gene 77, 61-68.

63   Wu, D.Y., Ugozzoli, L., Pal, B.K and Wallace, R.B. (1989) Proc. Nat. Acad. Sci. U.S.A. 86, 2757-2760.

64   Newton, C.R., Graham, A., Heptinstall, L.E., Powell, S.J., Summers, C., Kalssheker, N., Smith, J.C. and Markham, A.F. (1989) Nucl. Acids Res. 17, 2503-2516.

65   Gibbs, R.A., Nguyen, P.-N. and Caskey, C.T. (1989) Nucl. Acids Res. 17, 2437-2448.

66   Ehlen, T. and Dubeau, L. (1989) Biochem. Biophy. Res. Commun. 160, 441-447.

67   Petruska, J., Goodman, M.F., Boosalis, M.S., Sowers, L.C., Chaejoon, C. and Tinoco, I. (1988) Proc. Nat. Acad. Sci. U.S.A. 85, 6252-6256.

68   Kinzler, K.W. and Vogelstein, B. (1989) Nucl. Acids. Res. 17, 3645-3653.

69   Fried, M. and Crothers, D.M. (1981) Nucl. Acids Res. 9, 6505-6525.

70   Garner, M.M. and Revzin, A. (1981) Nucl. Acids Res. 9, 3047-3060.

71   Mueller, P.R. and Wold, B. (1989) Science 246, 780-786.

72   Church, G.M. and Gilbert, W. (1984) Proc. Nat. Acad. Sci. U.S.A. 81, 1991-1995.

73   Arnheim, N., White, T. and Rainey, M. Bioscience (in press).

74   Rappolee, D.A., Brenner, C.A., Schultz, R., Mark, D. and Werb, Z. (1988) Science 241, 1823-1825.

75   Cui, X., Li, H., Goradia, T.M., Lange, K., Kazazian, H.H., Galas, D. and Arnheim, N. (1989) Proc. Nat. Acad. Sci. U.S.A. 86, 9389-9393.

76   Rappolee, A.D., Mark, D., Banda, M.J. and Werb, Z. (1988) Science 241, 708-712.

77   Shibata, D., Martin, W.J. and Arnheim, N. (1988) Cancer Res. 48, 4564-4566.

78 Almoguera, C., Shibata, D., Forrester, K., Martin, J., Arnheim, N. and Perucho, M. (1988) Cell 53, 549-544.

79 Smit, V.T.B.M., Boot, A.J.M., Smits, A.M.M., Fleuren, G.J., Cornelisse, C.J. and Bos, J.L. (1988) Nucl. Acids Res. 16, 7773-7782.

REGULATION OF ALTERNATIVE SPLICING

Michael McKeown

Molecular Biology and Virology Laboratory
The Salk Institute, PO Box 85800
San Diego, CA 92138

INTRODUCTION

With careful study of the array of RNAs produced by many different genes, it has become clear that alternative splicing is an important mechanism for the regulation of the synthesis of various biologically important proteins. The simplest form of such regulation results from the tissue- or temporal-specific splicing events that lead alternatively to functional (protein coding) and nonfunctional (not protein coding) RNAs, but other events, such as the inclusion or exclusion of particular alternative coding sequences within an mRNA are also potential regulatory events in that they eliminate the production of one type of protein and allow the production of another. In addition the production of RNAs with alternative protein coding capacities can be used to greatly increase the potential array of products from any particular gene. One advantage of such regulation by alternative splicing, as opposed to having multiple copies of related genes each with its own different array of transcriptional regulatory elements, is that particularly useful groupings of tissue- and temporal-specific enhancer elements need not be remodeled to serve the multiple different functions of a gene.

This review attempts to discuss some of the factors which can have critical importance in the regulation of alternative splicing; to cover specific examples of alternative splicing which illustrate some of the potential mechanisms that are used to regulate certain kinds of splicing events; and to discuss some of the specific techniques that appear to be of value in experimental dissection and interpretation of alternative splicing. Much, but not all of the material concerns <u>in vivo</u> studies of alternative splicing and there is a bias toward systems in which

Genetic Engineering, Vol. 12
Edited by J.K. Setlow
Plenum Press, New York, 1990

it is possible to know, from other observations such as studies
of mutants, which factors are involved in the control of particular splicing events. This review does not attempt to catalogue
all of the different alternative splicing events that have been
observed, nor does it attempt to discuss in great detail the
latest advances in the study of the mechanism of splicing. For
other reviews covering partially overlapping topics, see the
papers by Breitbart et al. (1), Leff et al. (2) and Bingham et
al. (3).

MECHANISM OF SPLICING

Figure 1 briefly summarizes the intermediate steps in
splicing. This subject has been reviewed by Padgett et al. (4),
Green (5), Maniatis and Reed (6) and Sharp (7). As diagrammed,
introns contain three identified regions which are important for
the splicing reaction. The junction between the 3' end of the
first exon and the 5' end of the intron is commonly referred to
as the 5' splice site or splice donor site. Note that the 5'
splice site is identified relative to the intron and the 5'
splice sites occur at the 3' ends of exons. Examination of a
large number of introns from metazoans identifies a consensus
sequence for 5' splice sites of C/AAG↓GURAGU (R=purine), where the
arrow represents the boundary between the exon and the intron.
The junction between the 3' end of the intron and the 5' end of
the second exon is commonly referred to as the 3' splice site or
the splice acceptor site. Note that the 3' splice site is
identified relative to the 3' end of the intron and the 3' splice
sites occur at the 5' ends of exons. This region is generally
some variant of the consensus sequence Y_nNYAG↓ (Y=pyrimidine and
N=any base). Finally, at some point greater than about 17

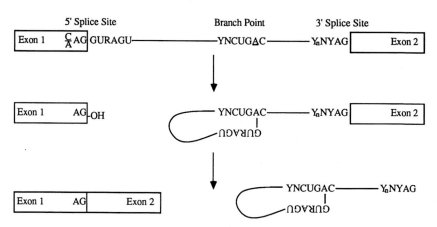

Figure 1. A schematic representation of splicing.

nucleotides upstream of the 3' splice site, there is a sequence
referred to as the branch point. This has the loose consensus of
YNCUGAC (8) in which the underlined A residue is involved in the
formation of the lariat intermediate of splicing.   Sequences
similar to these are found in yeast introns, although the
sequence requirements are more stringent for the 5' splice site
and branch point sequences and less stringent for the 3' splice
sites.

In the course of splicing, the 2' hydroxyl group of the A of
the branch point sequence becomes joined to the 5' phosphate
residue of the first G at the 5' end of the intron. As part of
this transesterification reaction the first exon is separated
from the rest of the pre-mRNA.   In the second step of splicing,
the now free 3' hydroxyl of the free first exon becomes joined to
the 5' phosphate of the first nucleotide of the second exon as
part of a transesterification reaction that joins the two exons
together and frees the intron.

Proper mRNA splicing depends upon the consensus sequences
being correctly recognized and acted upon by a set of ribonucleoprotein particles (snRNPs) containing small nuclear RNAs (snRNAs)
and proteins. These snRNPs are involved in the identification of
correct splice sites and in the formation of a high molecular
weight structure (the spliceosome) in which splicing occurs.
Regulation of splicing is likely to result from changes in the
way in which these snRNPs interact with pre-mRNAs.   These snRNPs
include the U1 snRNP, which is involved in the recognition of the
5' splice site as part of the earliest steps in splicing, the U2
snRNP, which becomes bound to the branch point sequence, the U5
snRNP, which is involved in a way which is not yet clear in the
proper identification of the 3' splice site, and the U4 and U6
snRNPs, which are necessary components for the assembly and
function of the spliceosome.   In addition to these particles,
there are other factors which have been implicated in identification of the 3' splice site, including specifically the U2
assembly factor (U2AF) (9) and an unnamed 70 to 100 kD protein
(10,11). See the reviews mentioned previously for references on
specific aspects of these processes.

TYPES OF ALTERNATIVE SPLICING

Figure 2 catalogues a series of possible alternative
splicing events and represents schematically a large fraction of
possible alternative splicing events.   This is similar to a
diagram presented by Breitbart et al. (1) and proceeds from
comparatively simple examples such as cases in which a potential
intron is included in only one version of a processed transcript
to much more complicated examples having to do with skipping of
multiple consecutive exons or the use of mutually exclusive
exons.

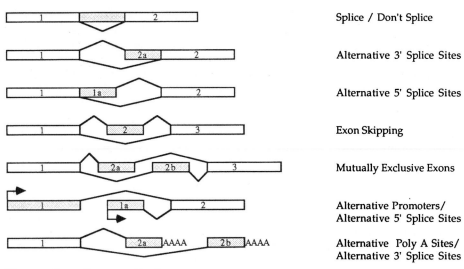

Splice / Don't Splice

Alternative 3' Splice Sites

Alternative 5' Splice Sites

Exon Skipping

Mutually Exclusive Exons

Alternative Promoters/
Alternative 5' Splice Sites

Alternative Poly A Sites/
Alternative 3' Splice Sites

Figure 2. Types of alternative splicing. Various simple types
of alternative splicing events are diagrammed such that only the
particular events are illustrated. Boxes represent sequences
that can be found in "mature" RNAs while thin lines represent
splicing out of intervening sequences. Alternative events are
represented above and below the line of the exons. Alternatively
used exons are shaded.

THEORETICAL CONSIDERATIONS

Use of the Ratios of Alternative Products to Estimate
Changes in Splicing Efficiency

In the course of this review, various experimental manipulations are described that will be used to try to dissect the
mechanisms by which alternative splicing events are regulated.
Prior to considering these different experiments, it is helpful
to have considered at least some of the theoretical considerations which may lead us to favor one experimental approach over
another.

As a result of such a theoretical analysis one wants to have
an initial approximation of the way in which specific regulatory
outcomes are likely to be affected by changes in regulatory
sequences or in regulatory factors, and to have a framework in
which to consider the way in which other processes, such as RNA
turnover, will alter experimental analysis. In addition, one
wants to have a sense of the experimentally measurable quantities
which are likely to yield the most information about changes in
splicing. In order to provide such a first approximation of the
consequences of changes in splicing regulation, consider the

simplified models of alternative splicing described in Figure 3.
Although these models cover only the simplest of alternative
splicing events (intron inclusion/exclusion and 5' or 3' competition) they lead to a pair of conclusions that are both experimentally and theoretically interesting and which help to establish a perspective for much of the rest of this review.   These
conclusions are 1) that the level of no single product of
alternative splicing is informative about key changes in regulation through the whole range of splicing efficiencies, i.e., just
measuring the steady state level of a certain mRNA or precursor
RNA is not necessarily informative about the efficiency of
production of that RNA and 2) that the ratio of the steady state
levels of the various RNA products of alternative splicing is
extremely informative about the relative efficiencies of particular events.   Thus the best experimental situation, being able to
compare signals relative to internal standards, is also the
theoretically most informative.

The particular analysis of alternative splicing events
presented in Figure 3 is derived from work initially done by
Pikielny and Rosbash (12). These authors considered the simplest
possible alterative event, the inclusion or exclusion of an
intron from a messenger RNA.   They considered two kinds of
theoretical models: 1) a simple situation in which precursor RNA
is either degraded in the nucleus or spliced and exported to the
cytoplasm (Figure 3A) and 2) a more complicated two compartment
model in which precursor can be degraded in the nucleus, spliced
and exported to the cytoplasm, or exported to the cytoplasm
without splicing (Figure 3B). It is also possible to extend this
analysis to the situation in which there are two 5' splice sites
competing for a single 3' site, or two 3' sites competing for a
single 5' site (Figure 3C).   These models, and the kinetic
analyses derived from them, make a number of simplifying assumptions that make the analyses easier to follow but which should be
made clear to the reader. First, they assume that the splicing
and RNA processing machinery is not limiting.   Second, they
assume that all molecules are at steady state levels.   Third,
they assume that transcription has a zero order rate constant, T,
and that all other processes occur with first order rate constants.   Fourth, splicing and export to the cytoplasm are viewed
as a single process with a single first order rate constant that
is specific for each potential splicing event.

Although the analysis leads to similar conclusions for all
three situations, consider the simplest (first) case as a way of
illustrating the reasons why the level of any particular RNA is
an insufficiently informative tool and why the ratio of accumulated RNAs is an extremely sensitive tool for the analysis of in
vivo effects on splicing.   A graph of relative mature and
precursor RNA concentrations as a consequence of changes in the
splicing rate constant (k_{sp}) is shown in Figure 4. As long as
the rate constant for splicing is higher than the rate constant

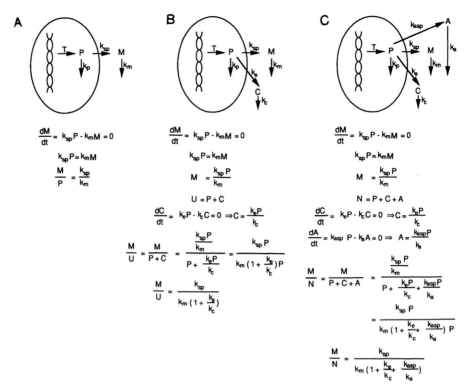

A

$$\frac{dM}{dt} = k_{sp}P - k_m M = 0$$

$$k_{sp}P = k_m M$$

$$\frac{M}{P} = \frac{k_{sp}}{k_m}$$

B

$$\frac{dM}{dt} = k_{sp}P - k_m M = 0$$

$$k_{sp}P = k_m M$$

$$M = \frac{k_{sp}P}{k_m}$$

$$U = P + C$$

$$\frac{dC}{dt} = k_e P - k_c C = 0 \Rightarrow C = \frac{k_e P}{k_c}$$

$$\frac{M}{U} = \frac{M}{P+C} = \frac{\dfrac{k_{sp}P}{k_m}}{P + \dfrac{k_e P}{k_c}} = \frac{k_{sp}P}{k_m\left(1 + \dfrac{k_e}{k_c}\right)P}$$

$$\frac{M}{U} = \frac{k_{sp}}{k_m\left(1 + \dfrac{k_e}{k_c}\right)}$$

C

$$\frac{dM}{dt} = k_{sp}P - k_m M = 0$$

$$k_{sp}P = k_m M$$

$$M = \frac{k_{sp}P}{k_m}$$

$$N = P + C + A$$

$$\frac{dC}{dt} = k_e P - k_c C = 0 \Rightarrow C = \frac{k_e P}{k_c}$$

$$\frac{dA}{dt} = k_{asp}P - k_a A = 0 \Rightarrow A = \frac{k_{asp}P}{k_a}$$

$$\frac{M}{N} = \frac{M}{P+C+A} = \frac{\dfrac{k_{sp}P}{k_m}}{P + \dfrac{k_e P}{k_c} + \dfrac{k_{asp}P}{k_a}}$$

$$= \frac{k_{sp}P}{k_m\left(1 + \dfrac{k_e}{k_c} + \dfrac{k_{asp}}{k_a}\right)P}$$

$$\frac{M}{N} = \frac{k_{sp}}{k_m\left(1 + \dfrac{k_e}{k_c} + \dfrac{k_{asp}}{k_a}\right)}$$

Figure 3. Simplified kinetic models of splicing events. Three simplified models of splicing are presented covering (A) the situation in which a premessage is either spliced or degraded in the nucleus, (B) in which a message is spliced, degraded in the nucleus or exported and degraded in the cytoplasm, or (C) spliced in one of two ways, as a result of having competing 5' or 3' splice sites, degraded in the nucleus, or exported and degraded in the cytoplasm. In all models the system is considered to be at steady state and it is assumed that the basic splicing machinery is not saturated by the RNAs of interest. In each diagram P represents nuclear precursor RNA and M represents cytoplasmic mRNA of the species of interest. The rate of transcription is represented by T. The rate constant for nuclear degradation of precursor is symbolized as k_p. In this simple model splicing and export to the cytoplasm are treated as a single event with a single first order rate constant k_{sp}. Messenger RNA is assumed to degrade with a rate constant k_m. In the more complicated models of (B) and (C), C represents the amount of unspliced RNA which has been exported to the cytoplasm. U is the total amount of unspliced RNA (P+C). Transport of unspliced RNA occurs with a rate constant of k_e. Degradation of unspliced cytoplasmic RNA occurs with rate constant k_c. Section (C) is additionally complicated by the presence of alternatively

(Figure 3. cont.) spliced cytoplasmic RNA, A, which is formed with rate constant k_{asp} and degraded with rate constant k_a. This RNA is assumed to result from the use of either an alternative 3' splice site or an alternative 5' splice site. The calculations below each model follow directly from the models themselves. In each case they lead to an equation for the ratio of the steady state amount of RNA of the type of interest (M) to the steady state amount of RNA resulting from all other possible events. In the case of (A), this alternative fate is P and the ratio is M/P. In the case of (B), the alternative fates are either P or C and their sum is the total amount of unspliced RNA, U. The relevant ratio is M/U. In the case of (C), there are three alternative RNA fates, P, C and A. Their sum is the total amount of RNA following "Not M" pathways, N. In each case, it can be seen that the ratio of M to alternative pathways is linearly related to the rate constant for splicing of P to M (as long as the other rate constants are viewed as constant). The model and calculation of (A) are directly from Pikielny and Rosbash (12). The model of (B) is described in the same paper and the conclusions are given, although the calculations are not presented. (C) is derived by proceeding in the same manner as in (B).

for precursor degradation (k_p), the steady state level of messenger RNA is within two-fold of the maximal observed level, i.e., for a wide range of splicing efficiencies, there is essentially no change in the steady state level of mRNA. In this same range of rate constants, the level of the precursor decreases nearly linearly with changes in the splicing rate. On the other hand, if the splicing rate constant is lower than the rate constant for precursor degradation, the level of precursor is within a factor of two of the maximal level while the mRNA level drops essentially linearly with changes in the splicing rate constant. Thus, over the full range of splicing rate constants neither mRNA levels nor precursor levels alone are sufficient to indicate changes in the efficiency of splicing.

Curves of similar shape to the M/M_{max} curve of Figure 4 are predicted for the amount of RNA spliced for both of the more complicated models shown in Figures 3B and 3C. In both of these models the level of spliced RNA is greater than half maximal as long as the rate constant for splicing is greater than the sum of the rate constants for all other processes that remove precursor RNA from the nucleus by degradation, export or alternative splicing.

Although the level of no single product is sufficient to judge changes in the efficiency of splicing, the calculations of Figure 3 show that for all three situations considered the ratio of the level of RNA that has been spliced to the sum of the levels of all other possible outcomes (the M/P ratio in A, the M/U ratio in B and the M/N ratio in C) varies in a linear manner

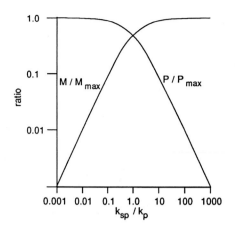

Figure 4.   Unsuitability of mature RNA or precursor levels alone
as indicators of splicing efficiency.   The amount of mature RNA
(as in Figure 3A) expressed relative to the maximum amount of
mature RNA (M/M_{max}) and the amount of unspliced precursor (as in
Figure 3A) expressed relative to the maximum amount of precursor
(P/P_{max}) are plotted as functions of the ratio of the splicing
rate constant (k_{sp}) to the degradation rate constant for precursor (k_p).   Since an assumption of the models in Figure 3 is
that k_p does not change appreciably with the changes leading to
changes in k_{sp}, k_{sp}/k_p is proportional to k_{sp}.   At the point at
which k_{sp} and k_p are equal, P/P_{max} and M/M_{max} are equal and are
half of the maximum observable levels.   Infinite increases in the
efficiency of splicing can only double the steady state amount of
message relative to this level and infinite decreases in the
efficiency of splicing can only increase the level of precursor
by a factor of two relative to this level.   Thus, the absolute
levels of M or P are poor indicators of changes in splicing
efficiency above or below the point where $k_{sp}=k_p$ respectively.
This diagram is based on calculations in Pikielny and Rosbash
(12) and is redrawn from a similar plot in that paper.

with the splicing rate constant throughout the entire range of
splicing rates.   Thus, in theory at least, it becomes possible to
use simple experimental assays to assess the relative rates of
splicing for different mutant RNAs under similar conditions or
for a single RNA under different experimental conditions (for
example, alternative tissue types or developmental times).   In
addition, such simple calculations allow us to predict how large
a change in the ratios of particular products is expected for a
given magnitude of change in the splicing rate constant.
Specifically, the model predicts that changes in the ratio of
accumulated RNAs should vary in a direct linear manner with
changes in the splicing rate constant, e.g., a change of two-fold

in the rate should result in a two-fold change in the ratio of spliced to unspliced RNA and a change of 50-fold in rate should result in a 50-fold change in the ratio of spliced to unspliced RNA.

As a test of these theoretical considerations, Pikielny and Rosbash (12) analyzed the levels of spliced and unspliced RNAs derived from both "wild type" and mutant versions of a specially constructed test gene. These experiments verify the broad predictions of the model. Specifically, they show that mRNA levels can remain at wild-type levels in sets of mutants that result in spliced to unspliced ratios ranging from near 30 (like wild type) to as low as 1.4. On the other hand, as predicted, a spliced-to-unspliced ratio of less than 1.0 was inevitably accompanied by a noticeable lowering in the level of spliced RNAs. Although the results were not calculated with specific regard to precursor levels, examination of the raw data indicates that, as predicted, once the level of splicing drops below a certain point, the concentration of unspliced RNA does not increase.

An obvious weak spot in experimental analyses based on the models is the specific assumption that changes brought about by any particular mutation in the mature or precursor RNA, or any particular changes in the cell types being tested, do not substantially alter any of the rate constants which are part of the denominator in these equations. For certain situations, such as when the mutational change is relatively small and affects only the precursor and not the mature RNA, there is not likely to be a problem with the above assumptions, but in other situations there might be a problem. Examination of the equations suggests the consequences of a breakdown in the assumptions. For all of these models, the experimentally measured ratio of RNAs varies inversely with changes in the rate constant for degradation of the mRNA of interest. Thus, changes that inadvertently alter mRNA degradation rates relative to related test constructs will alter the M/not M ratio with a magnitude proportional to the magnitude of the change in degradation constant. For models involving production and degradation of alternative cytoplasmic RNAs, the measured ratio varies inversely with $(1 + k_{synthesis}/k_{degradation})$ for these RNAs. As long as synthesis rate constants are high relative to degradation rate constants, then changes in either of these numbers will be reflected in an approximately linear manner in the experimentally measured ratio. One particular situation in which the degradation constants might vary substantially from construct to construct involves cytoplasmic degradation of translatable versus nontranslatable RNAs. If one of the measured RNAs is translatable in one construct and not in another, then its degradation constant is likely to be substantially different. Such situations seem to have been observed by Legrain and Rosbash (13) and Zachar et al. (14).

The problems of different turnover rates are most easily dealt with in situations in which RNAs which vary by only a few nucleotides are compared in a single cell type. On the other hand, it is also possible to study the action of tissue-specific splicing regulators by looking at a single construct in alternative host cells. The caveats about transport and degradation constants must be given particular consideration in such cases since these parameters might also vary in a tissue-specific manner. In addition, it is possible that in some situations, for example, cases of negative feedback regulation, specific regulatory factors may be limiting. In such cases, the ratio of alternative products continues to be a powerful tool in the detection of changes in the efficiency of splicing, but the apparent changes in splicing rates are not likely to be as simply interpretable as implied in the calculations of Figure 3. In some cases it may be possible to overcome these difficulties by overexpressing active regulatory factors or by varying the levels of transcription for the regulated RNAs.

The examples considered above cover only the simplest of alternative splicing events. It is theoretically more complicated to analyze other types of alternative events. Thus, for more complicated events, it is not clear what the exact relationship between the levels of alternative processing products means in terms of the relative efficiencies of different events. On the other hand, it is likely that many more complicated events involve primary regulatory mechanisms similar to those considered by the models of Figure 3. In addition, it seems intuitively clear that even for more complicated events, changes in the ratios among the different products will be a sensitive indicator of changes in the relative efficiencies of the different processing pathways and that this method gives a simple, internally controlled method of examining these changes in efficiency.

The above analyses were derived with reference to in vivo splicing, but many of the same considerations also apply to the analysis of in vitro splicing. Since there is often unspliced RNA present in the products of in vitro splicing, even after long times of incubation, it is not informative to calculate a spliced to unspliced ratio for such reactions. On the other hand, it is possible to analyze the time course of splicing in hopes of identifying mutations that result in slower initial rates of splicing. Alternatively, it is possible to establish a competitive situation similar to the in vivo situations of 3C above and to use the two possible splicing outcomes as a way of comparing the relative efficiencies of two events occurring on the same potential substrates. As will be described below, such an approach has great practical value and has been used to demonstrate the importance of certain regions that were previously thought to make little or no contribution to the efficiency of splicing (15,16).

Sites at which Competition Could be Altered

From the perspective of experimental dissection of alternative splicing, it is easiest to consider first the possibility that differences in splicing are mediated through changes in the activity of a single splice site and that all other changes in splicing patterns are consequences of the changes in the activity of this splice site. This, however, is not the only possibility for regulated alternative splicing. For example, one can imagine that multiple sites are each altered in their efficiency to a slight extent such that a larger overall change in splicing is observed. Even so, the working hypothesis of a single regulated site has the advantages that it is the simplest hypothesis consistent with alternative splicing and that it leads to clear predictions and tests for the importance of each individual site such that the need to invoke more complicated regulatory mechanisms (if necessary) will be made clear by experimental results.

Consider now the alternative splicing events shown in Figure 2 with specific regard to the sites that have their activity regulated. These considerations are trivial in the simple cases such as introns which alternatively splice or don't splice or for alternative 5' or 3' splice sites. In considering the other types of alternative splicing, the splicing patterns described below refer to the particular diagrams of Figure 2.

Exon skipping and mutually exclusive exons create more complicated situations. Exon skipping can be viewed as having one key regulated splicing event and then a second splicing event which occurs only if the proper first event has occurred. For example, it is possible to view exon skipping as a competition between the two alternative 3' splice sites for the 5' site of exon 1, just as in simple 3' splice-site competition. If the first 3' splice site is used, i.e., if exon 1 first splices to exon 2, then the RNA contains an additional pair of splice sites bounding the residual intron, and these are then used. If the 3' splice site at exon 3 first joins to exon 1, then exon 2 and its 5' and 3' splice sites are deleted from the pre-mRNA. Similarly, it is possible to view exon skipping solely as a competition between 5' splice sites for joining the 3' splice site of exon 3. If the 5' site on exon 2 first splices to exon 3, then the remaining 5' splice site splices to exon 2. On the other hand, just as for 3' competition, if the 5' splice site of exon 1 wins the competition for exon 3, then other splice sites become irrelevant.

It is important to note that both in nature and in the laboratory the level of activity of the "unregulated" sites in complicated alternative splicing situations such as exon exclusion is important. Mutations which make the secondary sites better substrates for splicing could drastically change the outcome of regulated splicing even in the absence of specific regulatory factors. Consider the exon skipping situation in which the critical regulated event is the competition between the 3' splice sites of exons 2 and 3 for splicing to exon 1. For this competition to be meaningful, the efficiency of the 5' splice site of exon 2 must be low relative to the splicing efficiency of exon 1, otherwise at least some of the time exon 2 will be joined to exon 3 without regard for the factors regulating competition between exons 2 and 3 for exon 1 and the regulatory system will be rendered partially nonfunctional.

The case of multiple exon skipping is perhaps the most complicated of the alternative splicing events. In some cases it is possible to imagine that the primary regulation of a single splicing event then establishes the context for all other events. For example, joining of exon 2 to exon 3 may create a context in which the 3' splice site of exon 2 and the 5' splice site of exon 3 now become active, but such models require an extremely complicated set of secondary consequences and don't work easily for situations involving even more exons. In such cases it may be necessary to consider the possibility that all, or many, of the exons, or their flanking introns, contain information involved in the regulation of their use.

Mutually exclusive exons can, as in exon skipping, be viewed as having a single primary regulatory site, but there is the additional complication that for at least some of the splicing patterns there is an additional potential alternative splice. For example, if the initial splicing event occurs from exon 1 to exon 2a, then some mechanism is needed to stop splicing from exon 2a to exon 2b. Similarly, if the initial event is splicing from exon 2b to exon 3, then some mechanism is needed to make sure that exon 1 splices to exon 2b and not to exon 2a.

The alternative promoter situation requires no specific regulation of splicing. If the second promoter is used, there is no potential alternative splicing event. If the first promoter is used, it is only necessary that the first 5' splice site have an endogenously greater splicing efficiency for the single 3' splice site than does the second 5' splice site. The situation is more complicated for alternative polyA sites. If the initial and critical event is in fact polyA site choice (either by termination of transcription prior to the second potential polyA site, or by polyA choice on a common precursor RNA) then the situation is formally similar to the alternative promoter situation. Alternatively, if the splicing event is the key regulatory event (as appears to be the case for calcitonin/CGRP (17)) then this situation is analogous to alternative 3' splice sites.

FACTORS INFLUENCING SPLICE SITE COMPETITION IN THE
ABSENCE OF SPECIFIC REGULATORS

One of the great difficulties in studying alternative
splicing is that splicing of any particular site is potentially
sensitive to changes in sequence context. Thus even small
changes in the region around one splice site can have unexpected
consequences for the use of other splice sites. As an illustration of this phenomenon, and as a way of characterizing some of
the factors which can be important for establishing the outcome
of basal splicing competition events, even in the absence of
specific regulatory factors, consider a set of <u>in vitro</u> splicing
experiments performed by Reed and Maniatis (15,16). The system
employed incorporated essentially identical splice sites in cis
competition for splicing from single 5' or 3' splice sites.
Specifically, the small first intron of β-globin was used to
supply the 5' and 3' splice sites. Test substrates contained one
or two copies of these sites with various amounts of spacer or
globin flanking sequence adjacent to the internal competing site
(see Figure 5). Note that in each case the outside exons were
wild type and that it was the internal sequences that varied,
thus allowing tests of the sensitivity of splicing involving the

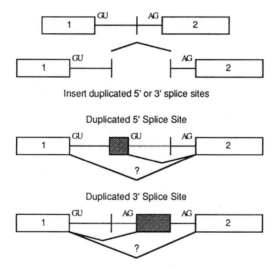

Figure 5. Test substrates for basal level competition experiments. The first intron of rabbit β-globin and its flanking
exons are diagrammed schematically. A convenient restriction
site in the intron allowed Reed and Maniatis (15) to insert
various portions of intronic and exonic material between the two
wild-type splice sites and exons, thus creating a system in which
wild-type splice sites at the outside are in competition with
mutant splice sites at internal positions.

internal site in the presence of various intron or flanking
sequence mutations. The key to these experiments is the presence
of competing splice sites on the same molecule. Many of the
mutant sequences tested in these assays and shown to be less
efficient than wild type had previously been characterized as
functional in the absence of cis competition, i.e., the presence
of a competing splice site revealed the loss of splicing efficiency in the mutants. These experiments involved tests of
duplicated 5' splice sites, duplicated 3' splice sites, duplicated 5' and 3' splice sites, and alterations of the internal
branch point sequence.

     The Reed and Maniatis results (15,16) demonstrate that small
regions of normal flanking exonic sequence (up to at least 55
bases) are not necessarily sufficient to cause the internal site
to be chosen over the wild-type external splice site. This is
true for both 5' and 3' sites and serves to indicate the importance of wild-type exonic regions in establishing a context for
fully efficient splicing. Regions of greater than about 90 to
115 bases of wild-type exonic sequences were sufficient to bring
about high level splicing to the internal sites at the expense of
splicing to the external sites. In these cases, since the
competing sites are identical in sequence over the regions which
are duplicated, at least some of the observed effect results from
splice site proximity. Thus with similar splice sites in
competition, the site closest to the single site is favored over
the more distant site. On the other hand, distance is not the
only factor, since the use of heterologous spacers to increase
the distance to the distal splice site did not necessarily lead
to the activation of otherwise unused internal splice sites.
Finally, changes in the branch point sequences can be shown to
alter the efficiency of splicing to a particular site in vitro
and to lead to use of a downstream 3' splice site in vivo.

     Having considered the various types of alternative splicing,
some ways to think about studying these alternative splicing
events and some of the factors that can alter the outcomes of
regulated splicing events, it is now possible to consider
specific examples of alternative splicing. These different
events will be considered in approximately the same order as
described in Figure 2.

INTRON INCLUSION/EXCLUSION

General Considerations

     Consider first the presence or absence of a specific intron
from an RNA. Although this is conceptually the simplest kind of
alternative splicing, it is one of the more difficult for which
to determine if regulation results from positive or negative
interactions at particular sites. The difficulty stems from the

fact that for particular mutations to be informative they must still allow function of the intron, i.e., a loss of splicing might either be the result of damage to a necessary site for positive regulation or merely disruption of a context that allows splicing. This is a particular problem in the mapping of sites at which positive regulation might occur. These difficulties are by and large alleviated in systems in which it is possible to identify and manipulate mutant or wild-type versions of the genes that control the alternative splicing events. Thus it is possible to know that a particular gene is necessary for splicing to occur in the tissue in which splicing normally occurs (implying a positive action of the gene product on the splicing event) or that a particular gene is necessary to keep splicing from occurring in the tissue in which splicing normally does not occur (implying a negative action of the gene product on the splicing event). Although these are not definitive conclusions (e.g., there could be negative regulation of an activator of splicing) they strongly implicate certain regulatory mechanisms, especially in the case of feedback of a product on its own synthesis.

Three situations will be considered in which it is possible to manipulate both the gene of interest and the cell type or genetic background in which the gene is being expressed (Figure 6). For two of these it has been possible to identify (by genetic means) a particular regulatory molecule that alters the splicing pattern and therefore to infer the likely type of regulation involved. In the third situation, it has been possible to identify tissues in which the alternative events occur, and to map regions sufficient for regulation, but, from the published data, it is not possible to infer positive or negative regulation.

RPL32

There are relatively few intron-containing genes in Saccharomyces cerevisiae, yet a large fraction of the genes coding for ribosomal proteins contain introns (18). The synthesis of ribosomal proteins must be coordinated with the synthesis of rRNAs and with the synthesis of other ribosomal proteins. If any product appears in excess of the amount needed its synthesis must be decreased or its degradation increased. In the course of a study of the homeostatic mechanisms controlling ribosomal protein synthesis, Warner et al. (19) observed that unspliced versions of the RPL32 and CYH2 RNAs appeared when the genes were over-expressed relative to the rest of the ribosomal protein genes. This raised the possibility that these proteins negatively regulate, directly or indirectly, the splicing of their own precursor RNAs. Dabeva et al. (20) specifically tested this hypothesis making use of the fact that the ratio of precursor to

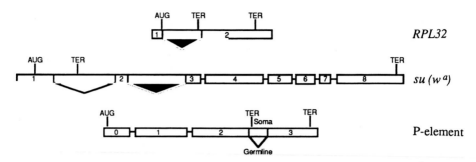

Figure 6. Examples of intron inclusion/exclusion. The RPL32, su(wᵃ) and P-element splicing patterns are diagrammed. Alternatively spliced introns are indicated by unnumbered boxes (unspliced) and by bent lines below the boxes (spliced). Constitutive splicing events are indicated by straight lines between the numbered boxes representing exons. The diagrams are drawn from information given in the references in the relevant portions of the text. TER indicates a stop codon.

mRNA levels is sensitive to changes in the efficiency of splicing.

These studies showed that cells carrying extra copies of the RPL32 gene accumulate more than 100-fold more unspliced RNA than do cells with only one copy of the gene, while the level of mRNA appears to increase no more than about 2- to 4-fold. Thus the ratio of M/U (see Figure 3) decreases as much as 50-fold under these conditions. Similar changes can also be brought about in the processing of RNA from the endogenous gene by the expression of multiple copies of cDNA for RPL32. This regulation is specific for RPL32; other unspliced RNAs do not accumulate as a result of overexpression of RPL32, and regulation is dependent upon the activity of the RPL32 protein since translationally blocked versions of the gene will not lead to a blockage of splicing. The authors were also able to replace the sequences downstream of the intron with lacZ coding sequences and to show that these hybrid genes also responded to expression of RPL32 protein (20). Additional fusion constructs involving the 5' end of the RPL32 intron and the 3' end of a different intron map the sequences sufficient for regulation to upstream of nucleotide 14 of the intron, i.e., to the first exon and the region immediately around the 5' splice site. Deletion/replacement and site-directed mutagenesis within the first exon identify sequences around both the 5' end of the RNA and the 5' splice site that are necessary for regulation. Theoretical structural analysis of these regions shows that they could form a stem loop structure that would block the U1 RNP binding site at the 5' splice site and suggests a model in which free RPL32 protein recognizes and stabilizes this structure, thereby blocking splicing (21).

Finally, the authors were able to apply the same analytical techniques to the expression of the CYH2 gene. These studies showed that cells overexpressing the RNA from this gene, or overexpressing a cDNA from the gene, did have a high level of precursor, and a low M/U ratio, but that this ratio was the same as observed for cells with only a single copy of the gene. They conclude that the RNA from CYH2 splices inefficiently, but that this inefficient splicing is unaffected by changing the levels of the mRNA for this gene (20).

su(w^a)

The suppressor of white apricot (su(w^a)) gene was identified by the fact that mutations in the gene result in a lessening of the severity of the white eye phenotype caused by the apricot allele of the white gene (w^a). This allele of white results from the insertion of a copia retroposon within one intron of the white gene. In a wild-type background, most w^a transcripts become polyadenylated within one or the other of the copia LTRs. Only a very small fraction of RNAs have the copia-containing intron removed. In the absence of su(w^a) function, the intron which contains the retroposon is spliced more efficiently and polyadenylation within the retroposon occurs less efficiently relative to su(w^a)$^+$, resulting in an increase in white mRNA (22). This implies that su(w^a)$^+$ function in some way alters these RNA processing events. When transcripts from the wild-type su(w^a) gene were examined (Figure 6), it was found that the major transcripts are partially unspliced nonfunctional versions of the fully spliced and functional mRNA. In contrast to the nonfunctional RNAs, the fully functional RNA is quite rare (14,23). Thus, in a manner similar to the RPL32 situation, the amount of functional RNA is controlled at the level of RNA splicing.

From the simple observation, it is not possible to distinguish if this ratio of spliced to unspliced products represents a specific regulatory event and if so if this event is blockage or activation of the splices needed for the production of functional RNA. Analysis of the splicing pattern of RNA from various su(w^a) point mutants shows that these mutant genes accumulate substantially less of the nonfunctional RNAs than wild type. Careful examination of the RNAs from these mutants shows that not only is there less of the normally nonfunctional RNAs, but there is an increase in the amount of RNA spliced in the pattern of functional message. One interpretation of these data is that the protein product derived from the functional su(w^a) RNA represses the splicing events that lead to the production of the functional mRNA. As a verification of this, Zachar et al. (14) showed that a wild-type allele of su(w^a) is capable of repressing splicing of the RNAs from both wild-type and mutant genes in the same individual. Taken together, these results show that su(w^a) activity negatively regulates the splicing of its own primary

transcript such that most of the primary transcripts are spliced
in a manner which lead to nonfunctional RNAs.

P-Element

The third example of intron inclusion/exclusion involves the
Drosophila P-element transposon (Figure 6). Wild-type P-elements
are stable in somatic tissues but, after appropriate crosses,
they transpose at a relatively high frequency in the germline
tissues. This germline transposition depends on the activity of
a P-element encoded transposase. When Laski, Rio and Rubin
examined the structure of potential P-element RNAs they found
sequences suggesting the presence of three introns, but they only
observed RNAs that had the first two of these removed when
examining RNA from whole flies (24,25). The fact that RNA from
whole flies is predominantly somatic RNA and the fact that the P
transposase is inactive in somatic tissues lead to the hypothesis
that incompletely spliced RNA (RNA with only the first two
introns removed) codes for a nonfunctional transposase and that
fully spliced RNAs code for the active transposase. Since P-
element transcripts are quite rare, it was not possible to test
this hypothesis by looking directly at germline RNAs. Instead
the authors created a version of the transposase gene in which
the putative third intron had been removed, thus leading to the
production of fully spliced RNAs in both the somatic and germline
tissues. Under such conditions P-elements transpose in both the
germline and the soma. Thus the authors concluded that P
transposase activity is controlled by germline-specific splicing
of the third P intron.

In principle this alternative splicing could be induced by
germline-specific factors or repressed by somatic factors.
Unfortunately, no genetic data point to potential factors
necessary for somatic repression or germline activation, and it
has been necessary to rely on experiments that attempt to
identify necessary cis-acting sequences and from a knowledge of
these to attempt to infer positive or negative regulation.

As a result of the rarity of the P-element RNAs, it has been
necessary to assay the alternative splicing events by assaying
enzymatic functions that occur if splicing has occurred. It has
not been possible to assay the ratios of spliced and unspliced
RNAs in the various tissues. Recent experiments involving in
vitro mutagenesis of the regulated intron and insertion of the
mutated introns into transgenic flies under conditions in which
the phenotypic effects of splicing of the intron can be assayed
have delimited the region(s) necessary for regulated splicing to
small portions of the intron and its flanking sequences (25a).
The published data do not yet distinguish between activation and
repression of the third intron splicing.

ALTERNATE 3' AND 5' SPLICE SITES

General Considerations

The formal properties of the regulation of alternative 5' or 3' splice sites are quite similar and can be considered together. These situations may offer the easiest chance to distinguish between positive and negative regulation and to map the sites involved in this regulation. This analysis is particularly aided by the fact that it is in principle always possible to assay splicing away from the site being mutated and by the fact that the ratio of the alternative products of splicing should be experimentally easy to determine and theoretically informative. These types of alternative splicing events are useful to consider because, as discussed earlier, more complicated events such as exon skipping and alternative 3' splice sites/alternative poly-adenylation are likely to have as their primary regulated events 3' or 5' splice site choice.

For simplicity of discussion consider 3' splice site choice. The same arguments apply to 5' splice site choice. As mentioned before, it is experimentally easiest to test the simple hypothesis of regulation occurring at a single site. In the case of competing 3' splice sites, regulation could result from repression of one splice site, leading to an increase in the fraction of RNA which is spliced in the alternative manner, or regulation could result from an activation of the otherwise less-used splice site without a specific repression of the normally more active splice site. Experiments to distinguish between these two possibilities require the ability to express mutant genes in tissues which normally exhibit one or the other type of regulation. This is a critical point because particular mutational changes can alter the efficiency of splicing even in the absence of specific regulatory factors (for example see the descriptions of experiments by Reed and Maniatis (15) above), making it absolutely necessary to determine the baseline splicing pattern for every construct in nonregulating tissues and then to compare this pattern to the pattern of splicing in the regulating tissue. In systems in which genetic information identifies the tissue in which active regulation occurs, it is easy to tell which tissue is supplying baseline information and which is supplying experimental information. In systems in which it is not previously established which is the regulating tissue, it is necessary to consider either tissue as baseline or regulated.

In principle it is straightforward to establish if only one of the competing splice sites is being regulated and which one. Mutations, such as small deletions, which inactivate either competing splice site are made and tested for activity in tissues showing the alternative splicing patterns. Since each construct contains only one splice site, the ratio of spliced to unspliced RNA must be calculated for the alternative tissues (i.e., the M/U

ratio must be calculated for each tissue (Figure 3)). The
mutation that disrupts the regulated splice site should lead to
identical RNA patterns in both tissues (M/U will be equal), since
the site through which regulation is manifested is missing, while
the mutation disrupting the nonregulated site should continue to
show regulation as a difference in the M/U ratio between the two
tissues (i.e., in one tissue splicing to the remaining splice
site will be less efficient than in the other and an increased
concentration of unspliced RNA will be present). It is important
to note that deletions of both splice sites must be performed to
serve as independent determinations of which site is regulated.
This is because it is possible, for example, that deletion of the
nonregulated splice site can alter the sequence context necessary
for proper regulation of the other site, leading to the incorrect
conclusion that the deleted site is really the site whose
activity is regulated. In this case the analysis of the second
deletion would yield contradictory information and indicate the
necessity of further experiments.

In systems in which it is not known which of the alternative
tissues regulates the splicing pattern, these results identify
the regulated splice site but do not establish positive or
negative regulation. For systems in which the genetic experiments identify the tissues in which active functions alter the
splicing patterns, the experiments not only establish the site
the activity of which is regulated, they establish if the
regulation is positive or negative. In the absence of genetic
data about which tissue supplies the putative regulatory factors
it is difficult to determine, by mutation of the regulated gene,
whether the regulated site is repressed or activated, although
certain mutations can lead to phenotypes consistent with alteration of sites involved in positive or negative regulation.

The following discussion of specific examples of alternative
3' and 5' splice sites describes one system in which the regulatory factors are known and in which the manipulations discussed
above can be performed. The two systems of 5' competition which
are described illustrate alternative ways in which the 5' choice
can be modulated, even in the absence of specific factors and, in
one case, suggest a surprising method for the alteration of 5'
splice site choice.

Drosophila Sex Determination and the transformer Locus

The transformer locus of Drosophila, one of a set of genes
controlling sexual differentiation, is a clear example of a gene
regulated by alternative 3' splice site choice. Since the
Drosophila sex differentiation hierarchy is well characterized
both genetically and molecularly, we can know a substantial
amount about the regulation of transformer splicing (26-37).

Postembryonic expression of the Drosophila sex differentiation regulatory cascade is controlled by a series of regulated

alternative splicing events. In each case, the splicing pattern observed in females is regulated by the action of upstream genes in the cascade while the splicing pattern observed in males represents the default pattern seen in the absence of specific regulators. The structure of this hierarchy has been defined by a combination of both genetic and molecular studies and is diagrammed in Figure 7. The Sex-lethal (Sxl) gene is at the top of the cascade. During very early development the initial activity state of this gene is set in response to the X chromosome to autosome ratio such that active Sxl protein is produced in females but not in males. Shortly after the blastoderm stage a non-sex-specific promoter becomes active. If Sxl protein is present an alternative splicing event occurs that results in exclusion of an exon that would otherwise have been incorporated into the RNA (Figure 8). The absence of this exon results in an RNA that codes for Sxl protein while the presence of the exon results in a nonfunctional RNA. Thus Sxl protein positively

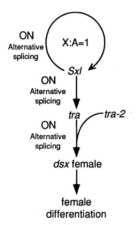

Figure 7. Drosophila sex differentiation cascade. Sex differentiation in Drosophila is controlled by a series of alternative splicing events. In females, (in which the ratio of X chromosomes to number of sets of autosomes (X:A) is 1) the Sex-lethal gene is active and continues to activate its own expression via female-specific alternative splicing of its own primary transcripts (33,38). In addition to controlling the splicing of its own RNA Sxl protein regulates female-specific splicing of RNA from the transformer locus (30,32,35,37,41). The tra product, in a process requiring a functional tra-2 gene, leads to female-specific splicing of RNA for the doublesex gene, leading to female differentiation (26,35,41,42). The interaction of tra and tra-2 occurs after transcription and processing of the tra-2 RNA and is quite possibly the result of protein-protein interactions (29,58,59).

regulates the production of its own mRNA at the level of RNA
splicing (33,38-40).

In addition to regulating Sxl splicing, Sxl protein also
regulates the splicing of RNA derived from the transformer (tra)
locus. In this case the presence of Sxl protein results in the
use of a female-specific 3' splice site that leads to the production of a functional tra mRNA (Figure 8). In the absence of Sxl
function, an alternative 3' splice site is used that results in
nonfunctional and unnecessary RNAs (29-32,35,41). The tra
protein, in a process requiring in addition the wild type
transformer-2 (tra-2) product, regulates the splicing of RNA from
the doublesex (dsx) gene. In the presence of tra protein,
splicing and polyA addition occur at a female-specific terminal
exon (Figure 14). In the absence of tra function, splicing and
polyadenylation occur at a more distal male-specific set of
exons. Each of the sex-specific dsx RNAs results in the production of a dsx protein with a sex-specific function (26,27,35,
41,42).

At each step in the cascade the genetic studies allow us to
know that the splicing pattern seen in males is the default
splicing pattern and the splicing pattern seen in females is the
result of the action of specific regulatory factors. In addition, the genetic and molecular studies allow us to know which
genes must function for these alternative splicing events to
occur. Finally, in the case of tra, tra-2 and dsx neither total
loss of function nor inappropriate gain of function alter
viability, allowing the testing of the action of various mutations altering these genes or their expression. Regulation of
Sxl and dsx will be covered in sections appropriate to their
particular splicing patterns. Since tra is an example of 3'
splice site choice, it will be considered here.

As discussed above, it is possible to infer the site through
which regulation occurs for alternative 3' splice sites by
creating mutations which block the use of one or the other site
and comparing the splicing patterns in the tissues which exhibit
alternative splicing. In the case of tra the alternative tissues
are the alternative sexes, with female being the sex in which the
regulatory gene Sxl is active and male being the sex in which Sxl
is not active. Deletion of the non-sex-specific splice site
leads to use of the normally female-specific splice site in both
males and females as well as the accumulation of some unspliced
RNA. The ratio of spliced to unspliced (M/U) is the same in both
sexes. This is exactly as predicted for deletion of the splice
site which undergoes regulation. Deletion of the female-specific
splice site has little effect on the use of the non-sex-specific
splice site in males, but leads to substantial accumulation of
unspliced RNA in females, i.e., the M/U ratio is high in males
and lower in females, indicating a substantial decrease in the
activity of the non-sex-specific splice site in the presence
of the female-specific regulatory factors. That the patterns

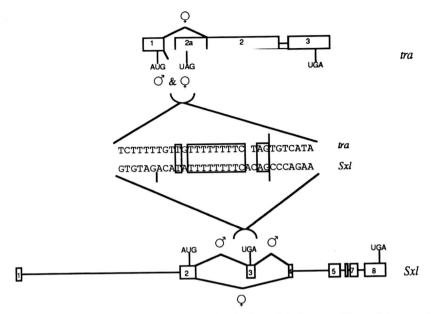

Figure 8. **Transformer** and **Sex-lethal** splicing. The alternative
splicing patterns of **tra** and **Sxl** are shown. Constitutive
splicing events are indicated by straight lines between exons.
The sequences in the center show a comparison of the non-sex-
specific 3' splice site of **tra** with the male-specific 3' splice
sites of **Sxl**. Boxes show a region of primary sequence similarity
between the two genes and vertical lines indicate the actual
splice junctions. Note that **Sxl** actually contains two splice
junctions in close proximity. The maps and sequences are derived
from data in Boggs et al. (30) and Bell et al. (38).

observed in males and females are different for this mutant is
exactly as predicted for deletion of the site which is not
directly regulated and independently confirms the results of the
other deletion indicating that the non-sex-specific site is the
site of regulation. In addition, genetic data indicating the
regulatory factors are acting in a female-specific manner and the
identification of the non-sex-specific splice site as the
regulated splice site indicate that the activity of the non-sex-
specific site is negatively regulated. This conclusion is
independently verified by the observation that the activity of
the non-sex-specific site is lowered in females relative to males
when the female-specific 3' splice site is deleted (37).
    The identification of the non-sex-specific splice site as a
negatively regulated splice site does not identify the exact
sequences at which this regulation occurs. Additional experiments have identified the sites which are sufficient for proper
regulation and some regions that are essential. By **in vitro**

mutagenesis it was possible to create a portable version of the regulated intron and to place this intron into other genes. In a context in which this intron exhibits splicing (the Herpes Simplex virus thymidine kinase gene), the splicing is properly sex-specific, indicating that the regulatory sequences are located within the intron itself. Deletions of the tra gene that remove various parts of the intron show that all of \the intron more than 38 nucleotides beyond the non-sex-specific splice site can be removed and yet maintain sex-specific regulation, thus limiting the regulatory sequences to the 111 nucleotides from the 5' splice site to 38 nucleotides past the non-sex-specific site.

Since tra and Sxl share a common regulatory factor, Sxl, it was possible to search the potential regulated splice sites of these genes for common sequence elements. One such element was found in the region and around the regulated site of tra and the likely regulated site of Sxl (see below) (Figure 8). Mutational alteration of this site such that it continues to function as a (slightly less efficient) splice site results in a total loss of regulated splicing. Specifically, mutation within the region suggested by homology to Sxl leads to a splice site that fails to be blocked by the female activity of Sxl and a failure to increase usage of the female splice site in response to Sxl protein. This thus localizes a necessary regulatory region to the immediate vicinity of the regulated 3' splice site (37).

It is interesting to note that the genetic data identify Sxl as a positive regulator of tra activity, since Sxl function is necessary for the production of the active tra mRNA. Yet when the regulation of tra is examined it becomes clear that the action of Sxl is negative, resulting in the blockage of a particular splice site which, if used, would result in a nonfunctional RNA. Thus the apparent positive action of Sxl actually results from repression of an event which inactivates tra expression.

Immunoglobulin κ Light-Chain Splicing

The germline copies of the genes coding for the immunoglobulin κ light-chain proteins in mice contain a number of variable (V) regions upstream of four potential joining (J) regions (Figure 9). In the course of B cell differentiation a single V region becomes joined to one of the J regions to produce a V + J exon. Transcripts containing the V + J exon splice from the 5' splice site of V + J to the 3' splice site bounding the constant region (C). This occurs even if residual potential J regions (and their potential splice sites) remain in the pre-mRNA. This splice site choice can be duplicated in multiple cell types and in heterologous organisms, and therefore appears to be a property of the response of the general splicing machinery to the specific RNAs involved. A number of investigators have examined the

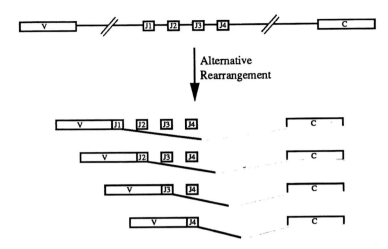

Figure 9.   Immunoglobulin κ light chain rearrangement and splicing. The germline structure of part of the immunoglobulin κ light
chain gene is shown at the top. In the course of B-cell development rearrangement occurs such that a single V region becomes
joined to one of the 4 J regions. The four possible VJ combinations are shown in the lower part of the figure. In each case,
as shown, splicing occurs from the 5' splice site bounding the J
region which is next to the V region.

factors involved in controlling this 5' splice site choice
(43,44).
     In the course of these studies, a series of constructs
containing multiple competing J regions were produced. These
included constructs in which the relative positions of the J
regions were switched or in which the upstream J regions were
either joined to a V region or not. The conclusions from these
studies are that 1) there is not a hierarchy of J region splice
sites, e.g., J3 does not carry with it an intrinsically more
efficient 5' splice site than J4, since if the order of the two
sites is switched splicing then occurs from J4 in preference to
J3.   2) The upstream J region does not require fusion to V to
function since constructs lacking V regions but containing
multiple J regions next to the sequences that flank them in
germline DNA still show splicing favoring the upstream J region.
3) This choice does not depend upon transcription since Xenopus
oocytes splice in vitro synthesized RNAs with the same preference
for the upstream of the competing J regions. The authors are thus
left to conclude that it is the most upstream of the J regions
which is spliced to the C region and that the key fact is that
the J region which is used is the first of the array.   The
mechanism for making this choice remains unclear.

SV40 T Antigen Splicing

The SV40 T antigen gene codes for both large-T and small-t antigens. This duel coding capacity is brought about via alternative 5' splice sites as diagrammed in Figure 10. A striking feature of the structure of the T antigen RNAs is the small size of the small-t intron, only 66 bases, which is very close to the minimum possible size. Recent work by Noble et al. (45) has shed some light on the differences in the splicing mechanism between large-T and small-t splicing. The surprising conclusion to come from this analysis is that the differences in 5' site utilization are accompanied by differences in branch point utilization. This suggests the possibility of regulating other systems of 5' splice site choice via selective blockage of the branch point sequences favored by one of the two 5' sites.

As part of an examination of T antigen splicing, branch point sequences were mapped for both large-T and small-t splicing. Large-T precursor RNAs can form lariats with nine different bases positioned from 18 to 32 bases upstream of the 3' splice site. Small-t precursors on the other hand form lariats only with bases -18 or -19 relative to the single 3' splice site. The small-t splice lariats are probably constrained by the small size of the intron, i.e., the bases farther than 19 from the 3' splice site are probably too close to the 5' splice site to be used in branch point formation (see also (46) and as discussed below). Mutations which alter the -18 or -19 branch point sequences drastically lower the amount of small-t type splicing relative to large-T whereas mutations which alter the major large-T-specific branch point bases lead to an increase in the relative amount of small-t splicing. Thus, although this is not a system in which tissue-specific splicing factors are involved, it is possible to see how specific factors could alter branch point choice and thereby alter 5' splice site choice.

EXON SKIPPING

General Considerations

Single exon skipping is a potentially more complicated system to study since there are four splice sites to consider and any one of these is a potential site for regulation. As discussed in an earlier section, it is simplest, as a first approximation, to consider this kind of regulation as an example of 5' or 3' splice site choice in which the regulated event leads to a defined set of secondary outcomes. Even in such a situation, study of the particular regulation can be difficult since alteration of normally unregulated splice site could change splicing efficiencies in a manner that obscures the normal regulatory pattern. As with other types of alternative splicing,

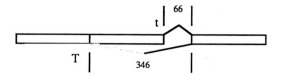

Figure 10.   SV40 T-antigen splicing.   The early RNAs from SV40
are represented schematically.   Large-T and small-T proteins are
produced from RNAs derived by alternative 5' splice site utilization.   The numbers by the introns represent the sizes of the
introns.

it is easiest to deal with a situation in which it is possible to
know, as a result of other studies, which tissue has the default
splicing pattern and which has the regulated splicing pattern.

Sex-lethal

As discussed above, the Sex-lethal gene of Drosophila is a
prime example of a gene controlled by exon exclusion (33,38).   As
in the case with tra we know both a specific regulatory factor
involved in regulating alternative splicing (Sxl) and the sex in
which this factor is active (female).   In addition, from analysis
of another gene controlled by Sxl (tra) it is possible to make an
educated guess as to the way in which Sxl is likely to be
regulated, i.e., by 3' splice site blockage.   Examination of the
Sxl splicing pattern suggests that if there is female-specific 3'
splice site blockage, the blocked site is the one that is
normally used only in males.   Comparison of the sequence around
this splice site with a sequence known to be necessary for proper
regulation of tra splicing reveals a clear primary sequence
similarity, consistent with the hypothesis that this is the site
of regulation (Figure 8) (30,37,38).

Studies of Sxl regulation have been complicated by the fact
that, in addition to controlling itself and tra, Sxl controls the
activity of the genes involved in X chromosome dosage compensation (37,47-51).   In the absence of Sxl function chromosomal
females die, and in the inappropriate presence of Sxl activity
chromosomal males die.   Thus, studies of important cis-acting
sequences within and around the regulated intron region of Sxl
will have to be carried out on modified reporter constructs that
will not interfere with normal growth and viability of males or
females.   Special care will need to be taken to allow an assess-
ment of the steady state levels of expression of RNAs representing all possible outcomes of splicing.

Even given these difficulties, there already exist some
published data consistent with the hypothesis that it is the
male-specific 3' splice site which is blocked by Sxl activity.
These data concern a set of dominant mutants which express the

female Sxl RNAs even in males and therefore behave genetically as
dominant male lethal mutations. Those that have been examined
have been shown to be the result of transposable element insertions near the location of the male-specific exon (38,52). The
simplest interpretation of these insertions is that they disrupt
the sequence context necessary for the use of the male-specific
splice site. As such they behave analogously to the tra deletion
mutants which remove the non-sex-specific splice site and result
in a constitutive and unregulated use of the female-specific
splice site. Unfortunately, the male lethality of these mutants
and the positive autoregulatory action of Sxl make it difficult
to examine directly the relative levels of splice site utilization in males and females carrying these mutations.

Note that this interpretation of Sxl regulation suggests
that, as in the case of tra, the genetically positive action of
Sxl is in fact the repression of the production of nonfunctional
RNAs rather than the direct activation of the production of
functional RNAs.

H-ras

Although the system is not fully characterized, a recent
paper on the alternative splicing of the H-ras oncogene illustrates both the biological importance of alternative splicing,
even in situations in which both alternative events occur in a
single cell, and the sensitivity of exon skipping systems to
changes in the efficiency of any one of a number of splice sites
(53).

The majority of transcripts derived from the H-ras cellular
oncogene are spliced in a manner such that the RNA turns over
rapidly and, if translated, gives rise to a protein of ~19 kD
with no identified function (Figure 11). A small fraction of the
transcripts are spliced such that an exon between exons 3 and 4
(labeled IDX, for Intron Derived Exon) is excluded from the
mature RNA. These RNAs code for the previously identified 21 kD
ras protein and appear to have longer half-life than RNAs which
contain the IDX. As a result of its longer half-life, the p21
RNA is normally the only H-ras RNA which is seen. In the context
of the wild type introns, even potentially transforming mutations
within the ras protein coding region are not transforming. Mutations which lower the efficiency of splicing to the IDX create a

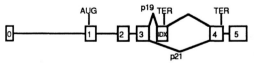

Figure 11. Alternative splicing of H-ras. The splicing pattern
of cellular H-ras is as presented in Cohen et al. (53).

situation in which the level of p21 RNA increases and in which ras protein coding sequence mutants are transforming. From these observations it is possible to infer that one consequence of the alternative splicing of ras is to decrease, posttranslationally, the level of RNA coding for p21 ras, and that this is a potentially meaningful level of regulation.

Studies of mutants which alter the splice sites bounding the IDX show how sensitive such an exon skipping system is to changes in the relative efficiencies of splice sites. Specifically, mutations which lower the match of either splice site to the consensus splice sequences have the effect of lowering the amount of splicing to the alternative exon and increasing exon skipping. This is true both for mutations of the splice site at the 5' end of the exon (a 3' splice site) and for mutations of the site at the 3' end of the exon (a 5' splice site).

The discovery of On/Off regulation of p21 ras production by alternative splicing calls our attention to the problem of ascertainment bias in the identification of genes regulated by alternative splicing. For genes such as RPL32, su(w[a]), P, tra and Sxl the relative stability of the nonfunctional RNAs makes them easy to detect. On the other hand, the apparently nonfunctional RNA from H-ras was not detected until its occurrence was inferred as the result of genetic data. This raises the possibility that there are substantially more genes regulated by alternative splicing than are currently identified as such for the simple reason that the nonfunctional RNAs are degraded fast enough that their steady state levels are quite low. (For an elaboration of this argument, and for a discussion of other reasons that regulation by alternative splicing might not be detected see the article by Bingham et al. (3).)

MULTIPLE EXON SKIPPING

General Considerations

Systems in which multiple consecutive exons are excluded from some of the RNAs derived from a transcription unit are potentially much more complicated than any of the systems considered so far. Even in cases in which only two exons are included or excluded together, thinking of a single primary regulatory site requires that after the primary choice is made it establishes a sequence context in which all other events follow by default.

The situation is further complicated by the fact that many multiple exon exclusion systems are part of a combinatorial exon usage system like troponin T, in which all, or nearly all possible combinations of introns can appear in mRNAs (see Breitbart et al. (1) for review). Both the general complications of multiple exon exclusion and the possibility of combinatorial

use of exons suggest the possibility that there may be specific
sequences involved in the inclusion or exclusion of many or all
of the alternatively used exons.

CD45

One system that exhibits the properties of multiple exon
exclusion as well as a combinatorial use of exons is the CD45
gene (also known as leukocyte common antigen (LCA), or T200, or
B220). This gene codes for a family of cell surface glycopro-
teins of molecular weight approximately 180,000 to 220,000 which
appear to be protein tyrosine phosphatases (briefly reviewed in
Streuli and Saito (54)). As diagrammed in Figure 12, exons 4, 5,
and 6 are included in some B cell RNAs, resulting in the
production of the B cell-biased 220 kD form of CD45. These three
exons are excluded from some T cell RNAs, resulting in the
expression of the T cell-biased 180 kD form of the protein. In
addition, at least 4 of the other 6 possible patterns of
inclusion or exclusion of these introns have been observed (for
references see (54)).

In order to test the possibility that individual exons
contain the information for their inclusion or exclusion from the
CD45 RNA, Streuli and Saito (54) created a series of mini-genes
which contained either exon 4 or exon 6 flanked by various other
CD45 exons. Their results show that the ability to make the
correct B cell or T cell choice of exon usage is a property of
the exons themselves (and the flanking splice sites included with
them in the constructs) and does not depend upon the particular
pair of flanking exons. This is particularly clear for exons 4
and 6 and suggests that each exon contains sequences involved in
determining its utilization. Exon 5 did not exhibit a clear
pattern of tissue-specific usage, consistent with previous in
vivo data suggesting that the use of this exon is less clearly
tissue-specific than the use of exons 4 and 6.

Additional experiments map three discontinuous regions
within exon 4 which are necessary for proper regulation.
Alteration of any of these regions results in the failure to
exclude exon 4 from the RNAs produced in T cells. The simplest
explanation of the data is that these are sites which are in-

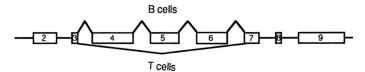

Figure 12. Multiple exon skipping in CD45. Alternative splicing
patterns of CD45 are drawn to emphasize the cell type specific
skipping of multiple exons. The use or skipping of exons 4, 5
and 6 is not fully correlated (see (54) for references).

volved in the negative regulation of exon 4 splicing in T cells
(and other cell types as well).

MUTUALLY EXCLUSIVE EXONS

General Considerations

As discussed previously, it is possible to consider mutually
exclusive exon utilization as under the control of a single
primary event, but there is the additional complicating factor
that for at least certain primary splicing outcomes additional
splice sites remain. For example, in the diagram of Figure 2, if
splicing occurs from exon 1 to exon 2a, then some additional
mechanism must function to ensure that exon 2a is spliced to exon
3 and not to exon 2b, and, in addition, to insure that exon 2b
does not then splice to exon 3. This suggests that to understand
specific pairs of mutually exclusive exons it will be necessary
both to define the primary splicing decision and to characterize
the regulated secondary decision. Two examples of such multiple
levels of regulation will serve to illustrate the possibilities.

β-Tropomyosin

A single gene in rat gives rise to both fibroblast TM1 and
skeletal muscle β-tropomyosin mRNAs. This dual coding occurs in
part by the use of one or the other of a pair of mutually
exclusive exons (called 6 and 7) between exons 5 and 8 (Figure
13). This splicing pattern has been examined by Helfman et al.
(55). Schematically expressed, TM1 is 5, 6, 8 and β-tropomyosin
is 5, 7, 8. When mini-gene constructs containing exons 5-9 and
their introns are expressed in HeLa cells, or when similar
constructs are examined in vitro in splicing extracts derived
from HeLa and containing high concentrations of $MgCl_2$, the
splicing pattern is of the TM1 type. The in vitro pattern is
informative in that the first splicing events observed occur in
the pathway for joining of exon 6 to exon 8. Only later, and at
a substantially slower rate, is exon 5 observed to splice to exon
6. (As an aside, at low $MgCl_2$ concentrations splicing occurs
from exon 5 to exon 8 with the exclusion of both exons 6 and 7.
With substrates containing exon 5 to 8 and the introns separating
them, splicing of exon 7 was not observed under any salt conditions.) The failure to observe splicing of 5 to 6 unless
splicing first occurred from 6 to 8 suggests that splicing of 6
to 8 is the primary regulated event and that splicing from 5 to 6
then follows as a consequence of this first event having hap-
pened. If splicing of 6 to 8 is the first event of TM1 splicing,
it suggests that splicing of 7 to 8 may be the first critical
event in the splicing of β-tropomyosin RNA. As a test of this
hypothesis, mini-genes in which exon 6 was fused to exon 8 or in

β–Tropomyosin/ TM1

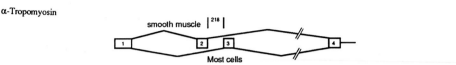

α-Tropomyosin

Figure 13. Mutually exclusive exon use. Both α- and β-tropomyosin genes show mutually exclusive exon use. In both cases shown
the mutually exclusive exons have similar sizes and appear to be
alternative variants of similar sequences. The β-tropomyosin map
is redrawn from data in Helfman et al. (55) and is not to scale.
The map of α-tropomyosin is drawn from data in Smith and Nadal-
Ginard (46). The 218 between exons 2 and 3 indicates the numbers
of bases separating these exons and illustrates how close
together they are.

which exon 7 was fused to exon 8 were constructed and tested for
splicing both in vivo (in HeLa cells) and in vitro. In all cases
the exon which had been "prespliced" to exon 8 was used as a
substrate for splicing to exon 5, consistent with the idea that
the first key event is the splicing to exon 8. This conclusion
is further strengthened by the observation that exon 6 fused to
exon 8 is only an efficient substrate for splicing of exons 5 and
6 if there are greater than about 26 nucleotides of exon 8
present in the fusion to exon 6.

     From all of these data it is possible to derive a plausible
model of the events that lead to mutually exclusive use of exons
6 and 7. The key regulatory event appears to be the competition
between the 5' splice sites at the ends of exons 6 and 7 for the
single 3' splice site bounding exon 8. Splicing from exon 6 to 7
is not a possibility because, in the absence of exon 8, the
sequence context of the 3' splice site at exon 7 is not favorable
for splicing. Similarly, exon 6 is not a substrate for splicing
to exon 5 because in the absence of exon 8 sequences the 3'
splice site of exon 6 is not a substrate for splicing. After
splicing of either exon 6 or exon 7 to exon 8, then the exon
joined to exon 8 becomes functional as a 3' splice site.
Splicing of this 3' splice site to the 5' splice site at exon 5
then occurs. The inability to splice from 5 to 6 or 7, or from 6
to 7, in the absence of exon 8 need not be regulated and can
possibly represent the basal response of the basic splicing
machinery to such substrates. Similarly, the splicing from 5 to
7 rather than from 6 to 7 in the production of 5, 7, 8 can easily

be envisioned as a response of the basal splicing machinery to a
simple 5' splice site competition between the 5' splice sites
bounding exons 5 and 6. The data do not rule out the possibility
of additional mechanisms for inhibiting the splicing of 6 to 7/8
(see below and as discussed in Smith and Nadal-Ginard (46)).

α-Tropomyosin

The specific question of the mechanism of inhibiting
splicing between the two mutually exclusive exons has been
considered for the α-tropomyosin gene by Smith and Nadal-Ginard
(46). In the case of α-tropomyosin exon 1 is spliced to exon 3
which is spliced to exon 4 (1, 3, 4) in most cell types while in
muscle the pattern is exon 1, exon 2, exon 4 (1, 2, 4) (Figure
13). Splicing from exon 2 to exon 3 is not observed. As
discussed above, one possibility for the failure to splice
between exons 2 and 3 is that the splicing event between them may
be relatively inefficient when compared to splicing from exon 1
to 3 or from 2 to 4. Alternatively, splicing from 2 to 3 could
be absolutely inefficient, even in the absence of competing
splice sites.

As a test of these hypotheses, exons 2 and 3 were expressed
in COS cells under conditions in which there were no alternative
splice sites in cis. In vitro experiments were also performed in
HeLa cell splicing extracts with substrates similar to the RNAs
tested in vivo. In both cases, RNA from these constructs, or
from constructs in which the 5' splice site was mutated to the
consensus sequence, showed no splicing from exon 2 to exon 3.
Examination of the sequence of the unspliced intron revealed a
possible reason for this incompatibility: the most likely site
for branch point formation is located only 42 nucleotides away
from exon 2 and almost 180 nucleotides from exon 3. This spacing
places the branch point sequence and the 5' splice site of exon 2
too close for them to function together. Insertion of as few as
17 bases between these sites allows splicing of exon 2 to exon 3,
indicating that there is no absolute incompatibility between the
5' and 3' sites of the intron between exons 2 and 3 and completely consistent with the idea that branch point to splice site
spacing normally prohibits the splicing of exon 2 to exon 3.

As part of the discussion of this phenomenon, the authors
note that it may not be necessary for a branch point to be within
a range that absolutely inhibits splicing as long as the branch
point is sufficiently close to the 5' splice site that it renders
splicing of those particular sites relatively less efficient than
the splicing from an upstream 5' splice site. In addition this
observation of a remote (~180 nt) branch point sequence raises
the possibility that such a phenomenon will be observed as a
mechanism for decreasing the splicing rate for other regulated
splicing events.

ALTERNATIVE 3' EXONS

General Considerations

As noted in the discussion of Figure 2, alternative 3' exon choice can be generated by either a primary polyadenylation choice or by a primary choice between competing 3' splice sites. If the primary event is the choice of polyA sites, then the regulation of splicing becomes a relatively trivial matter. If the first polyA site is chosen, then only a single pair of 5' and 3' splice sites remains. If the second polyA site is chosen, then the observed skipping of the first potential 3' exon is reasonably viewed as the result of an endogenously higher affinity of the second 3' splice site for the single 5' site. If on the other hand polyA site choice is dependent on the splicing pathway of a single set of primary transcripts, the problem reduces to 3' splice site competition as was discussed before. As such, all of the same experimental considerations should apply.

Calcitonin/CGRP

The well characterized calcitonin and calcitonin gene related peptide (CGRP) gene exhibits alternative 3' exon use. The structures of the gene and of the two RNAs derived from it are represented in Figure 14. Calcitonin RNA has the structure 1, 2, 3, 4 while CGRP RNA has the structure 1, 2, 3, 5, 6. If RNA from the calcitonin/CGRP gene is expressed throughout the body of transgenic mice, most cells splice it in the calcitonin manner. Most neurons on the other hand splice the precursor RNA into CGRP mRNA (56). These choices can be mimicked in culture by HeLa cells (calcitonin) and F9 teratocarcinoma cells (CGRP) (57). Experiments specifically designed to test for the importance of polyA site sequences in these terminal exon choices indicate that polyA site choice is not the event regulating splicing (17,57). Thus the analysis of this alternative splicing can be considered as similar to the analysis of alternative 3' splice site choice.

As with the analysis of transformer splicing, the key experiments involved deletion of one or the other of the competing 3' splice sites, in this case the sites bounding exons 4 and 5, followed by comparison of the splicing patterns of these RNAs in HeLa cells or in F9 cells (57). In the case of the deletion of the 3' splice site bounding exon 4, the splicing pattern in HeLa looks very similar to the splicing pattern observed for wild-type RNAs in F9 cells. If anything the splicing to exon 5 is more efficient (as judged by the ratio of 3 to 5 spliced RNAs relative to unspliced exon 4 or exon 5 RNAs) for mutant RNAs in HeLa than for wild-type RNAs in F9. In addition, the splicing pattern of this deletion is similar in F9 to the pattern observed in HeLa (M.G. Rosenfeld, personal communication). As discussed

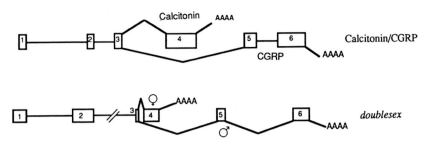

Figure 14.   Alternative polyadenylation/alternative 3' splice
sites.   Both the calcitonin/CGRP and doublesex genes undergo
specific regulated splicing and polyadenylation (17,42,57).

in the consideration of 3' splice site choice, these results are
all as predicted if the 3' splice site near exon 4 is the key
regulated site and not as predicted for regulation of the 3'
splice site near exon 5.

The analysis of a deletion mutation removing the 3' splice
site bounding exon 5 independently supports the conclusion that
use of the exon 4 splice site is the regulated event. When such
a deletion mutation is expressed in HeLa cells, the major amount
of RNA results from splicing of exon 3 to exon 4 (M/U is high).
When this same construct is expressed in F9 cells, there is no
splicing to exon 4, but there is a substantial increase in the
amount of unspliced RNA (M/U decreases substantially as a result
of changes in both M and U). This is as expected for a situation
in which the activity of the exon 4 splice site is being regulated. Deletions which remove major portions of the central parts
of the introns upstream of exon 4 or exon 5 do not alter the
tissue-specific regulation of splicing and suggest that the
critical regulatory sequences are not found in these regions.

The deletion mapping experiments suggest that regulation is
mediated through sequences around the 3' splice site bounding
exon 4. The use of exon 4 in nearly all tissues suggests, but
does not prove, that regulation occurs in the cells that use exon
5. If so, regulation in those cells is a negative regulation of
the use of exon 4. Deletion and substitution mutations map the
sequences likely to be involved in this regulation to the region
very near the exon 4 splice site with particular emphasis on
those sequences between 38 bases upstream of the splice site and
11 bases downstream of it. As with tra, a substitution mutation
replacing the wild-type splice site with a functioning splice
site of different sequence (in this case replacing the bases from
-58 to +11 with a region from -38 to +8 of another 3' splice
site) leads to a dramatic increase in the use of the exon 4
splice site in F9 cells. This suggests that this region is
involved in regulation and is consistent with the idea that
regulation of this site is negative. In this case the site of

negative regulation does not appear to have been completely
defined in that even the large substitution still undergoes some
regulation as judged by a substantial increase in the M/U ratio
in the utilization of exon 4 in F9 relative to HeLa.

In summary, the calcitonin/CGRP gene appears to be a case in
which alternative 3' exon choice is mediated by cell type-
specific splice site blockage, similar, in a formal sense, to the
regulatory mechanisms acting at Sxl and tra. As with these other
systems, the blocked 3' splice site appears to be the site
closest to the relevant 5' splice site. Thus for three different
forms of alternative splicing events it is plausible that there
is a single underlying mechanism of control.

doublesex

The doublesex gene (dsx) is the last defined regulatory gene
in the sex determination hierarchy of Drosophila. As diagrammed
in Figure 14, the splicing pattern of dsx RNAs is very similar to
that of calcitonin/CGRP RNAs (41,42). In the absence of the
action of the regulatory factors tra and tra-2, as occurs in
males, splicing proceeds from exon 3 to 5 and polyadenylation
occurs after exon 6 (41). The protein product of these male RNAs
has a male-specific function in sexual differentiation. In the
presence of the action of both tra and tra-2, as occurs in
females, splicing of dsx RNAs joins exon 3 to 4 and polyadenylation occurs at the end of exon 4. The protein product of the
female-specific RNAs has a female-specific function in sexual
differentiation (26,35,41,42).

The fact that dsx gain or loss of function has no effect on
viability makes dsx regulation an amenable system for study of
the mechanism of alternative splicing. The current published
data do not definitely establish if the regulation of dsx
splicing is based on polyA site choice or on splice site choice,
nor do they as yet establish, if splice site choice is regulated,
which particular site is regulated. Given the above caveats,
there are a number of pieces of evidence which fit easily with a
model in which the female-specific 3' splice site of exon 4 is
positively activated by the action of tra and tra-2. If so, this
is clearly a different situation from that observed for Sxl, tra,
or calcitonin/CGRP.

The argument that dsx is regulated by activation of exon 4
goes as follows. First of all, the intron separating exon 3 from
exon 4 is only 114 nucleotides, while the distance from exon 3 to
exon 5 is 4.1 kb. From other studies of splice site position, it
would normally be expected that the closer 3' splice site would
be favored over the farther. The failure of the female site to
be used in the absence of regulatory factors then suggests that
this site is not an active site in the absence of additional
factors. Examination of the sequences around this site shows
them to be a poor match to consensus with a comparatively purine

rich region right before the splice site. Second, there are
dominant mutants of dsx which express the male RNAs without
regard for the action of the upstream genes tra and tra-2. All
of these are insertion or deletion mutants mapping near the
female-specific 3' splice site, consistent with the idea that
they interrupt potential activating sequences for the female
splice site. This idea is supported by the observation that in
the region near the site of the insertion mutations is a sequence
of 13 nucleotides which is repeated six times. These sequences
are just downstream of the female-specific protein coding
sequences, with the first such repeat less than 300 bases from
the female splice site. The last of these repeats is more than
590 bases from the site of polyA addition. This is consistent
with the idea that these sites are used in the activation of the
female splice site. Finally, the S1 nuclease data of Burtis and
Baker (42) indicate that RNA containing the unspliced female exon
accumulates in a male-biased fashion. This is not easily
consistent with a model involving female-specific blockage of the
male-specific splice site.

    As noted above, most of these observations are consistent
with the activation of the female splice site, and they can
easily be joined into a consistent whole. On the other hand, it
should be noted that there are other interpretations of these
observations that also make a consistent whole and that are
completely consistent with the idea that regulation is a blockage
of the male-specific splice site. This interpretation, while
being similar to what happens with Sxl, tra, calcitonin/CGRP,
RPL32 and su(w^a), in the sense of involving negative control of a
particular splice site, is different from these in that it
involves negative control of a more distal site rather than of
the site most proximal to the nearest 5' or 3' site.

GENETICALLY IDENTIFIED SPLICING REGULATORY FACTORS AND THEIR STRUCTURE

Identified Splicing Regulators

    The ability to analyze a particular alternative splicing
pathway genetically not only has the advantage that it allows us
to know which splicing pathway occurs by default and which occurs
as a result of the action of specific regulatory factors, but it
also allows us to identify likely candidates for proteins which
are specifically involved in the regulation of particular
splices. In the course of this review, specific candidates for
regulatory proteins have been the protein products of the RPL32,
su(w^a), Sxl, tra and tra-2 genes. In the case of each of these
proteins, a direct biochemical involvement in regulated splicing
has not been shown, but in each case, the genetic evidence,
coupled with certain aspects of the protein structure or function

(as revealed by sequence analysis or by knowledge of other activities of the gene product) suggest a direct involvement of these proteins in the regulation of splicing.

The ribosomal protein, RPL32, is a clear example of a protein known to be involved in formation of a complex structure with RNA and known to be involved in splicing regulation. In addition there are specific secondary structures that could be formed by the pre-mRNA for this protein that are similar to certain possible structures of the rRNA near where RPL32 is likely to bind. Thus it is easy to see how RPL32 might regulate the splicing of its own RNA. Unfortunately, as in the case of other ribosomal proteins, there is no clear set of sequence motifs that help us to identify how this protein might bind RNA or to identify genes coding for proteins of similar function.

In the case of the proteins coded for by the other genes listed above, there are sequence motifs found in more than one of these and also found in other proteins involved in RNA binding. From this information it may be possible to search for other potential regulators of splicing and to infer some aspects of how these known regulators function.

RNP Consensus

Sex-lethal protein and transformer-2 protein (probably in conjunction with the transformer protein) regulate at least three different splicing events. Examination of the inferred protein sequences of these genes shows that they contain a sequence motif found on other proteins known to be involved in RNA binding (38,58,59) (reviewed in Bandziulis et al. (60)). The Sxl and tra-2 versions of this conserved motif are shown in Figure 15. This region is 80 to 100 amino acids in length and contains within it two highly conserved regions. The entire 80 to 100 amino acid region has been dubbed the RNA binding domain, while the more highly conserved internal regions have been referred to as RNP2 and RNP1. RNP1 has also been referred to as the ribonucleoprotein consensus sequence (RNP-CS). In the case of U1 associated 70 kD protein, recent work has shown that the RNA binding domain, plus 20 additional amino acids from the region just C-terminal to the RNA binding domain, are sufficient to bind to the same portion of the U1 snRNA as does the full length protein (61). Many of the proteins observed to have such motifs contain multiple copies of this motif. This is the case with Sxl which has two such motifs.

Most proteins which contain the RNA binding domain are not involved with the regulation of splicing, but the fact that many proteins containing this sequence are known RNA binding proteins strengthens the argument that Sxl and tra-2 exert their regulatory effects through direct RNA binding, possibly to the genetically identified target RNAs.

```
Sex-lethal-1  MNDPRASNTN LIVNYL .PQDMTDRELYALFRA.IGPINTCRIM.....RDY KTGYSFGY AFVDFTSEMDSQRAIKVLNGITVRNKRLKV
Sex-lethal-2  PGGESIKDTN LVVTNL .PRTITDDQLDTIFGK.YGSIVQKNILRDKLTGRP. RGVAFVRY NKREEAQEAISALNNVIPEGGSQPLSVRLA
tra-2         SREHPQASRC IGVFGL N.TNTSQHKVRELFNK.YGPIERIQMVIDAQTQRS. RGFCFIYF EKLSDARAAKDSCSGIEVDGRRIRVDFSIT

RNP Consensus LVVGNL     E L F FG I    K    KGFGFVXF    A    L G
              IYIKG      D   Y V       R    R RYA Y          I

              RNP 2                          RNP 1
```

Figure 15. RNP consensus domains in Sex-lethal and transformer-2. The two RNP domains of Sxl and the single RNP domain of tra-2 are aligned. The RNP consensus sequence beneath them is from Bandziulis et al. (60). The RNP 1 and RNP 2 regions are in boldface in the sequences and in the consensus.

RSRDRE

The products of the su(w^a) and tra genes both contain domains composed largely, but not exclusively, of serine and arginine residues (SR domains). A comparison of the primary sequence of these regions shows that, in spite of having similar amino acid composition, they have different primary sequences (23,30). In addition to being found on splicing regulatory molecules such as tra and su(w^a), SR domains can be found on another molecule involved in splicing, the U1 associated 70 kD protein (61,62). In the case of the U1 70K protein, the SR domain is also accompanied by sequences composed of arginine interspersed with the acidic amino acids aspartic acid and glutamic acid, thus creating domains of RSRDRE. If the serine residues of either SR or RSRDRE domains were to be phosphorylated, it would create arginine-acid domains rather than the more basic domains containing unphosphorylated serines.

The similarity in amino acid composition, and the difference in primary sequence, between the SR domains of different proteins raises the question of whether they perform specific sequence recognition functions or if they are involved in interactions other than primary sequence recognition. In the case of the U1 70 kD protein the RSRDRE domains have been shown not to be involved in primary sequence recognition, which as described above, is performed instead by the RNP-CS type region. This suggests that the SR domains of su(w^a) and tra may also be involved in something other than direct RNA sequence recognition, and that they may in fact be necessary for interaction with other components of the splicing machinery. As noted above, the function of su(w^a) appears to be negative, since it blocks splicing events from occurring, while, if the interpretation of dsx regulation is correct, tra acts positively by activating female-type splicing of dsx. Thus, if the SR domains of su(w^a) and tra act in a similar manner, this function can be used in either positive or negative regulation of splicing.

Acknowledgments: Special thanks to Russell Boggs for multiple helpful discussions during the course of the writing of this review. In addition, thanks to Paul Bingham, Michael Rosbash, Michael G. Rosenfeld and Jonathan Warner for helpful discussions about part of their work discussed in this review. Work done in the author's lab has been supported by funds from the NIH. The author is a Pew Scholar in the Biomedical Sciences.

REFERENCES

1   Breibart, R.E., Andreadis, A. and Nadal-Ginard, B. (1987) Annu. Rev. Biochem. 56, 467-495.

2   Leff, F.E., Rosenfeld, M.G. and Evans, R.M. (1986) Annu. Rev. Biochem. 55, 1091-1118.
3   Bingham, P.M., Chou, T.-B., Mims, I. and Zachar, Z. (1988) Trends Genet. 4, 134-138.
4   Padgett, R.A., Grabowski, P.J., Konarska, M.M., Seiler, S. and Sharp, P.A. (1986) Annu. Rev. Biochem. 55, 1119-1150.
5   Green, M.R. (1986) Annu. Rev. Genet. 671-708
6   Maniatis, T. and Reed, R. (1987) Nature 325, 673-678.
7   Sharp, P.A. (1987) Science 235, 766-771.
8   Keller, E.B. and Noon, W.A. (1984) Proc. Nat. Acad. Sci. U.S.A. 81, 7417-7420.
9   Ruskin, B., Zamore, P.D. and Green, M.R. (1988) Cell 52, 207-219.
10 Tazi, J., Albert, C., Temsamani, J., Reveillaud, I., Cathala, G., Brunel, C. and Jeanteur, P. (1986) Cell 47, 755-766.
11 Gerke, V. and Seitz, J.A. (1986) Cell 47, 973-984.
12 Pikielny, C.L. and Rosbash, M. (1985) Cell 41, 119-126.
13 Legrain, P. and Rosbash, M. (1989) Cell 57, 573-583.
14 Zachar, Z., Chou, T.-C. and Bingham, P. (1987) EMJO J. 6, 4105-4111.
15 Reed, R. and Maniatis, T. (1986) Cell 46, 681-690.
16 Reed, R. and Maniatis, T. (1988) Genes and Dev. 2, 1268-1276.
17 Leff, S.E., Evans, R.E. and Rosenfeld, M.G. (1987) Cell 48, 517-524.
18 Leer, R.J., van-Raamsdonk-Duin, M.M., Hagendoorn, N.J., Mager, W.H. and Planta, R. (1984) Nucl. Acids Res. 12, 6685-6700.
19 Warner, J.R., Mitra, G., Schwindinger, W.F., Studeny, M. and Fried, H.M. (1985) Mol. Cell. Biol. 5, 1512-1521.
20 Dabeva, M.D., Post-Beittenmiller, M.A. and Warner, J.R. (1986) Proc. Nat. Acad. Sci. U.S.A. 83, 5854-5857.
21 Eng, F.J., Johnson, S.P. and Warner, J.R. (unpublished data).
22 Levis, R., O'Hare, K. and Rubin, G.M. (1984) Cell 38, 471-481.
23 Chou, T.-C., Zachar, Z. and Bingham, P. (1987) EMBO J. 4095-4104.
24 Laski, F.A., Rio, D.C. and Rubin, G.M. (1986) Cell 44, 7-19.
25 Rio, D.C., Laski, F.A. and Rubin, G.M. (1986) Cell 44, 21-32.
25a Laski, F.A. and Rubin, G.M. (1989) Genes and Dev. 3, 720-728.
26 Baker, B.S. and Ridge, K. (1980) Genetics 94, 383-423.
27 Baker, B.S. and Belote, J.M. (1983) Annu. Rev. Genet. 17, 345-379.
28 Baker, B.S. (1989) Nature 340, 521-524.
29 Belote, J.M., McKeown, M., Boggs, R.T., Ohkawa, R. and Sosnowski, B.A. (1984) Dev. Genetics 10, 143-154.

30   Boggs, R.T., Gregor, P., Idriss, S., Belote, J.M. and
      McKeown, M. (1987) Cell 50, 739-747.
31   Cline, T.W. (1987) Genetics 90, 683-698.
32   Cline, T.W. (1979) Dev. Biol. 72, 266-275.
33   Cline, T.W. (1984) Genetics 107, 231-277.
34   McKeown, M., Belote, J.M. and Baker, B.S. (1987) Cell 48,
      489-499.
35   McKeown, M., Belote, J.M. and Boggs, R.T. (1988) Cell 53,
      887-895.
36   McKeown, M., Boggs, R.T., Nash, K., Ohkawa, R., Manly, A.,
      Sosnowski, B.A. and Belote, J.M. (1989) in Gene Transfer in
      Animals (Beaudet, A.L., Mulligan, R. and Verma, I.N., eds.)
      pp. 1-8, Alan Liss, New York, NY.
37   Sosnowski, B.A., Belote, J.M. and McKeown, M. (1989) Cell
      58, 449-459.
38   Bell, L.R., Maine, E.M., Schedl, P. and Cline, T.W. (1988)
      Cell 55, 1037-1046.
39   Nicklas, J.A. and Cline, T.W. (1983) Genetics 103, 617-631.
40   Salz, H.K., Cline, T.W. and Schedl, P. (1987) Genetics 117,
      221-231.
41   Nagoshi, R., McKeown, M., Burtis, K., Belote, J.M. and
      Baker, B.S. (1988) Cell 53, 229-236.
42   Burtis, K. and Baker, B.S. (1989) Cell 56, 997-1010.
43   Lowery, D.E. and Van Ness, B.G. (1988) Mol. Cell. Biol. 8,
      2610-2619.
44   Kedes, D.H. and Steitz, J.A. (1988) Genes and Dev. 2, 1448-
      1459.
45   Noble, J.C.S., Prives, C. and Manley, J.L. (1988) Genes and
      Dev. 2, 1460-1475.
46   Smith, C.W.J. and Nadal-Ginard, B. (1989) Cell 56, 749-758.
47   Cline, T.W. (1983) Dev. Biol. 95, 260-274.
48   Gergen, J.P. (1987) Genetics 117, 477-495.
49   Lucchesi, J.C. and Skripsky, T. (1981) Chromosoma 82, 217-
      227.
50   Lucchesi, J.C. and Manning, J.E. (1987) Adv. Genet. 24, 371-
      430.
51   Sanchez, L. and Nöthiger, R. (1983) EMBO J. 2, 485-491.
52   Maine, E.M., Salz, H.K., Cline, T.W. and Schedl, P. (1985)
      Cell 43, 521-529.
53   Cohen, J.B., Broz, S.D. and Levinson, A.D. (1989) Cell 56,
      461-472.
54   Streuli, M. and Saito, H. (1989) EMBO J. 8, 787-796.
55   Helfman, D.M., Ricci, W.M. and Finn,L.A. (1988) Genes and
      Dev. 2, 1627-1638.
56   Crenshaw, E.B., Russo, A.F., Swanson, L.W. and Rosenfeld,
      M.G. (1987) Cell 49, 389-398.
57   Emeson, R.B., Hedjran, F., Yeakly, J.M., Guise, J.W. and
      Rosenfeld, M.G. (1989) Nature 341, 76-80.
58   Amrein, H., Gorman, M. and Nöthiger, R. (1988) Cell 55,
      1025-1035.

59   Goralski, T.J., Edström, J.-E. and Baker, B.S. (1989) Cell
      56, 1011-1018.
60   Bandziulis, R.J., Swanson, M.S. and Dreyfus, G. (1989) Genes
      and Dev. 3, 431-437.
61   Query, C.C., Bentley, R.C. and Keene, J.D. (1989) Cell 57,
      89-101.
62   Theissen, H., Etzerodt, M., Reuter, R., Schneider, C.,
      Lottspeich, F., Argos, P., Lührmann, R. and Philipson, L.
      (1986) EMBO J. 5, 3209-3217.

STRUCTURE AND FUNCTION OF THE NUCLEAR RECEPTOR SUPERFAMILY FOR

STEROID, THYROID HORMONE AND RETINOIC ACID

Vincent Giguère

Division of Endocrinology
Hospital for Sick Children
555 University Ave.
Toronto, Canada, M5G 1X8
and
Department of Medical Genetics
University of Toronto

INTRODUCTION

A central problem in eukaryotic molecular biology is the elucidation of molecules and mechanisms that mediate specific gene regulation in response to exogenous inducers such as hormones and growth factors. Nuclear receptors are intracellular proteins that mediate complex effects on development, growth and physiological homeostasis by selective modulation of gene expression in a ligand-dependent manner (1). During the last five years, studies in a number of laboratories have led to the characterization by molecular cloning of a superfamily of genes that encode nuclear receptors for steroid and thyroid hormones, Vitamin D3 and retinoic acid, an active metabolite of Vitamin A. In the course of these studies, the application of recombinant DNA technology allowed the creation of mutant and chimeric receptor that led to a better understanding of the molecular mechanisms involved in regulation of gene expression by this family of trans-activating factors. In this chapter, I do not intend to undertake a complete review of the literature but instead to use specific examples to demonstrate the application of genetic engineering to the study of the mechanism of action of the nuclear receptors. I will first review the strategies used to clone, identify and resolve the structure and function of the steroid hormone receptors, with an emphasis on the human gluco-corticoid receptor (hGR), then describe the methodology used to identify a novel receptor, the retinoic acid receptor.

Genetic Engineering, Vol. 12
Edited by J.K. Setlow
Plenum Press, New York, 1990

MOLECULAR ANALYSIS OF THE GLUCOCORTICOID RECEPTOR

Cloning of Glucocorticoid Receptor cDNAs

The primary structure of the human glucocorticoid receptor (hGR) was the first to be elucidated by sequencing of a cloned complementary DNA (2). The availability of receptor-specific antibodies developed against the human glucocorticoid receptor (hGR) (3) made possible the expression cloning of hGR cDNAs. Using these receptor-specific antibodies as molecular probes, several cDNA clones encoding the immunogenic region of the hGR were isolated (4). The demonstration that these cDNA clones encoded the hGR was provided by the generation of β-galactosidase fusion proteins that were used for affinity-purification of receptor-epitope antibodies, which were subsequently recovered and identified by binding to protein blots of cellular extracts (4). Once the identity of one of the immunopositive cDNAs was established as encoding part of the hGR, this cDNA insert was in turn used as a probe to isolate full-length clones from a human fibroblast cDNA library. The deduced nucleotide sequence of the full-length cDNA revealed an open reading frame 777 amino acids long. To provide additional evidence that the cloned receptor was functional, an in vitro transcription system was used to generate mRNA from the hGR cDNA which in turn was in vitro-translated with rabbit reticulocyte lysate to generate the hGR protein. The in vitro-translated products were then incubated with a radiolabeled synthetic glucocorticoid analog and shown to acquire specific steroid-binding capacity. This series of experiments also revealed that the in vitro-translated product was 1) identical in size to the native hGR (94 kD), 2) immunolog-ically reactive to hGR-specific antibodies and 3) that the acquisition of steroid-binding properties did not require any specific post-translational modification other than the ones, if any, generated by the rabbit reticulocyte lysate.

During the same period of time, cDNA clones encoding the rat GR were also isolated. These clones were obtained by partial purification of receptor mRNA by polysome immunoenrichment, followed by construction of a cDNA library and differential screening with probes generated from immunoenriched mRNA and total mRNA (5). A composite rat GR cDNA revealed a protein of 795 amino acids, slightly longer than the hGR. The difference between the human and rat receptor is mainly accounted for by a stretch of 19 glutamine residues located in the aminoterminus region of the rat GR but absent in the hGR. Miesfeld and colleagues (6) demonstrated the full biological activity of the cloned rat GR by complementation of a receptor defect in a hormone-nonresponsive cell line containing low levels of receptor protein. This cell line, derived from the rat hepatoma cell line HTC, was stably transfected with a rat GR expression plasmid and the hormonal induction of tyrosine aminotransferase (TAT) activity and MMTV RNA levels were used as a measure of receptor function. Dexamethasone treatment showed that both inducibility of TAT activity and expression of MMTV RNA were restored in the transfected cell line.

A Superfamily of Ligand-Dependent Regulatory Proteins

The cloning of the glucocorticoid receptor was soon followed by the molecular cloning of a large number of cDNAs encoding nuclear receptors for estradiol, progesterone, aldosterone, androgene, Vitamin D3, thyroid hormones and retinoic acid (see below) from a variety of species (Figure 1, for Refs. see 1). Amino acid sequence comparison of homologous receptors from different species revealed five regions with varying degree of homology. The regions encoding the DNA- and ligand-binding domains are the most conserved, reflecting their relative importance for receptor function. Both the amino- and carboxy-terminal end of the proteins together with a region located between the DNA- and ligand-binding show the lowest level of conservation between homologous receptors from different species and show no amino acid sequence identity between heterologous receptors.

The high degree of identity in the DNA-binding domain allowed the isolation of cDNAs encoding other members of this superfamily by screening cDNA libraries under low stringency hybridization condition. For example, the mineralocorticoid receptor (MR) was isolated using a hGR hybridization probe (7). However, as discussed below, a number of cDNAs isolated by this method encode receptor-like proteins, termed orphan receptors, for which the putative ligands remain unknown.

This receptor superfamily can be further subdivided into subfamilies according to the varying degree of homology between their respective amino acid sequences and the relatedness of their ligand. Receptors for cortisol, aldosterone, progesterone and testosterone can be classified as one subfamily, while the estrogen receptor belongs to a subfamily that also includes four orphan receptors (see below). The thyroid hormone and retinoic acid receptors are in subfamilies of their own but each have the particularity to include several receptor isoforms for the same ligand. The physiological significance of the existence of multiple receptors for one ligand remains to be elucidated but the differential distribution of expression of each receptor isoform for T3 and retinoic acid suggests that each isoform may play a specific role in controlling development and homeostasis.

Mapping the Functional Domains in the Glucocorticoid Receptor

Understanding the mechanism by which steroid hormone receptors regulate gene transcription first required the characterization of their functional domains. The classic model of

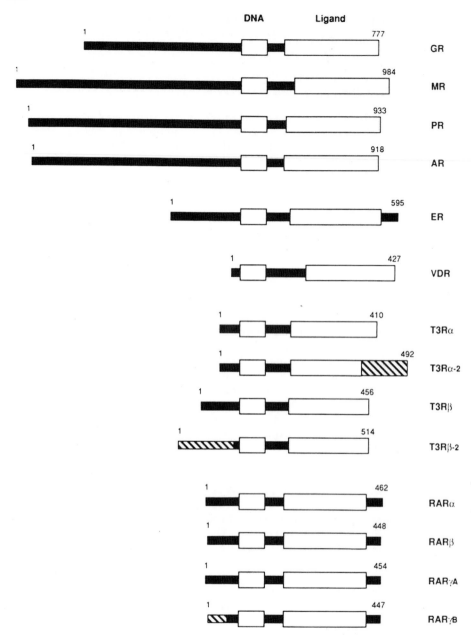

Figure 1. Schematic representation for the steroid, thyroid hormone and retinoic acid receptor superfamily. Each receptor is included into a subfamily according to the degree of homology of the receptor proteins and the structure of their respective ligand. The conserved DNA- and ligand-binding domains are represented by open boxes and the less conserved regions by the

steroid hormone action proposed three major domains necessary for receptor function: a ligand-binding domain, a DNA-binding domain and at least one domain for transcriptional activation. The strategy designed to identify these functional domains is shown in Figure 2.   In these experiments, a hormone-responsive enhancer/promoter element linked to a reporter gene (CAT or luciferase) is introduced into receptor-negative cells and this gene is therefore transcriptionally inactive. Contransfection of the reporter plasmid with a receptor expression vector then provides functional receptors that allow induction of the reporter gene activity upon treatment of the transfected cells with the natural hormone or a synthetic analog. The level of reporter enzyme activity is then directly proportional to the ability of the wild-type or mutagenized receptors to activate gene expression. In addition, biochemical studies such as Western blot analysis, steroid- and DNA-binding activity of the expressed receptor can be performed simultaneously. This assay, first developed for the study of the human glucocorticoid receptor (8), was also used subsequently to study the structure and function of all other members of the steroid/thyroid hormone receptor family for which a responsive promoter had been characterized.

The first genetically engineered mutant receptors were generated using a linker-insertion scanning approach (8).   To generate linker-insertion mutants of the hGR, the expression vector containing the hGR is first linearized by partial cleavage using a restriction enzyme that cuts DNA molecules with high frequency (a four base-pair cutter, for example). The DNA is then repaired with the Klenow fragment of DNA polymerase to yield blunt ends, the linear form of the plasmid is isolated and a synthetic linker (8 to 12 base pair) is added to restore the open reading frame encoding the hGR. The resulting mutants carry three or four additional amino acids which disrupt the natural sequence of the protein. One major advantage to creation of linker-insertion mutation is that this experimental design allows for the convenient generation of any desired set of small or large deletional mutants and the ability to add or switch domains between related molecules to study function. Because each insertional mutant is defined by a unique in-frame restriction enzyme site, deletion of sequences between any two linker positions can be conveniently achieved. This approach gave rise to double deletion mutants that are useful in characterizing interaction between two domains. The result from the study of a large number (>100) of hGR mutants led to the proposal that receptors are composed of a melange of regulatory domains.

(Figure 1. cont.) thin filled boxes. Striped boxes for T3Rα-2,T3Rβ-2 and RARγB indicate the region of the receptor isoform generated by alternative RNA processing.

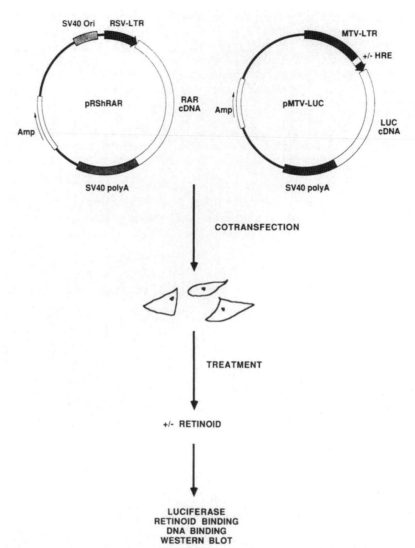

Figure 2. The contransfection assay. Cultured cells are transfected with a receptor cDNA in an expression vector (top left), encoding a retinoic acid receptor, and a plasmid containing a reporter gene linked to a promoter containing a hormone-responsive element (HRE) (top right). Transfected cells can be used to monitor induction of luciferase activity upon addition of the ligand, receptor binding to the DNA, ligand binding and expression of receptor protein.

Figure 3 shows a schematic representation of the hGR with the location of the functional domains.

The Ligand-Binding Domain

The ligand-binding domain encompasses a large portion of the hGR including the last 220 amino acid residues of the C-terminus. This domain is extremely sensitive to changes in the primary structure of the receptor since every insertion mutant in this region is inactive. However, expression of receptors deleted of their aminoterminus or/and their ligand-binding domain demonstrates normal steroid binding activity. Thus, the steroid binding domain can assume a proper tertiary structure independently of the presence of other functional domains. As expected, deletion of a large portion of the carboxyterminal half of the receptor eliminates all steroid binding capacities. Unexpectedly, this type of deletion gives rise to a constitutively active receptor, e.g., this mutant hGR was found to activate transcription in the presence or absence of hormone (9,10). These results demonstrate that the steroid-binding domain plays no major mechanistic role in transcriptional enhancement by the hGR, but rather exerts an overall negative effect on the remainder of the protein.

There is also evidence that this region of the protein harbors many other functions. A ligand-dependent nuclear translocation signal that directs the ligand/receptor complex to the nucleus has been mapped within the hormone-binding domain in both the rat GR (11) and the progesterone receptor (PR) (12). In addition, a segment of high amino acid sequence homology between all members of the steroid hormone receptor family present in the ligand-binding domain was proposed to interact with the heat shock protein 90 (HSP90) (13). Finally, the hormone-binding domain of both GR and estrogen receptor (ER) contains transactivation functions (see below).

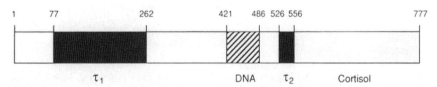

Figure 3.   Linear representation for the human glucocorticoid receptor with the location of the DNA- and cortisol-binding domains and the two acidic transactivation domains τ_1 and τ_2. Not shown are putative nuclear translocation functions located in the vicinity of the DNA-binding domain and inside the cortisol-binding domain.

The DNA-Binding Domain

Steroid hormone receptors selectively activate gene transcription by recognizing and binding to short cis-acting DNA
sequences present in the promoter of hormone-responsive genes.
These sequences, referred to as hormone-response elements (HREs),
are 15 to 20 base pair palindromes composed of two inverted
repeats separated by a spacer up to 6 base pairs in length
(Figure 4). The HREs function in a position- and orientation-
independent manner and thus behave like transcriptional enhanc-

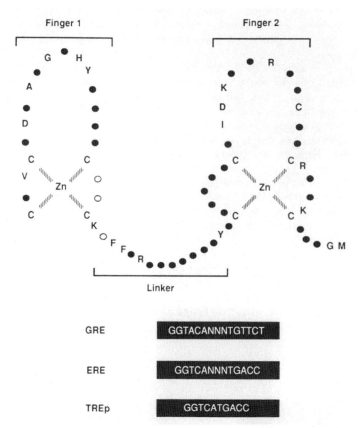

Figure 4. Nuclear receptor DNA binding domain and HREs. The
hypothetical zinc fingers are represented by the one-letter amino
acid code for the conserved residues and by filled circles for
the variable residues. The three open circles at the base of the
first finger represent amino acid residues that dictate target
gene specificity. The consensus sequences for the glucocorti-
coid- (GRE) and estrogen- (ERE) responsive elements and the
idealized palindromic structure of a thyroid hormone-responsive
element (TREp) are shown within the black boxes.

ers.   The apparent dyad symmetry of HREs suggests that they interact with receptor dimers with each receptor molecule binding to a half site of the HRE.   Comparison of several HREs has revealed that glucocorticoid-, estrogen- and T3-responsive elements are very similar to each other.   In fact, two point mutations are sufficient for converting an estrogen-responsive element to a glucocorticoid-responsive element (14,15), while the functional distinction between T3 and estrogen receptor binding sites resides in the presence or absence of the spacer between the two half sites (16).   Moreover, the GR, MR, progesterone (PR) and androgene (AR) receptors all recognize the same HRE (7,17,18), while the retinoic acid and thyroid hormone receptors can induce gene expression through a common responsive element (see below) (19).   Thus, the sequence and functional similarities of the HREs suggest that the DNA-binding domains of the steroid/thyroid hormone receptors must be encoded in a region of high homology between all the nuclear receptors.

Early amino acid comparison analyses of steroid and thyroid hormone receptors revealed a highly conserved region composed of 66 to 68 amino acids which contains nine cysteine residues whose spacing was reminiscent of the DNA-binding protein TFIIIA (20) and other DNA-binding proteins (21).   This DNA-binding structure is now referred to as the "zinc fingers motif" in which a zinc ion interacts with four cysteine and/or histidine residues to form a loop (Figure 4).   Using EXAFS (extended X-ray absorption fine structure) and visible light spectroscopy, Freedman et al. (22) indeed showed the presence of zinc ion in the DNA-binding domain of the glucocorticoid receptor.   The DNA-binding domain of steroid hormone receptors has two "zinc fingers", and intensive mutational analysis of the GR showed that receptors mutated in either finger were defective in DNA-binding activities (9,23,24).   However, the definitive proof that this segment of the protein is conferring DNA-binding specificity was provided by a "domain swapping" experiment.   With the use of in vitro mutagenesis techniques, chimeric receptors were constructed in which the DNA-binding domain of the estrogen receptor was replaced with the DNA-binding domain of the glucocorticoid receptor (25,26).   This chimeric receptor was then shown to be able to activate transcription of the glucocorticoid-responsive promoter in the presence of estradiol, thus providing evidence that the two putative "zinc fingers" were sufficient to recognize a specific DNA sequence.   Further "swapping" experiments involving whole or subregions of DNA-binding fingers demonstrated that the first finger is responsible for sequence-specific recognition while the second finger is required for discrimination of hormone response element half-site spacing (27,28).   The most unexpected result from these studies was the fact that the loops themselves can be interchanged without affecting binding specificity and that the target gene specificity of the hGR can be converted to that of the estrogen receptor by changing three amino acid

residues clustered at the base (knuckle) of the first zinc finger
(28-30).   In the course of these studies, it was found that a
single amino acid change in the DNA-binding domain of the hGR
produces a receptor that recognizes both glucocorticoid and
estrogen response elements, therefore suggesting a simple pathway
for the coevolution of receptor DNA-binding domains and HREs
(28).

Multiple and Cooperative Trans-Activation (τ) Domains

Trans-activating domains can be described as regions of a
transcription factor that increase effective initiation of
transcription by RNA polymerases.   The trans-activation domains
can be identified either by loss or gain of function.   Deletion
of these domains decreases activity of the transcription factor
while duplication increases it.   With insertion and deletion
analysis, activation properties have been found in both the
amino- and carboxyterminal regions of the hGR and were referred
to as τ_1 and τ_2 (8,9).   Duplication of either of these two
transcriptional activating regions increases receptor activity
and, remarkably, the position of the duplicated regions within
the hGR is not critical (23).
A definitive proof of function can be obtained by linking a
putative trans-activating region of a protein to a DNA-binding
domain of another.   Therefore, by fusing various regions of a
receptor to the DNA-binding domain of an unrelated trans-acting
factor, one can assess the ability of a particular region to
induce transcription.   Using this approach with chimeric proteins
engineered by linking regions of the GR or ER to the DNA-binding
domain of the yeast transcription factor GAL4 or the bacterial
repressor LexA, three groups (23,31,32) have shown that the
trans-activating function embodied in both the amino- and
carboxyterminal regions of steroid hormone receptors can confer
activation properties to the GAL4 DNA-binding domain.   In the
hGR, the two transcriptional activating regions are unrelated in
their primary structure but are both acidic in character.
Similar acidic regions possessing trans-activating functions have
been located in the yeast transcription factors GAL4 and GCN4
(for review, see 33) and thus providing evidence that unrelated
transcription factors may activate through common mechanisms.   In
contrast, the hER contains two independent transcriptional
activation functions that are nonacidic (34) and therefore
distinct from those of acidic activators.
Deletional mutagenesis also suggests that the trans-activating function present in the aminoterminus of steroid hormone
receptors can dictate target gene specificity.   As expected,
deletion of the aminoterminal region of the hGR results in an
important decrease in transcription activation function (8).
However, studies with the progesterone and estrogen receptors
have shown that the trans-activating domain in the aminoterminal

region can function in a cell and target gene specificity.
Deletion of their respective aminoterminal region can have either
no effect or lead to dramatic reduction of the ability to
activate transcription depending on the target cell or reporter
gene (26,35). The precise mechanism by which the aminoterminal
region may determine target gene specificity is unknown but is
likely to involve differential interaction with other transcription factors which bind to the promoter of the target gene.
Supporting this argument is the demonstration that steroid
hormone receptors compete for functionally limiting transcription
factors that mediate their enhancer function (36). Both the
isolated N-terminal region and ligand-binding domain of the hER
can inhibit transcription induced by the progesterone receptor.
Moreover, steroid hormone receptor-mediated inhibition of the rat
prolactin gene expression has been shown not to require the
receptor DNA-binding domain, again suggesting that protein-
protein interactions between steroid hormone receptors and other
transcription factor(s) mediate inhibition (37).

Direct Applications of the Modular Structure of the Steroid Hormone Receptors

The study of the functional domains of steroid hormone
receptors has revealed a modular structure for these proteins.
Each domain can function independently and retains its function
when grafted on another unrelated protein. In particular, hybrid
proteins containing the ligand-binding domain of steroid hormone
receptors acquire a ligand-dependent nuclear translocation
function which can be used to control the access of this chimeric
protein to the nucleus. Exploiting this strategy, Yamamoto and
colleagues have created chimeric proteins between the myc
oncoprotein and the hER and between the adenovirus E1A protein
and the rat GR. In both cases, the function of the targeted
protein remained intact but became hormone-inducible. The
myc/hER chimera was shown to transform a rat fibroblast cell line
in a tightly estrogen-dependent manner (38) while the E1A/rGR
hybrid activated transcription in a manner characteristic of E1A,
but in a glucocorticoid-dependent fashion (39). This type of
chimeric protein will provide useful tools for studying not only
the mechanisms of oncogenic transformation but also a wide
variety of other physiological and pathological processes under
the strict control of regulatory molecules.

IDENTIFICATION OF NOVEL RECEPTORS

Orphan Receptors

The expression cloning of the human glucocorticoid receptor
(hGR) described above not only provided the first completed

structure of a steroid receptor but also revealed amino acid sequence identity with the oncogene erbA (40). The relationship between hGR and erbA was confirmed by the molecular cloning of the receptors for estradiol, progesterone, aldosterone and Vitamin D_3 (Figure 1, for Refs. see 1). The characterization of the erbA proto-oncogene product led to its identification as the thyroid hormone receptor (41,42), thus demonstrating for the first time that receptors for unrelated ligands were members of a unique superfamily of ligand-dependent transcription factors. This remarkable discovery led to the suggestion that previously unrecognized ligand-inducible nuclear transcription factors might be associated with this family and that the presence of a segment common to all members of this family might be used to identify related cDNAs. This segment, the cysteine-rich region encoding the "zinc fingers" which function as the DNA-binding domain, shows a high degree of conservation in both the amino acid and nucleotide sequences. Therefore, a direct approach to search for unrecognized hormone response systems is systematically to employ reduced stringency hybridization conditions to screen cDNA libraries made from mRNAs isolated from a variety of target tissues. The hybridization probe can either be a DNA fragment encoding the DNA-binding domain of a particular receptor or an oligonucleotide corresponding to a consensus sequence derived from the DNA sequences encoding the DNA-binding domain of a number of nuclear receptors. Phage plaques or colonies giving positive signals are then rescreened under high stringency conditions with cDNA probes derived from the DNA-binding domain of cloned receptors to eliminate clones of these receptors. The cDNA inserts of the remaining positive clones are then analyzed further by restriction enzyme mapping and DNA sequencing. With the use of this strategy, a large number of cDNAs encoding putative ligand-dependent transcription regulators displaying all the structural features of steroid hormone receptors have been identified from organisms ranging from Drosophila to man (Figure 5) (43-51).

Identification of the Specific Ligand Associated with Orphan Receptors

With the exception of COUP-TF that is essential for expression of the chicken ovalbumin gene (52) and the Drosophila gap gene kni which is involved in the organization of the posterior segment pattern (50), the physiological functions of these orphan receptors are totally unknown, and for all of them, the identity of their putative ligand(s) remains to be elucidated. Therefore, one of the major steps toward the characterization of the physiological role of orphan receptors will be to identify the putative ligand(s) that activates those receptors. However, because the DNA sequences recognized by these novel receptors are not characterized, it is not possible to use a direct transcrip-

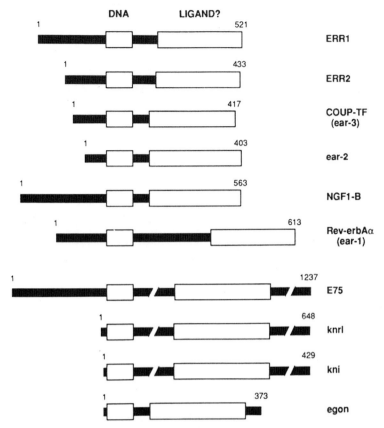

Figure 5.   The orphan receptors.   Linear representation of
identified gene products with high level of homology with other
members of the steroid hormone receptor superfamily.   The DNA-
and putative ligand-binding domain are represented by open boxes.
The location of the ligand-binding domain is hypothetical at this
point since no ligand has been identified for these proteins.
The first six gene products are deduced from human cDNAs with the
exception of NGF1-B which was cloned from the rat PC12 cell line
while the last four orphan receptors were identified from the
fruit fly Drosophila melanogaster.

tional assay such as the one described above for the study of the
hGR.   Instead, a general assay that takes advantage of the
modular structure of the steroid hormone receptors and the
finding that functional domains may be interchangeable (see
above) was developed to reveal the identity of these ligands.   In
this procedure, the DNA-binding domain of the novel receptor is
replaced by the well-characterized DNA-binding domain of the
human glucocorticoid receptor (the DNA-binding domain of any

receptor for which the DNA recognition element has been elucidated can be used for the domain switch). To achieve the domain
switch, the respective cDNAs must first be mutagenized in vitro
to introduce common restriction enzyme sites on each border of
the DNA-binding domains. The choice of the novel restriction
enzyme sites is governed by two factors. First, for added
convenience, it is preferable that these sites not be present in
any of the two cDNAs or the expression vector. Second, one must
try to minimalize the changes resulting from the mutagenesis in
the amino acid sequences encoded by the two cDNAs since drastic
changes in amino acid sequence (a proline residue for a glycine
residue, for example) could engender a non-functional hybrid
receptor. Once the mutagenesis is completed, DNA fragments
encoding the respective DNA-binding domain are interchanged to
yield chimeric receptors bearing novel ligand specificity.

The assay system is established by transfecting recipient
cells (CV-1 or HeLa cells are suitable) with the hybrid receptor
cDNA and glucocorticoid-responsive reporter gene. Transfected
cells are then systematically challenged with candidate ligands
and hybrid receptor activation is monitored by changes in
reporter enzyme activity. The chloramphenicol acetyltransferase
(CAT) or the luciferase (LUC) gene linked to the Mouse Mammary
Tumor Virus long terminal repeat (MMTV-CAT or MMTV-LUC) have been
successfully used in this assay.

Discovery of the Retinoic Acid Receptors

With the use of the "domain swap" assay, retinoic acid was
found to activate such a chimeric receptor. Retinoic acid is an
active metabolite of Vitamin A essential for normal growth and
development and has been implicated as a potential morphogen
responsible for pattern formation in the developing limb (see 53
for review). The discovery that a receptor for retinoic acid
belongs to the superfamily of ligand-dependent transcription
factors led to the suggestion that the putative morphogen
controls limb formation by regulating a set of developmental
control genes. Three distinct receptors for retinoic acid have
been identified to date and are now referred to as RARα (54,55),
RARβ (56,57) and RARγ (58,59). In addition, a novel RARγ isoform
that possesses a different amino terminal domain generated by the
usage of a different promoter has recently been characterized
(60). As expected, the three RAR proteins show a high degree of
amino acid identity in the DNA- and hormone-binding domains. The
region located between the DNA- and hormone-binding domains,
which probably functions as a hinge, shows moderate amino acid
homology. However, the aminoterminal regions of these four
receptors are totally unrelated in their amino acid sequence.
Knowing that the aminoterminal regions of steroid hormone
receptors contain transcriptional activation functions that
dictate target gene and cell specificities, we can speculate that

these distinct aminoterminal domains may contribute to important functional differences between RAR isoforms.

Assay for Transcriptional Regulation by the RARs

Because the DNA sequences conferring retinoic acid responsiveness have not been identified, a direct functional assay for transcriptional regulation by the RARs is not possible at present. Once again, the homology in the amino acid sequence of the DNA-binding domains between various members of the steroid and thyroid hormone receptor family suggests a strategy to identify a retinoic acid-response element (RARE). Because the DNA-binding domains of the retinoic acid and thyroid hormone receptor are highly related, the possibility that the retinoic acid receptor could activate gene expression through a thyroid hormone response element (TRE) was explored (19). Novel thyroid hormone responsive promoters were first constructed by replacing the glucocorticoid responsive elements present in the MMTV-LTR with an oligonucleotide encoding a palindromic TRE (TREp). These promoters were then fused to the LUC gene to generate reporter plasmids such as ΔMTV-TREp-LUC. With the use of the contransfection assay described above (Figure 2), receptorless COS-1 cells were cotransfected with the reporter plasmid and the expression vector containing the RARα, RARβ or RARγ and γB. The transfected cells were then challenged with retinoids. This assay revealed that the concentration of retinoic acid leading to 50% of maximum LUC activity (ED$_{50}$) is significantly different between RARα, RARβ and RARγ, ranging from 20 to 0.5 nM respectively. Thus, this system offers a rapid and sensitive assay to test the relative potency of a large number of natural and synthetic retinoids for their ability to activate receptor function. The availability of a reporter gene responsive to retinoic acid also provides a direct assay to test for the presence of, and the transcriptional response to endogenous RAR(s) in a particular cell line. For example, endogenous RAR(s) present in the F9 teratocarcinoma cells can lead to high levels of LUC activity when transfected with the reporter plasmid ΔMTV-TREp-LUC and treated with retinoic acid (19). TREp-containing vectors provide a simple and quantitative assay for retinoic acid responsiveness in cultured cells.

CONCLUDING REMARKS

In this chapter, I have demonstrated that the practical use of genetic engineering technology has been central to the enhancement of our understanding of the structure and function of the steroid hormone receptors. In addition, the ability to create and express chimeric proteins directly led to the discovery of previously unrecognized hormonal and morphogenetic systems and provided tools to study the molecular mechanism of action of a variety of regulatory proteins, such as nuclear oncogenes. Finally, the elucidation of novel ligand-responsive systems and the development of functional assays to study them not only increase our current knowledge of development and homeostatic processes but are likely to provide opportunities to create new drugs for the treatment of various diseases, including cancer.

Acknowledgments: I wish to thank Dr. Ronald M. Evans for the support he provided me during my postdoctoral studies in his laboratory. This research is supported by the Medical Research Council and the National Cancer Institute of Canada.

REFERENCES

1   Evans, R.M (1988) Science 249, 899-895.
2   Hollenberg, S.H., Weinberger, C., Ong, E.S., Cerelli, G., Oro, A., Lebo, R., Thompson, E.B., Rosenfeld, M.G. and Evans, R.M. (1985) Nature 318, 635-641.
3   Harmon, J.M., Eisen, H.L., Brower, S.T., Simons, S.S., Langley, C.L. and Thompson, E.B. (1984) Cancer Res. 44, 4540-4547.
4   Weinberger, C., Hollenberg, S.H., Ong, E.S., Harmon, J.M., Brower, S.T., Cidlowski, J., Thompson, E.B., Rosenfeld, M.G. and Evans, R.M. (1985) Science 228, 740-742.
5   Miesfeld, R., Okret, S., Wilkstrom, A.C., Wrange, O., Gustafsson, J.A. and Yamamoto, K.R. (1984) Nature 312, 779-781.
6   Miesfeld, R., Rusconi, S., Godowski, P., Maler, B.A., Okret, S., Wilkstrom, A.C., Gustafsson, J.A. and Yamamoto, K.R. (1986) Cell 46, 389-399.
7   Arriza, J.L., Weinberger, C., Cerelli, G., Glaser, T.M., Handelin, B.L., Housman, D.E. and Evans, R.M. (1987) Science 237, 268-275.
8   Giguère, V., Hollenberg, S.H., Rosenfeld, M.G. and Evans, R.M. (1986) Cell 46, 465-652.
9   Hollenberg, S.H., Giguère, V., Segui, P. and Evans, R.M. (1987) Cell 49, 39-46.
10 Godowski, P., Rusconi, S., Miesfeld, R. and Yamamoto, K.R. (1987) Nature 325, 365-368.
11 Picard, D. and Yamamoto, K.R. (1987) EMBO J. 6, 333-340.
12 Guiochon-Mantel, A., Loosfelt, H., Lescop, P., Sokhavuth, S., Atger, M., Perrot-Applanat, M. and Milgrom, E. (1989) Cell 57, 1147-1154.
13 Pratt, W.B., Jolly, D.J., Pratt, D.V., Hollenberg, S.H., Giguère, V., Cadepond, F., Schweizer-Groyer, G., Catelli, M.-G., Evans, R.M. and Baulieu, E.-E. (1988) J. Biol. Chem. 263, 267-273.
14 Klock, G., Strähle, U. and Schütz, G. (1987) Nature 329, 734-736.

15 Martinez, E., Givel, F. and Wahli, W. (1987) EMBO J. 6, 3719-3727.
16 Glass, C.K., Holloway, J.M., Devary, O.V. and Rosenfeld, M.G. (1988) Cell 54, 313-323.
17 Strahle, U., Klock, G. and Schütz, G. (1987) Proc. Nat. Acad. Sci. U.S.A. 84, 7871-7875.
18 Ham, J., Thonson, A., Needham, M., Webb, P. and Parker, M. (1988) Nucl. Acids Res. 16, 5263-5276.
19 Umesono, K., Giguère, V., Glass, C.K., Rosenfeld, M.G. and Evans, R.M (1988) Nature 336, 262-265.
20 Miller, J., McLachlan, A.D. and Klug, A. (1985) EMBO J. 4, 1606-1614.
21 Evans, R.M. and Hollenberg, S.H. (1988) Cell 52, 1-3.
22 Freedman, L.P., Luisi, B.F., Korszun, Z.R., Basavappa, R., Sigler, P.B. and Yamamoto, K.R. (1988) Nature 334, 543-546.
23 Hollenberg, S.H. and Evans, R.M. (1988) Cell 55, 899-906.
24 Severne, Y., Weiland, S., Schaffner, W. and Rusconi, S. (1988) EMBO J. 7, 2503-2508.
25 Green, S. and Chambon, P. (1987) Nature 325, 75-78.
26 Kumar, V., Green, S., Stack, G., Berry, M., Jin, J.R. and Chambon, P. (1987) Cell 51, 941-951.
27 Green, S., Kumar, V., Theulaz, I., Whali, W. and Chambon, P. (1988) EMBO J. 7, 3037-3044.
28 Umesono, K. and Evans, R.M. (1989) Cell 57, 1139-1146.
29 Mader, S., Kumar, V., de Verneuil, B. and Chambon, P. (1989) Nature 338, 271-274.
30 Danielsen, M., Hink, L. and Ringold G.M. (1989) Cell 57, 1131-1138.
31 Webster, N.J.G., Green, S., Jin, J.R. and Chambon P. (1988) Cell 54, 199-207.
32 Godowski, P., Picard, D. and Yamamoto, K.R. (1988) Science 241, 812-816.
33 Sigler, P. (1988) Nature 333, 210-212.
34 Tora, L., White,, J., Brou, C., Tasset, D., Webster, N., Scheer, E. and Chambon, P. (1989) Cell 59, 477-487.
35 Tora, L., Gronemeyer, H., Turcotte, B., Gaub, M.P. and Chambon, P. (1988) Nature 333, 185-188.
36 Meyer, M.-E., Gronemeyer, H., Turcotte, B., Bocquel, M.-T., Tasset, D. and Chambon, P. (1989) Cell 57, 433-442.
37 Adler, S., Waterman, M.L., He, X. and Rosenfeld, M.G. (1988) Cell 52, 685-695.
38 Ellers, M., Picard, D., Yamamoto, K.R. and Bishop, M.J. (1989) Nature 340, 66-68.
39 Picard, D., Salser, S.J. and Yamamoto, K.R. (1988) Cell 54, 1073-1080.
40 Weinberger, C., Hollenberg, S.H., Rosenfeld, M.G. and Evans, R.M. (1985) Nature 318, 670-672.
41 Weinberger, C., Thompson, C.C., Ong, E.S., Lebo, R., Gruol, D.J. and Evans, R.M. (1986) Nature 324, 641-646.

42   Sap, J., Muñoz, A., Damm, K., Goldberg, Y., Ghysdael, J.,
     Leutz, A., Beug, H. and Vennström, B. (1986) Nature 324,
     635-640.

43   Giguère, V., Yang, N., Sequi, P.and Evans, R.M. (1988)
     Nature 331, 91-94.

44   Miyajima, N., Kadowaki, Y., Fukushige, S.-I., Shimizu, S.-
     I., Semba, K., Yamanashi, Y., Matsubara, K.-I., Toyoshima,
     K. and Yamamoto, T. (1988) Nucl. Acids Res. 16, 11057-11074.

45   Milbrandt, J. (1988) Neuron 1, 183-188.

46   Hazel, T.G., Nathans, D. and Lau, L.F. (1988) Proc. Nat.
     Acad. Sci. U.S.A. 85, 8444-8448.

47   Lazar, M.A., Hodin, R.A., Darling, D.S. and Chin, W.W.
     (1989) Mol. Cell. Biol. 9, 1128-1136.

48   Miyajima, N., Horiuchi, R., Shibuya, Y., Fukushige, S.-I.,
     Matsubara, K.-I., Toyoshima, K. and Yamamoto, T. (1989) Cell
     57, 31-39.

49   Oro, A., Ong, E.S., Margolis, J.S., Posakony, J.W., McKeown,
     M. and Evans, R.M. (1988) Nature 336, 439-496.

50   Nauber, U., Pankratz, M.J., Kienlin, A., Seifert, E., Klemm,
     U. and Jackle, H. (1988) Nature 336, 489-492.

51   Rothe, M., Nauber, U. and Jackle, H. (1989) EMBO J. 8, 3087-
     3094.

52   Wang, L.-H., Tsai, S.Y., Cook, R.G., Beattie, W.G., Tsai,
     M.-J. and O'Malley, B.W. (1989) Nature 340, 163-166.

53   Brockes, J.P. (1989) Neuron 2, 1285-1294.

54   Giguère, V., Ong, E.S., Segui, P. and Evans, R.M. (1987)
     Nature 330, 624-629.

55   Petkovitch, M., Brand, N., Krust, A. and Chambon, P. (1987)
     Nature 330, 444-450.

56   Brand, N., Petkovitch, M., Krust, A, Chambon, P., de Thé,
     H., Marchio, A. Tiollais, P. and Dejean, A. (1988) Nature
     332, 850-853.

57   Benbrook, D., Lenhardt, E. and Pfahl, M. (1988) Nature 333,
     669-672.

58   Zelent, A., Krust, A., Petkovitch, M., Kastner, P. and
     Chambon, P. (1989) Nature 339, 714-717.

59   Kurst, A., Kastner, P, Petkovitch, M., Zelent, A. and
     Chambon, P. (1989) Proc. Nat. Acad. Sci. U.S.A. 86, 5310-
     5314.

60   Giguère, V., Shago, M., Zirngibl, R., Tate, T., Rossant, J.
     and Varmuza, S. Mol. Cell. Biol. (in press).

IDENTIFICATION AND FUNCTIONAL ANALYSIS OF

MAMMALIAN SPLICING FACTORS

Albrecht Bindereif[1] and Michael R. Green[2]

[1]Max-Planck-Institut für Molekulare Genetik
Otto-Warburg-Laboratorium
Ihnestrasse 73, D-1000 Berlin 33, Germany
[2]Department of Biochemistry and Molecular Biology
Harvard University, 7 Divinity Avenue,
Cambridge, MA 02138

INTRODUCTION

In eukaryotic cells most primary transcripts of protein-coding genes are processed through a variety of post-transcriptional modification reactions to yield mature mRNA. One of these processing reactions termed pre-mRNA splicing involves the accurate excision of intron RNA sequences and the coordinate ligation of exon RNA sequences. In addition, there are post-transcriptional modifications at the 5' end (capping) and at the 3' end (polyadenylation). A systematic biochemical study of pre-mRNA splicing began with the development of efficient in vitro splicing systems [reviewed in (1)] that accurately splice synthetic pre-mRNAs in nuclear extracts prepared from cultured mammalian cells. In most cases, relatively short model substrates consisting of two exons and one intron were used for these studies (see Figure 1 for a schematic representation). The biochemical analysis of this highly complex RNA processing reaction has since made rapid progress. Initial studies revealed a two-step mechanism involving the three conserved splicing signals (Figure 1). First, cleavage at the 5' splice site occurs concomitantly with lariat formation at the intron branch point. Second, the RNA intermediates, exon 1 and intron lariat-exon 2, are cleaved at the 3' splice site, and the two exon RNA sequences are ligated together. The intron RNA sequences are thereby released in the form of a lariat. Soon after the establishment of in vitro splicing systems it became clear that this complex series of reactions occurs in large ribonucleoprotein (RNP)

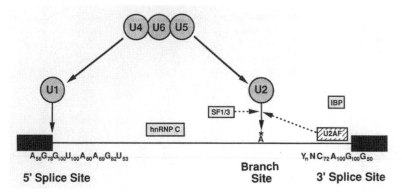

Figure 1: Interactions of snRNP and non-snRNP splicing factors with the pre-mRNA. The consensus sequences for eukaryotic 5' and 3' splice site regions, $A_{56}G_{78}-G_{100}U_{100}A_{60}A_{69}G_{82}U_{53}$ and $Y_{n}NC_{72}A_{100}G_{100}-G_{50}$, respectively [the splice junctions indicated by a hyphen, the polypyrimidine tract by Y_{n} (Y=pyrimidine), and with the actual nucleotide frequencies] are based on analysis of the GENBANK databank [for a recent analysis, see (150)]. The branch point is indicated by A^{*}, since in most cases branch formation occurs at the adenosine residue of a weakly conserved mammalian consensus sequence, $UNCURA^{*}C$ (R=purine), which is located between 18 and 38 nucleotides upstream of the 3' splice site [see, for example, (117,120-123,152); for a review, see (1)]. Splicing factors are included, for which pre-mRNA interactions have been analyzed (snRNP splicing factors indicated by circles, non-snRNP factors by boxes).

structures of the pre-mRNA called splicing complexes or spliceosomes (2-7). The ordered assembly of splicing complexes is a prerequisite of the pre-mRNA splicing reaction and requires the coordinate action of small nuclear ribonucleoproteins (snRNPs) and of additional proteins, so-called non-snRNP splicing factors.

In recent years it has become evident that splicing is not only required for the accurate processing of pre-mRNAs, but that it also provides many targets for regulating gene expression. Differentially spliced mRNAs can be generated starting from a single primary transcript by alternative splice site selection. We know of a number of examples where this selection is regulated in a temporal or tissue-specific manner [for a review, see (8,9)]. To understand the molecular mechanisms how splicing events can be regulated, we first need to acquire a thorough knowledge of the basic splicing machinery.

This review emphasizes recent advances in the identification and functional analysis of mammalian splicing factors; to previous work we refer only when necessary as a basis for understanding recent studies or sometimes for reasons of completeness. Most of the conclusions have been drawn from work on mammalian systems; however, we will also discuss a few results derived from two other metazoan systems, Xenopus and Drosophila, which exhibit a pre-mRNA splicing mechanism very similar to the mammalian system. Several reviews of the past few years have summarized research on pre-mRNA splicing (1,10-12), snRNPs (13-16), hnRNPs (17), alternative splicing (8,9), RNA catalysis and group I intron self splicing (18).

snRNP-SPLICING FACTORS

Identification of snRNPs as Splicing Factors

snRNPs (small nuclear ribonucleoprotein particles) are abundant, nuclear RNP-particles consisting of an snRNA and multiple protein components [reviewed recently in (19,20)]. The biochemical composition of the snRNPs involved in splicing (U1, U2, U4/U6 and U5 snRNPs) has been analyzed over the last ten years, yet the exact stoichiometry of the components and, for some snRNPs, the complete list of their protein constituents are still not fully established. And only in the last few years have approaches been developed to characterize the functions of individual snRNPs in the splicing mechanism.

Several approaches have been taken to demonstrate the requirement of snRNPs in the pre-mRNA splicing reaction. First, snRNP-specific antibodies have been used to inactivate snRNPs in unfractionated nuclear extracts and to deplete extracts of snRNPs. Antibody-inhibition experiments proved very useful in the initial studies that established the requirement of snRNPs of the Sm-type (U1, U2, U4/U6 and U5 snRNPs) and specifically of U1 snRNP in the in vitro splicing reaction. Antibody-depletion/ complementation studies provided another approach to identify snRNP splicing factors. Nuclear extracts could be depleted of snRNPs with anti-Sm antibodies as well as with antibodies directed against the m3G-cap structure characteristic of U1, U2, U4 and U5 snRNA. Splicing activity in these snRNP-depleted and splicing-inactive extracts could be complemented by an immuno-purified fraction of snRNPs (21,22). With the use of immunodepleted nuclear extracts and highly purified 20S U5 snRNP, direct evidence for U5 snRNP being a splicing factor and required for both steps of the pre-mRNA splicing reaction has recently been obtained (22).

Second, a general approach to inactivate specific snRNPs in a complex mixture was introduced with the oligonucleotide-directed RNase H cleavage reaction. This approach was first applied in the in vitro splicing system to assess the requirement of an intact 5' end of U1 snRNA (23). Subsequently, U2 snRNP was shown to be a necessary splicing factor in the same way (24). Furthermore, oligonucleotide-directed RNase H cleavage of U4 and U6 snRNAs, which are present within a single snRNP particle, the U4/U6 snRNP (25-27), inhibited the splicing activity in nuclear extract (28,29), strongly arguing for U4/U6 snRNP acting as a splicing factor. U5 snRNP was the only abundant snRNP that could not be efficiently inactivated by U5 snRNA-specific oligonucleotides and RNase H, presumably because of its compact organization in an RNP structure (30,31). Recently, site-directed RNase H cleavage was also applied in the Xenopus oocyte system to demonstrate the splicing activity of human-Xenopus hybrid and of Xenopus U1 and U2 snRNPs (32,33). In these studies the endogenous snRNPs were first inactivated by oligonucleotide-directed RNase H cleavage followed by the microinjection of snRNA genes. Transcribed snRNAs assembled with endogenous Xenopus snRNP proteins and rescued splicing activity.

A fourth approach to assess a splicing function of specific snRNPs was introduced by Lamond and coworkers (34,35) who used modified anti-sense RNA oligonucleotides that stably bound to certain regions of U2, U4, or U6 snRNAs thereby inhibiting pre-mRNA splicing.

Finally, a more indirect line of evidence arguing for snRNP-splicing factors came from the analysis of affinity-purified splicing complexes. All the major snRNPs (U1, U2, U4/U6 and U5 snRNPs) were detected as specific components of the 60S splicing complex (36,37), the formation of which is a prerequisite of the pre-mRNA splicing reaction.

Purification and Structural Characterization of snRNPs

The major mammalian snRNPs (U1, U2, U4/U6 and U5 snRNPs) were first purified by chromatographic fractionation procedures resulting in preparations of pure U1 snRNP and highly enriched U2 snRNP (38-40). Since the fractionation of individual snRNPs proved very difficult, a second approach, CsCl density gradient centrifugation in the presence of $MgCl_2$, was introduced by Jeanteur and his coworkers. At least core structures of snRNPs are surprisingly stable under these conditions and could be partially fractionated according to their different densities. In combination with ion-exchange chromatography pure preparations of U1 and U5 snRNPs were obtained (41). Third, immune affinity chromatography provides probably the most gentle method to fractionate and purify native snRNPs (42). Monoclonal antibodies against snRNP proteins were used as well as antibodies directed against modified nucleosides, in particular against the snRNA-specific m_3G-cap structure and against 6-methyladenosine, m^6A [reviewed in (20)]. For future work, affinity purification of snRNPs based on anti-sense oligonucleotides [see, for example, (34)] may become an important tool.

The protein composition of HeLa-snRNPs is summarized in Table 1. These proteins can be divided into two classes, a set of seven common proteins (B'BDD'EFG) shared by all major snRNPs and specific proteins, which have until now been detected for U1, U2 and U5 snRNPs.

Table 1
snRNP Splicing Factors

snRNA[a]	Common Proteins[b]	Specific Proteins[b]
U1 (164 nt)		70K 70 kD A 34 kD C 22 kD
	B' 29 kD B 28 kD D 16 kD D' 15.5 kD E 12 kD F 11 kD G 9 kD	
U2 (187 nt)		A' 33 kD B'' 28.5 kD
U4/U6 (141/106 nt)		?
U5 (116 nt)		25 kD[c] 40 kD[d] 52 kD[d] 100 kD[d] 102 kD[d] 116 kD[d] 200 kD[d]

[a]The major mammalian snRNAs and their sizes are listed [for a review, see (19), and references therein].
[b]Common and specific snRNP proteins from HeLa cells and their sizes are given [for review, see (20), and references therein].
[c](41)
[d](97)

The Sm Core Structure of snRNPs

Several protection studies using purified snRNPs [for example (41)] had found as the most stably protected snRNA region the so-called Sm binding site, $A(U)_nG$, which is a conserved sequence element of U1, U2, U4 and U5 snRNA and consists of a single-stranded region flanked by two double-stranded stems (43). A core RNP structure containing the common Sm polypeptides could be isolated by sucrose gradient centrifugation of snRNPs in the presence of sarcosyl (44).

A multi-step snRNP assembly has been suggested based on pulse-chase studies of the in vivo synthesis and assembly of snRNPs (45). An RNA-free protein complex of 6S containing the DEFG polypeptides is formed first, followed by its binding to the Sm binding site of snRNAs and further general and snRNP-specific assembly steps. More recently, by in vivo assembly in Xenopus oocytes of mutated U2 snRNAs and of an artificial RNA, the Sm binding site has been shown to be necessary, but not sufficient for binding of the common Sm proteins to U2 snRNA (46-48). By UV-crosslinking of purified, [^{32}P]-labeled snRNPs only the F protein, one of the common snRNP proteins, was detected in close contact with U1 snRNA (49), suggesting that this protein recognizes the Sm binding site directly.

Several of the common Sm proteins have been studied by cDNA cloning and analysis: B'B, D and E (see below). The primary sequence of the B'B proteins has recently been determined from cDNA cloning (50-52). There is evidence that the B' and B proteins arise by alternative splicing of one pre-mRNA and differ only at their carboxyterminus, consistent with earlier protein mapping data (53). The cDNA analysis identified also a sequence motif, which is arranged in repeated units (for example, Pro-Pro-Pro-Gly) interspersed with hydrophobic and basic residues. It is also present in the U1 snRNP-specific A and C proteins.

A non-specific endoribonuclease activity of snRNP core particles residing in the aminoterminal half of B'B was recently discovered (54), which is activated only after conversion of snRNPs to core particles. The functional significance of this nuclease activity in the splicing mechanism remains speculative, and perhaps it is involved in snRNA 3' processing or turnover. Of great interest in this context is also a brain-specific cDNA, which encodes a snRNP-associated and Sm epitope-bearing polypeptide called N (55-57). The B'B proteins have an amino acid sequence very homologous with the N protein, except for a region of 50 amino acids in B'B inserted towards the carboxyterminus. The additional finding that N appears to replace B in a brain-specific manner suggests that switching of snRNP components can function as a mechanism for tissue-specific RNA processing. Furthermore, cDNAs have been isolated for the D and E polypeptides (58-60). In the D protein sequence the most striking features are nine repeats of a glycine-arginine stretch of

unknown function, but no RNP-consensus sequence (RNP-CS) was
found (58). The RNP-consensus sequence consists of an octapep-
tide region, which has recently been discovered as a highly
conserved domain in many RNA binding proteins including a number
of snRNP polypeptides [see the following two sections; for a
review, see (61)]. Interestingly, the RNA-consensus sequence was
absent not only in the D protein, but also in the B'B and E
proteins (50,60).

It remains to be investigated whether the B'BDD'EFG protein
complex bound at the Sm binding site functions in the actual
splicing mechanism beyond being a potential nuclear transport
signal (46) and being required for cytoplasmic cap trimethylation
(47).

U1 snRNP

Through the application of a wide range of approaches and
the efforts of several groups much progress has recently been
made in the structural and functional characterization of the U1
snRNP. U1 snRNP was the first snRNP that was systematically
studied by in vitro reconstitution (62-66). With the use of
various crude extracts containing free snRNP polypeptides and
assay by immunoprecipitation, nuclease protection, buoyant
density, and sedimentation, efficient reconstitution of a
structurally correct U1 snRNP was achieved.

A coupled immunoprecipitation-nuclease protection assay was
used to map RNA-protein interactions in an in vitro reconstituted
U1 snRNP particle. The binding site of the common Sm protein
complex was mapped to the Sm consensus sequence between the two
3' terminal stems, and the binding site of the U1 snRNP-specific
70K protein to stem-loop I (63,64). Based on co-immunoprecipita-
tion of the two protected U1 snRNA regions with either antibody,
an interaction between 70K and the Sm proteins was also suggested
(64). As extensive mutagenesis revealed and in agreement with
the high evolutionary conservation of the loop I region, almost
all positions within this region were found to be essential for
70K binding (67).

This U1 snRNA binding analysis was recently extended to the
U1 snRNP-specific A and C proteins. UV crosslinking revealed
that the A protein contacts U1 snRNA directly (65). Mattaj and
coworkers studied the assembly of U1 snRNP both in vitro and in
vivo using Xenopus oocytes and synthetic wild-type and mutant U1
snRNAs (68,69) and came to similar conclusions as studies using
mammalian extracts. Binding of the U1 snRNP-specific 70K protein
requires only stem-loop I (68), and essential contacts for this
binding reaction reside in the loop sequence (69).

With protein blots of partially purified snRNP proteins and
probes with ^{32}P-labeled U1 snRNA fragments, specific binding of
the A protein to stem-loop II of U1 snRNA was demonstrated. This
recognition requires most of stem-loop II sequences and no other

proteins (70). Scherly et al. further demonstrated that only the evolutionarily conserved loop II sequence is required for binding of the A protein (71). By immune electron microscopy of purified U1 snRNPs the m₃G-cap structure was recently mapped to the base of U1 snRNP that appears as a body with two characteristic protuberances (72). New insight in the structure of U1 snRNP and in the specific binding of U1 snRNP proteins has in recent years been gained by cDNA cloning of all the U1 snRNP-specific proteins. Human cDNAs have been isolated for all the U1 snRNP-specific proteins: 70K (73-75), C (76,77) and A (78).

Analysis of the 70K cDNA revealed three regions of homology with known nucleic acid binding proteins. These are two arginine-rich regions in the carboxyterminal half and one copy of the so-called RNP consensus sequence (79) in the middle of the protein sequence. The deviation of the predicted molecular size of 52 kD from the apparent electrophoretic mobility of 70 kD is caused by an unusual charge distribution within the protein, mainly by the two arginine-rich regions near the carboxyterminus (80). Post-translational modifications such as serine phosphorylation (81) may play a functional role in this unusual protein. In addition, the analysis of heterogenous cDNAs and the detection of two mRNA species present in different ratios in different cell lines suggest that multiple forms of this protein might arise by alternative RNA processing and might provide a means to regulate splicing patterns (73,74). Finally, in vitro translated deletion derivatives of 70K protein were analyzed for U1 snRNA binding by immunoprecipitation of ^{32}P-labeled U1 snRNA and by native RNP gel electrophoresis with ^{35}S-labeled proteins (80). That way a 111-amino acid region of 70K was identified that retained specific U1 snRNA binding. This RNA binding domain encompasses the highly conserved RNP octamer sequence within a larger, 80-amino acid region conserved among many RNA-binding proteins. The two arginine-rich stretches are not required for U1 snRNA-specific binding. These conclusions were confirmed by another analysis with 70K-β-galactosidase fusion proteins and mutant derivatives expressed in E. coli (67).

A human cDNA for the second U1 snRNP-specific polypeptide, the A protein, has also been isolated and analyzed recently (78). A striking feature is the high homology between the U1 snRNP-specific A protein and the U2 snRNP-specific B'' protein (82) located in two sequence blocks covering the aminoterminal and carboxyterminal thirds of the protein. Both of these regions contain an RNP consensus sequence (70,78). Additional evidence for the relatedness of these two proteins has been obtained through immunological studies and peptide mapping (53,83,84). The A protein requires the intact RNP-consensus sequence for its specific binding to loop II (71).

For the third U1 snRNP-specific polypeptide, protein C, a human cDNA has been isolated as well (76,77). In contrast to the 70K and A proteins, the C protein contains no RNP consensus sequence. Post-translational modifications of the C protein had previously been demonstrated by a pulse-chase analysis (45).

U2 snRNP

Both of the U2 snRNP-specific polypeptides, the A' and the B'' proteins, have been analyzed by cDNA cloning (82,85). The most striking features of the A' protein are a leucine-rich region in the aminoterminal half and the absence of an RNP-consensus sequence (85). The close relatedness between the U1 and U2 snRNP-specific proteins A and B'', respectively, has been discussed in the context of U1 snRNP (see above). This parallel between these two proteins may further extend to the level of protein binding sites. A strong sequence similarity has been detected between the binding site of protein A in stem-loop II of U1 snRNA and a sequence element in stem-loop III of U2 snRNA (70). Binding of protein A to stem-loop III of U2 snRNA had previously been suggested (47).

Structurally correct assembly of U2 snRNP by in vitro reconstitution in extracts has been achieved with natural U2 snRNA (66) as well as with synthetic U2 snRNA (86). In another recent study the function of U2 snRNP in splicing was investigated with 2'-O-methylated RNA anti-sense oligonucleotides (34,87). Such modified oligonucleotides were directed against various regions of U2 snRNA where they bound in a very stable manner thereby blocking specific snRNP functions. First, a requirement of the U2 snRNA region between stem-loop I and II for U2 snRNP-pre-mRNA binding (complex A formation) was demonstrated that way. Second, U2 snRNA sequences from the 5' end and stem-loop I were not required for complex A formation, but for assembly of the spliceosome (complex B). These conclusions differ from what had previously been found on the basis of RNase H-cleavage studies (88-90).

U2 snRNP assembly has also been studied in vivo with Xenopus oocytes and RNase H-mediated snRNP inactivation (32). This system has recently been used to assay mutant U2 snRNAs for splicing complementation activity (33). Surprisingly, U2 snRNA regions essential for binding of U2 snRNP-specific proteins were not required for splicing complex formation and splicing.

U4/U6 snRNP

In mammalian extracts, most of the U4 and U6 is present in the form of the U4/U6 snRNP, which is unique in that it contains two snRNAs, U4 and U6, joined by intermolecular base pairing (25-27). U4 snRNA, but not U6 snRNA, contains a binding site for the Sm protein complex. Furthermore, U6 snRNA is the only spliceosomal snRNA that lacks the trimethylguanosine cap structure. Instead, it has a unique cap structure, which has been recently characterized as a 5' guanosine triphosphate methylated at the γ-

position (91). Based on extensive phylogenetic evidence, a
secondary-structure model of U4 and U6 snRNAs was recently
proposed that is strongly conserved from yeast to mammals
(15,92). According to this model, the base pairing interaction
between U4 and U6 snRNAs is organized in form of a so-called Y
structure, which consists of two intermolecular helices separated
by an intramolecular helix of U4 snRNA. That this proposed U4-U6
snRNA base pairing interaction exists in the U4/U6 snRNP was
recently confirmed by in vitro reconstitution of synthetic U6
snRNA mutant derivatives and by U6 snRNA interference analysis
(93). Both a U4 and a U4/U6 snRNP could be obtained by in vitro
reconstitution that are by several criteria structurally and
functionally indistinguishable from the native particles (66).

In Xenopus oocytes, in contrast, most of the U6 snRNA is not
in form of the U4/U6 snRNP, but in form of a U6 snRNP (94). An
RNA region near the 5' end of U6 snRNA was identified that is
required for migration of U6 snRNA from the cytoplasm to the
nucleus.

U5 snRNP

U5 snRNP has several unique features. Of the abundant
snRNPs, it is most resistant to nucleases and oligonucleotide-
directed RNase H cleavage, indicating its highly compact organization in an RNP complex (30,31). As the other snRNPs, U5 snRNP
contains the Sm proteins, B'BDD'EFG (38,41,42). Several candidates for U5 snRNP-specific polypeptides have been described: a
25K protein (41) a 70K protein (95) and the 100K IBP [intron
binding protein (96,116); also see below in the section Splice
Site and Branch Point Recognition of snRNPs]. Very recently,
immunoaffinity purification of U5 snRNP from splicing-active HeLa
nuclear extracts revealed a surprising complexity of several
polypeptides of molecular weights 40K, 52K, 100K, 102K, 116K and
200K, which all copurified with U5 snRNA through glycerol
gradient centrifugation (97). At least the 100K and the 200K
proteins are immunologically distinct. These U5 snRNP-specific
proteins may be involved in the formation of multi-snRNP complexes containing U5 snRNA (see below in the following section).
In summary, the U5 snRNP appears to be the largest (20S) and most
complex snRNP particle.

The 200K protein of human U5 snRNP was also identified with
an antibody against the U5 snRNP-specific Prp8 protein of yeast
(98). The human 200K protein was detected not only in U5 snRNP,
but also in the U4/U5/U6 snRNP complex and in affinity-purified
spliceosomes.

U4/U5/U6 Multi-snRNP Complex

By Northern analysis and native RNP gel electrophoresis of
HeLa nuclear extracts U5 snRNA was detected not only as individ-

ual U5 snRNP, but also in the form of a 25S multi-snRNP complex containing U4, U5 and U6 snRNAs (99). Its formation from U4/U6 and U5 snRNPs requires ATP (31). Although there is no direct evidence for this, the multi-snRNP complex most likely represents an early intermediate in spliceosome assembly (see below). Since by chemical modification no differences could be detected between free and multi-snRNP bound U5 snRNP, U5 snRNP probably interacts with U4/U6 snRNP more through protein-protein contacts than by RNA-RNA base pairing, perhaps involving some of the recently discovered U5 snRNP-specific polypeptides (97).

An even larger multi-snRNP complex containing U2, U4, U5 and U6 snRNAs, the so-called pseudospliceosome, forms under certain ionic conditions in the absence of pre-mRNA (100). Its relevance for the splicing reaction is unclear, yet its formation suggests that snRNP components, through their wide potential of interactions, may build up a backbone for the spliceosome.

Pre-mRNA Binding of snRNPs and Assembly of Splicing Complexes

Pre-mRNA binding of snRNPs and the order of assembly of snRNPs into splicing complexes have been studied by RNase protection assays, affinity purification of splicing complexes, in vitro snRNP reconstitution, and by native RNP gel electrophoresis.

Affinity purification was introduced to study the snRNP composition of splicing complexes, with biotinylated splicing substrates and streptavidin agarose (36,37). U2, U4/U6 and U5 snRNPs were identified that way as stably bound components present in approximately equimolar amounts in the 60S splicing complex. The presence of U1 snRNP in the 60S splicing complex has first been a controversial issue, since conflicting results were obtained from analyses with the use of immunoprecipitation, affinity purification and native gel electrophoresis (3,36,37, 89,90,101,102). Most evidence, however, is in favor of U1 snRNP being a component of the spliceosome and not only transiently associated with it. The apparent discrepancy was most likely caused by the relative lability of U1 snRNP binding, particularly in the presence of heparin (37).

By RNase protection assays, binding of U1 snRNP at the 5' splice site was detected as an initial step of splicing complex assembly followed by the ATP-dependent, stable association of U2 snRNP with sequences around the branch point (24,103,104). In the 60S splicing complex, binding of U1 and U2 snRNP was also detected, the 5' splice site protection by U1 sRNP being extended (37,89). This may reflect interactions of U1 snRNP with other spliceosome components, for example with U2 snRNP for which a mutational analysis provided some evidence (105), or with U4/U5/U6 snRNP (see below).

The question where on the pre-mRNA U4/U6 and U5 snRNPs bind has been addressed by a combination of RNase protection assays,

affinity purification of splicing complexes, and by studying the
kinetics of snRNP binding and the effects of splice site mutations. These data are all in support of an assembly model where
U4, U5 and U6 snRNAs bind together in form of the U4/U5/U6 multi-
snRNP complex. U4/U5/U6 snRNP most likely does not contact the
pre-mRNA directly but rather interacts with pre-mRNA bound U1 and
U2 snRNPs (37). Whereas binding of U4, U5 and U6 snRNAs depends
on ATP, there may be a different, ATP-independent mode of U5
snRNP binding at the 3' splice site early in the splicing
reaction (30).

The U4/U6 snRNP is so far the only snRNP which can be
reconstituted in vitro in a functional form. After reconstitution both synthetic U4 and U6 snRNAs are assembled under splicing
conditions into splicing complexes as shown by native RNP gel
electrophoresis (66). Assembly of U6 snRNA into spliceosome
complexes requires in addition to the U4/U6 snRNA interaction
domain (see above the section U4/U6 snRNP) two so-called spliceosome assembly domains that flank the base pairing region (93).
These two spliceosome assembly domains may function as binding
sites for splicing factors or may facilitate the formation of an
alternative U6 snRNA secondary structure during spliceosome
assembly.

After spliceosome formation, the U4/U6 snRNP undergoes a
major conformational change as evidenced by the dissociation of
U4 snRNA, but not of U6 snRNA, from spliceosomal complexes
resolved on non-denaturing gels (106). Since this conformational
change is concomitant with the appearance of splicing intermediates, dynamic interactions within the U4/U5/U6 snRNP complex
are probably directly involved in the catalytic activation of the
spliceosome. More recently, biotinylated anti-sense RNA
oligonucleotides were used to affinity-select snRNPs and snRNP-
containing splicing complexes (35). It was demonstrated that U4
snRNA is retained as a component of splicing complexes throughout
spliceosome assembly and the two steps of the splicing reaction.

The formation of splicing complexes has also been investigated using multi-intron pre-mRNAs (107). This study indicated
that multiple spliceosome units can assemble independently and
simultaneously on such more complex splicing substrates.

Finally, the biochemical composition and overall morphology
of splicing complexes that were purified by gel filtration in a
functional state have been studied by electron microscopy (108).

NON-snRNP SPLICING FACTORS

Identification and Functional Characterization

Soon after efficient in vitro splicing systems had been
established, it became evident that multiple factors are required
for this complex RNA processing reaction. Non-snRNP splicing

factors can in principle be distinguished from snRNPs by their resistance to micrococcal nuclease. An essential nucleic acid component might, however, be strongly resistant to nucleases as, for example, in the case of U5 snRNP (see the section U5 snRNP). Therefore the absence of a nucleic acid component has to be confirmed ultimately by purification of each non-snRNP splicing

Table 2
Non-snRNP Splicing Factors

	Function and Size	References
Fraction Ia	2nd step of splicing reaction	6,109
Fraction Ib		6,109
SF (splicing factor)2	1st step	110
SF3	2nd step	110
SF4A	2nd step	110
SF4B	1st step	110
SF (splicing factor)1	U2 snRNP binding	111,119
SF2	50 kD	111,112
SF3	U2 snRNP binding	111,119
SF4		111
U2AF (auxiliary factor)	U2 snRNP binding: 65 kD polypyrimidine tract/3' splice site recognition) and 35 kD	114,118
IBP (intron binding protein)	polypyrimidine tract/3' splice site recognition: 100 kD (U5 snRNP associated?)	95
70 kD protein	polypyrimidine tract/3' splice site recognition (U5 snRNP associated?)	116
Non-snRNP factor(s)	U1 snRNP binding at the 5' splice site	127,128
hnRNP protein C1 41 kD C2 43 kD	polypyrimidine tract recognition	115,129
SCF (splicing complementing factor)		113

factor to homogeneity. One should also keep in mind that this distinction between snRNP and non-snRNP splicing factors can become obscured, because at least some non-snRNP factors can associate with snRNPs. Most non-snRNP factors are biochemically still less well characterized than snRNPs, and only a little is known about their mechanism of action. Therefore the known non-snRNP factors will be grouped in the following according to how they have been identified (for a summary, see Table 2). A number of studies used complementation of splicing activity as an assay to fractionate non-snRNP factors (see the next section). Other assays include formation of splicing complexes, splice site and branch point recognition of snRNPs and pre-mRNA binding all described below. Separately discussed will be the activities and RNA binding factors that may be indirectly involved in splicing (see the section PERSPECTIVES).

Complementation of Splicing Activity

Furneaux et al. demonstrated the requirement for non-snRNP factors by fractionation of the splicing activity in HeLa nuclear extract into two fractions (I and II), one of which(I) was resistant to micrococcal nuclease treatment (109). Later, fraction I was further separated into two components (Ia and b). Fractions Ib and II are required for the first step of the splicing reaction and Ia in addition to the others for the second step (6).

Krainer and Maniatis developed a fractionation scheme to purify snRNPs and non-snRNP splicing factors (110). By selective inactivation and by complementation assays they identified four activities called splicing factors (SF) 2, 3, 4A and 4B. SF3 and 4A are required for the second step, SF 2 and 4B for the first step.

Krämer et al. also used complementation of splicing activity as an assay to fractionate and purify splicing factors (111). In addition to two snRNP fractions, four fractions were identified containing non-snRNP splicing factors [SF 1-4; nomenclature different from that of (110)]. One of them, SF2, has been extensively purified, is required early in the splicing reaction, and has a molecular weight of 50K (112).

A different splicing complementation assay was developed with the use of splicing complexes purified by gel filtration (102,108). Purified 60S splicing complexes underwent splicing once complemented with fractions containing non-snRNP factors.

A splicing complementation approach has also been used in the detection of a non-snRNP factor (splicing-complementing factor, SCF) in extracts from nuclear matrix preparations (113). Since different fractionation schemes were used in these studies, a definitive correlation between the non-snRNP splicing factors must await further purification and protein identification.

Formation of Splicing Complexes

That snRNPs are required for the assembly of splicing complexes and for the splicing reaction, has been established (see the section Identification of snRNPs as Splicing Factors). The requirement of purified non-snRNP factors in the formation of splicing complexes has also been analyzed by sucrose gradient centrifugation. Whereas fraction II containing non-snRNP factors alone gave a 30S complex of the pre-mRNA, only its combination with fraction Ib resulted in the assembly of 55S complexes containing pre-mRNA and splicing intermediates (6).

Splice Site and Branch Point Recognition of snRNPs

There are several lines of evidence that established that non-snRNP factors are required for snRNPs to recognize splice sites and the branch point. The branch point-U2 snRNP interaction has probably been studied in most detail both by biochemical fractionation of the components, by nuclease protection and native RNP gel electrophoresis.

Using a protection assay, Ruskin et al. found that partially purified U2 snRNP could not bind to its target sequence unless complemented with a micrococcal nuclease resistant fraction from nuclear extract (114). The complementing activity was called U2AF (U2 snRNP auxiliary factor) and shown to differ from the known U2 snRNP polypeptides and from other non-snRNP factors, hnRNP C protein (115) and the 70K 3' splice site binding protein (95,116). Competition assays showed that U2AF binding required an intact 3' splice site region including the polypyrimidine tract and the 3' splice site. A model for the selection of mammalian branch sites by U2 snRNP and U2AF was recently proposed (114,117). According to this model, U2AF first recognizes the polypyrimidine tract-3' splice site region, followed by U2AF-U2 snRNP interaction, which directs U2 snRNP to the best branch site available within the distance constraints of this interaction.

U2AF has recently been purified to homogeneity (118). It consists of two polypeptides of molecular weights 35,000 and 65,000, the latter of which binds in the absence of ATP specifically to the polypyrimidine tract-3' splice site region. After the subsequent ATP-dependent binding of U2 snRNP, U2AF remains associated with the pre-mRNA.

Using an RNP gel assay, Krämer also reported an activity that is required for partially purified U2 snRNP to form the presplicing complex (complex A) (119). This activity is present in nuclear extract and could be chromatographically fractionated into two components, SF1 and SF3 (see also the section Formation of Splicing Complexes).

Mammalian branch point recognition by U2 snRNP most likely requires base pairing between the branch point sequence of the pre-mRNA and U2 snRNA. This is based on convincing genetic evidence that branch point mutations could be efficiently suppressed by mutant U2 snRNAs carrying compensatory base changes (120,121). A weak consensus sequence can be derived for the mammalian branch point sequence, UNCURA*C, within which branch formation occurs, mostly at the adenosine (A*) residue at a distance of between 18 and 38 nucleotides from the 3' splice site [see, for example (120,122,123)]. In most cases the branch site could be predicted correctly by this consensus sequence as a recent compilation of all mapped mammalian branch site sequences revealed (117). In contrast to the branch point sequence, the putative base-pairing region of U2 snRNA is strongly conserved and located between stem-loops I and II (15).

Non-snRNP factors may also be involved in binding of U5 snRNP at the 3' splice site. This notion is based on two independent approaches with the use of RNase T1 protection-immunoprecipitation and Northwestern blotting (95,116). In one study, a 70K protein was identified after fractionation of nuclear extract that interacts with the polypyrimidine-3' splice site region, is immunoprecipitable with anti-Sm antibodies, and reversibly associates with an snRNA, most likely U5 snRNA (95). In another study a similar activity that binds to the same pre-mRNA region was identified by Northwestern blotting as a 100K protein called IBP [intron binding protein (116)]. These two proteins, which copurified with U5 snRNP, may be closely related with each other; their binding may represent one of the earliest steps in the splicing reaction.

Finally, there is evidence from mutational studies [for example (124-126)] that for U1 snRNP recognizing the 5' splice site RNA-RNA base pairing is necessary but not sufficient. In agreement with these observations, initial fractionation of splicing extracts resulted in non-snRNP fractions that enhanced U1 snRNP binding or complemented a U1 snRNP fraction deficient in binding (127,128).

Pre-mRNA Binding

The RNA-binding properties of hnRNP proteins have been studied for a long time [for a recent review, see (17)], but only recently direct evidence was obtained for their involvement in the pre-mRNA splicing reaction (115,129). With the use of immunodepletion with a monoclonal antibody, of the major hnRNP proteins only the C proteins (C1, 41 kD, and C2, 43 kD) were found to be required as splicing factors. Furthermore, the C proteins are a component of the 60S splicing complex (115). Stable binding of hnRNP proteins to synthetic pre-mRNA has also been assessed by UV-crosslinking (130). More recently, the RNA binding properties of the C proteins have been studied in detail (131). With synthetic RNAs and purified hnRNP proteins, RNP particles could be reconstituted in the absence of ATP on RNAs of a minimum length of about 700 nucleotides (132). The resulting

40S particles probably represent the repeating unit of hnRNPs. For a long time this pre-mRNA packaging was considered to occur in a nonspecific manner; recently, however, with RNase T1 protection-immunoprecipitation assays, only a subset of hnRNP proteins including the C proteins, A1 and D, was found to bind specifically to the polypyrimidine tract near the 3' end of the intron (133). Considering the abundance of the hnRNP proteins in the nucleus, they may function both in a nonspecific and sequence-specific manner in pre-mRNA packaging and pre-mRNA processing, respectively.

Two cDNAs for the C proteins have been cloned, which differ in their 3' untranslated region, but have the same coding capacity (134,135). Therefore C1 and C2 proteins, which are both phosphorylated, probably arise by post-translational modification. The sequence analysis revealed two distinct domains: an RNP consensus sequence near the aminoterminus, and an acidic-hydrophilic carboxyterminus with a putative nucleoside triphosphate (NTP)-binding fold and a protein kinase phosphorylation site.

SPLICING-RELATED FACTORS

An RNA unwinding (helicase) activity may have an auxiliary function in splicing by melting secondary structure elements in the pre-mRNA such as stem-loop structures, thereby exposing essential RNA regions during splicing complex assembly. In addition, more specific RNA unwinding activities may be involved in regulating RNA-RNA base pairing interactions of snRNPs with each other and with the pre-mRNA. In fact, an RNA unwinding activity has been detected in mammalian cell extracts (136,137) that appears to involve an adenosine to inosine modifying activity (138). Based on antibody-inhibition experiments (139), a 68 kD nuclear phosphoprotein with RNA helicase activity (140,141) is also a potential splicing factor.

A lariat debranching activity that specifically cleaves the 2'-5' phosphodiester bond of lariat RNAs and is present in nuclear and cytoplasmic fractions of mammalian cells (151) has been purified away from known splicing factors (112,142). Probably it has a function in the degradation pathway of lariat RNAs in vivo (143).

The requirement of a 5' terminal cap structure for pre-mRNA splicing and splicing complex formation is controversial (compare, for example, 144-146). Its effect appears to differ in nuclear and whole-cell extracts and may also be related to the stability of the pre-mRNA under the in vitro conditions. By UV-crosslinking nuclear cap binding proteins of 20 and 115 kD have been identified that may have a function in the splicing reaction (147).

PERSPECTIVES

Although we do not yet completely understand the molecular mechanism of the splicing reaction, it is evident that in many cases gene expression is regulated at the level of pre-mRNA splicing. Differentially spliced mRNAs can be generated from a single primary transcript by alternative splice site selection. We know of a number of examples from mammalian systems where this selection is regulated in a temporal or tissue-specific manner [for reviews, see (8,9)]. In Drosophila, alternative splicing processes have recently been characterized that involve the regulatory genes of sex determination [for a review, see (148)].

There are many potential mechanisms of action for transacting regulatory splicing factors and many potential targets for regulation at the level of spliceosome assembly. First, proteins may block splice site utilization by competing with snRNPs or non-snRNP factors for specific pre-mRNA sequences. Second, specific RNA-binding proteins could stabilize certain secondary structures of the pre-mRNA and thereby alter the splicing pattern. A third, more indirect splicing regulatory mechanism may act at the level of export of unspliced precursor RNA, for example by accelerating the rate of nuclear export. Recent studies of HIV gene expression suggest a function for the HIV regulatory protein Rev in this type of regulation [for a short review, see (149)].

Further targets for regulating splicing events are on the level of snRNA gene expression, of assembly of snRNPs and multi-snRNP complexes, and of their cytoplasmic-nuclear transport. For example, regulatory proteins may associate with snRNPs or may exchange with certain snRNP polypeptides thus switching perhaps an entire class of splicing events (for an example, see the section The Sm Core Structure of snRNPs). A similar switch may be triggered by altered expression of snRNA genes.

Most likely, more cases will be discovered where gene expression is regulated on the level of pre-mRNA splicing, and novel types of regulation on this level will probably be added to the presently known repertoire. Differential splicing has so far been studied mostly in vivo, and in some cases sequence requirements are known. The development of faithful in vitro systems will certainly facilitate the identification of positively or negatively acting factors involved in differential splicing and the study of their mechanism of action.

REFERENCES

1 Green, M.R. (1986) Annu. Rev. Genet. 20, 671-708.
2 Brody, E. and Abelson, J. (1985) Science 228, 963-967.
3 Grabowski, P.J., Seiler, S.R. and Sharp, P.A. (1985) Cell 42, 345-353.

4 Frendewey, D. and Keller, W. (1985) Cell 42, 355-367.
5 Bindereif, A. and Green, M.R. (1986) Mol. Cell. Biol. 6, 2582-2592.
6 Perkins, K.K., Furneaux, H.M. and Hurwitz, J. (1986) Cell 45, 869-877.
7 Kaltwasser, E., Spitzer, S.G. and Goldenberg, C.J. (1986) Nucl. Acids Res. 14, 3687-3701.
8 Leff, S.E., Rosenfeld, M.G. and Evans, R.M. (1986) Annu. Rev. Biochem. 55, 1091-1117.
9 Breitbart, R.E., Andreadis, A. and Nadal-Ginard, B. (1987) Annu. Rev. Biochem. 56, 467-495.
10 Padgett, R.A., Grabowski, P.J., Konarska, M.M., Seiler, S.R. and Sharp, P.A. (1986) Annu. Rev. Biochem. 55, 1019-1050.
11 Sharp, P.A. (1987) Science 235, 766-771.
12 Krainer, A.R. and Maniatis, T. (1988) in Transcription and Splicing (Hames, B.D. and Glover, D.M., eds.) pp. 131-206, IRC Press, Oxford, Washington, DC.
13 Brunel, C., Sri-Widada, J. and Jeanteur, P. (1985) Prog. Mol. Subcell. Biol. 9, 1-52.
14 Maniatis, T. and Reed, R. (1987) Nature 325, 673-678.
15 Guthrie, C. and Patterson, B. (1988) Annu. Rev. Genet. 22, 387-419.
16 Steitz, J.A., Black, D.L., Gerke, V., Parker, K.A., Krämer, A., Frendewey, D. and Keller, W. (1988) in Structure and Function of Major and Minor Small Nuclear Ribonucleoprotein Particles (Birnstiel, M.L., ed.) pp. 115-154, Springer Verlag, Berlin.
17 Dreyfuss, G. (1986) Annu. Rev. Cell. Biol. 2, 459-498.
18 Cech, T.R. and Bass, B.L. (1986) Annu. Rev. Biochem. 55, 599-629.
19 Reddy, R. and Busch, H. (1988) in Structure and Function of Major and Minor Small Nuclear Ribonucleoprotein Particles (Birnstiel, M.L., ed.) pp. 1-37, Springer Verlag, Berlin.
20 Lührmann, R. (1988) in Structure and Function of Major and Minor Small Nuclear Ribonucleoprotein Particles (Birnstiel, M.L., ed.) pp. 71-99, Springer Verlag, Berlin.
21 Krainer, A.R. (1988) Nucl. Acids Res. 16, 9415-9429.
22 Winkelmann, G., Bach, M. and Lührmann, R. (1989) EMBO J. 8, 3105-3112.
23 Krämer, A., Keller, W., Appel, B. and Lührmann, R. (1984) Cell 38, 299-307.
24 Black, D.L., Chabot, B. and Steitz, J.A. (1985) Cell 42, 737-750.
25 Bringmann, P., Appel, B., Rinke, J., Reuter, R., Theissen, H. and Lührmann, R. (1984) EMBO J. 3, 1357-1363.
26 Hashimoto, C. and Steitz, J.A. (1984) Nucl. Acids Res. 12, 3283-3293.
27 Rinke, J., Appel, B., Digweed, M. and Lührmann, R. (1985) J. Mol. Biol. 185, 721-731.
28 Black, D.L. and Steitz, J.A. (1986) Cell 46, 697-704.

29 Berget, S.M. and Robberson, B.L. (1986) Cell 46, 691-
    696.
30 Chabot, B., Black, D.L., LeMaster, D.M. and Steitz, J.A.
    (1985) Science 230, 1344-1349.
31 Black, D.L. and Pinto, A.L. (1989) Mol. Cell. Biol. 9, 3350-
    3359.
32 Pan, Z.-Q. and Prives, C. (1988) Science 241, 1328-1331.
33 Hamm, J., Dathan, N.A. and Mattaj, I.W. (1989) Cell 59, 159-
    169.
34 Lamond, A.I., Sproat, B., Ryder, U. and Hamm, J. (1989) Cell
    58, 383-390.
35 Blencowe, B.J., Sproat, B.S., Ryder, U., Barabino, S. and
    Lamond, A.I. (1989) Cell 59, 531-539.
36 Grabowski, P.J. and Sharp, P.A. (1986) Science 233, 1294-
    1299.
37 Bindereif, A. and Green, M.R. (1987) EMBO J. 6, 2415-2424.
38 Hinterberger, M., Pettersson, I. and Steitz, J.A. (1983) J.
    Biol. Chem. 258, 2604-2613.
39 Kinlaw, C.S., Dusing-Swartz, S.K. and Berget, S.M. (1982)
    Mol. Cell. Biol. 2, 1159-1166.
40 Kinlaw, C.S., Robberson, B.L. and Berget, S.M. (1983) J.
    Biol. Chem. 258, 7181-7189.
41 Lelay-Taha, M.-N., Reveillaud, I., Sri-Widada, J., Brunel,
    C. and Jeanteur, P. (1986) J. Mol. Biol. 189, 519-532.
42 Bringmann, P. and Lührmann, R. (1986) EMBO J. 5, 3509-3516.
43 Branlant, C., Krol, A., Ebel, J.-P., Lazar, E., Haendler, B.
    and Jacob, M. (1982) EMBO J. 1, 1259-1265.
44 Brunel, C., Sri-Widada, J., Lelay, M.-N., Jeanteur, P. and
    Liautard, J.-P. (1981) Nucl. Acids Res. 9, 815-830.
45 Fisher, D.E., Conner, G.E., Reeves, W.H., Wisniewolski, R.
    and Blobel, G. (1985) Cell 42, 751-758.
46 Mattaj, I.W. and De Robertis, E. (1985) Cell 40, 111-118.
47 Mattaj, I.W. (1986) Cell 46, 905-911.
48 Konings, D.A.M. and Mattaj, I.W. (1987) Exp. Cell Res. 172,
    329-339.
49 Woppmann, A., Rinke, J. and Lührmann, R. (1988) Nucl. Acids
    Res. 16, 10985-11004.
50 Rokeach, L.A., Jannatipour, M., Haselby, J.A. and Hoch, S.O.
    (1989) J. Biol. Chem. 264, 5024-5030.
51 Ohosone, Y., Mimori, T., Griffith, A., Akizuki, M., Homma,
    M., Craft, J. and Hardin, J.A. (1989) Proc. Nat. Acad. Sci.
    U.S.A. 86, 4249-4253.
52 Van Dam, A., Winkel, I., Zijlstra-Baalbergen, J., Smeeuk, R.
    and Cuypers, H.T. (1989) EMBO J. 8, 3853-3860.
53 Reuter, R., Rothe, S. and Lührmann, R. (1987) Nucl. Acids
    Res. 15, 4021-4034.
54 Temsamani, J., Alibert, C., Tazi, J., Rucheton, M., Capony,
    J.-P., Jeanteur, P., Cathala, G. and Brunel, C. (1989) J.
    Mol. Biol. 206, 439-449.

55 McAllister, G., Amara, S.G. and Lerner, M.R. (1988) Proc. Nat. Acad. Sci. U.S.A. 85, 5296-5300.
56 McAllister, G., Roby-Shemkovitz, A., Amara, S.G. and Lerner, M.R. (1989) EMBO J. 8, 1177-1181.
57 Schmauss, C., McAllister, G., Ohosone, Y., Hardin, J.A. and Lerner, M.R. (1989) Nucl. Acids Res. 17, 1733-1743.
58 Rokeach, L.A., Haselby, J.A. and Hoch, S.O. (1988) Proc. Nat. Acad. Sci. U.S.A. 85, 4832-4836.
59 Wieben, E.D., Rohleder, A.M., Neuninger, J.M. and Pederson, T. (1985) Proc. Nat. Acad. Sci. U.S.A. 82, 7914-7918.
60 Stanford, D.R., Kehl, M., Perry, C.A., Holicky, E.L., Harvey, S.E., Rohleder, A.M., Rehder, K. Jr., Lührmann, R. and Wieben, E.D. (1988) Nucl. Acids Res. 16, 10593-10605.
61 Mattaj, I.W. (1989) Cell 57, 1-3.
62 Wieben, E.D., Madore, S. and Pederson, T. (1983) J. Cell. Biol. 96, 1751-1755.
63 Patton, J.R., Patterson, R.J. and Pederson, T. (1987) Mol. Cell. Biol. 7, 4030-4037.
64 Patton, J.R. and Pederson, T. (1988) Proc. Nat. Acad. Sci. U.S.A. 85, 747-751.
65 Patton, J.R., Habets, W., van Venrooij, W.J. and Pederson, T. (1989) Mol. Cell. Biol. 9, 3360-3368.
66 Pikielny, C.W., Bindereif, A. and Green, M.R. (1989) Genes Dev. 3, 479-487.
67 Surowy, C.S., van Santen, V.L., Scheib-Wixted, S.M. and Spritz, R.A. (1989) Mol. Cell. Biol. 9, 4179-4186.
68 Hamm, J., Kazmaier, M. and Mattaj, I.W. (1987) EMBO J. 6, 3479-3485.
69 Hamm, J., van Santen, V.L., Spritz, R.A. and Mattaj, I.W. (1988) Mol. Cell. Biol. 8, 4787-4791.
70 Lutz-Freyermuth, C. and Keene, J.D. (1989) Mol. Cell. Biol. 9, 2975-2982.
71 Scherly, D., Boelens, W., van Venrooij, W.J., Dathan, N.A., Hamm, J. and Mattaj, I.W. (1989) EMBO J. 8, 4163-4170.
72 Kastner, B. and Lührmann, R. (1989) EMBO J. 8, 277-286.
73 Thiessen, H., Etzerodt, M., Reuter, R., Schneider, C., Lottspeich, F., Argos, P., Lührmann, R. and Philipson, L. (1986) EMBO J. 5, 3209-3217.
74 Spritz, R.A., Strunk, K., Surowy, C.S., Hoch, S.O., Barton, D.E. and Francke, U. (1987) Nucl. Acids Res. 15, 10373-10391.
75 Query, C.C. and Keene, J.D. (1987) Cell 51, 211-220.
76 Sillekens, P.T.G., Beijer, R.P., Habets, W.J. and van Venrooij, W.J. (1988) Nucl. Acids Res. 16, 8307-8321.
77 Yamamoto, K., Miura, H., Moroi, Y., Yoshinoya, S., Goto, M., Nishioka, K. and Miyamoto (1988) J. Immunol. 140, 311-317.
78 Sillekens, P.T.G., Habets, W.J., Beijer, R.P. and van Venrooij, W.J. (1987) EMBO J. 6, 3841-3848.
79 Adam, S.A., Nakagawa, T., Swanson, M.S., Woodruff, T.K. and Dreyfuss, G. (1986) Mol. Cell. Biol. 6, 2932-2943.

80 Query, C.C., Bentley, R.C. and Keene, J.D. (1989) Cell 57, 89-101.
81 Wooley, J.C., Zukerberg, L.R. and Chung, S.-Y. (1983) Proc. Nat. Acad. Sci. U.S.A., 80, 5208-5212.
82 Habets, W.J., Sillekens, P.T.G., Hoet, M.H., Schalken, J.A., Roebroek, A.J.M., Leunissen, J.A.M., van de Ven, W.J.M. and van Venrooij, W.J. (1987) Proc. Nat. Acad. Sci. U.S.A. 84, 2421-2425.
83 Reuter, R. and Lührmann, R. (1986) Proc. Nat. Acad. Sci. U.S.A. 83, 8689-8693.
84 Habets, W.J., Sillekens, P.T.G., Hoet, M.H., McAllister, G., Lerner, M.R. and van Venrooij, W.J. (1989) Proc. Nat. Acad. Sci. U.S.A. 86, 4674-4678.
85 Sillekens, P.T.G., Beijer, R.P., Habets, W.J. and van Venrooij, W.J. (1989) Nucl. Acids Res. 17, 1893-1906.
86 Kleinschmidt, A.M., Patton, J.R. and Pederson, T. (1989) Nucl. Acids Res. 17, 4817-4828.
87 Barabino, S., Sproat, B.S., Ryder, U., Blencowe, B.J. and Lamond, A.I. (1989) EMBO J. 8, 4171-4178.
88 Frendewey, D., Krämer, A. and Keller, W. (1987) Cold Spring Harbor Symp. Quant. Biol. 53, 287-298.
89 Chabot, B. and Steitz, J.A. (1987) Mol. Cell. Biol. 7, 281-293.
90 Zillmann, M., Zapp, M.L. and Berget, S.M. (1988) Mol. Cell. Biol. 8, 814-821.
91 Singh, R. and Reddy, R. (1989) Proc. Nat. Acad. Sci. U.S.A. 86, 8280-8283.
92 Brow, D.A. and Guthrie, C. (1988) Nature 334, 213-218.
93 Bindereif, A., Wolff, T. and Green, M.R. (1990) EMBO J. (in press).
94 Hamm, J. and Mattaj, I.W. (1989) EMBO J. 8, 4179-4187.
95 Gerke, V. and Steitz, J.A. (1986) Cell 47, 973-984.
96 Tazi, J., Temsamani, J., Alibert, C., Rhead, W., Khellil, S., Cathala, G., Brunel, C. and Jeanteur, P. (1989) Nucl. Acids Res. 17, 5223-5243.
97 Bach, M., Winkelmann, G. and Lührmann, R. (1989) Proc. Nat. Acad. Sci. U.S.A. 86, 6038-6042.
98 Pinto, A.L. and Steitz, J.A. (1989) Proc. Nat. Acad. Sci. U.S.A. 86, 8742-8746.
99 Konarska, M.M. and Sharp, P.A. (1987) Cell 49, 763-774.
100 Konarska, M.M. and Sharp, P.A. (1988) Proc. Nat. Acad. Sci. U.S.A. 85, 5459-5462.
101 Konarska, M.M. and Sharp, P.A. (1986) Cell 46, 845-855.
102 Abmayr, S.M., Reed, R. and Maniatis, T. (1988) Proc. Nat. Acad. Sci. U.S.A. 85, 7216-7220.
103 Ruskin, B. and Green, M.R. (1985) Cell 43, 131-142.
104 Krämer, A. (1987) J. Mol. Biol. 196, 559-573.
105 Lamond, A.I., Konarska, M.M. and Sharp, P.A. (1987) Genes Dev. 1, 532-543.

106 Lamond, A.I., Konarska, M.M., Grabowski, P.J. and Sharp, P.A. (1988) Proc. Nat. Acad. Sci. U.S.A. 85, 411-415.
107 Christofori, G., Frendewey, D. and Keller, W. (1987) EMBO J. 6, 1747-1755.
108 Reed, R., Griffith, J. and Maniatis, T. (1988) Cell 53, 949-961.
109 Furneaux, H.M., Perkins, K.K., Freyer, G.A., Arenas, J. and Hurwitz, J. (1985) Proc. Nat. Acad. Sci. U.S.A. 82, 4351-4355.
110 Krainer, A.R. and Maniatis, T. (1985) Cell 42, 725-736.
111 Krämer, A., Frick, M. and Keller, W. (1987) J. Biol. Chem. 262, 17630-17640.
112 Krämer, A. and Keller, W. (1985) EMBO J. 4, 3571-3581.
113 Zeitlin, S., Wilson, R.C. and Efstratiadis, A. (1989) J. Cell. Biol. 108, 765-777.
114 Ruskin, B., Zamore, P.D. and Green, M.R. (1988) Cell 52, 207-219.
115 Choi, Y.D., Grabowski, P.J., Sharp, P.A. and Dreyfuss, G. (1986) Science 231, 1534-1539.
116 Tazi, J., Alibert, C., Temsamani, J., Reveillaud, I., Cathala, G., Brunel, C. and Jeanteur, P. (1986) Cell 47, 755-766.
117 Nelson, K.K. and Green, M.R. (1989) Genes Dev. 3, 1562-1571.
118 Zamore, P.D. and Green, M.R. (1989) Proc. Nat. Acad. Sci. U.S.A. 86, 9243-9247.
119 Krämer, A. (1988) Genes Dev. 2, 1155-1167.
120 Zhuang, Y. and Weiner, A.M. (1989) Genes Dev. 3, 1545-1552.
121 Wu, J. and Manley, J.L. (1989) Genes Dev. 3, 1553-1561.
122 Ruskin, B., Krainer, A.R., Maniatis, T. and Green, M.R. (1984) Cell 38, 317-331.
123 Reed, R. and Maniatis, T. (1988) Genes Dev. 2, 1268-1276.
124 Zhuang, Y. and Weiner, A.M. (1986) Cell 46, 827-835.
125 Aebi, M., Hornig, H. and Weissmann, C. (1987) Cell 50, 237-246.
126 Nelson, K.K. and Green, M.R. (1988) Genes Dev. 2, 319-329.
127 Mayeda, A., Tatli, K., Kitayama, H., Takemura, K. and Oshima, Y. (1986) Nucl. Acids Res. 14, 3045-3057.
128 Zapp, M.L. and Berget, S.M. (1989) Nucl. Acids Res. 17, 2655-2674.
129 Sierakowska, I., Szer, W., Furdon, P.J. and Kole, R. (1989) Nucl. Acids Res. 14, 5241-5254.
130 Mayrand, S.H., Pedersen, N. and Pederson, T. (1986) Proc. Nat. Acad. Sci. U.S.A. 83, 3718-3722.
131 Kumar, A., Sierakowska, H. and Szer, W. (1987) J. Biol. Chem. 262, 17126-17137.
132 Conway, G., Wooley, J., Bibring, T. and Le Sturgeon, W.M. (1988) Mol. Cell. Biol. 8, 2884-2895.
133 Swanson, M.S. and Dreyfuss, G. (1988) EMBO J. 7, 3519-3529.
134 Nakagawa, T.Y., Swanson, M.S., Wold, B.J. and Dreyfuss, G. (1986) Proc. Nat. Acad. Sci. U.S.A. 83, 2007-2011.

135 Swanson, M.S., Nakagawa, T.Y., LeVan, K. and Dreyfuss, G. (1987) Mol. Cell. Biol. 7, 1731-1739.

136 Konarska, M.M., Padgett, R.A. and Sharp, P.A. (1985) Cell 42, 165-171.

137 Wagner, R.W. and Nishikura, K. (1988) Mol. Cell. Biol. 8, 770-777.

138 Wagner, R.W., Smith, J.E., Cooperman, B.S. and Nishikura, K. (1989) Proc. Nat. Acad. Sci. U.S.A. 86, 2647-2651.

139 Liew, C.C. and Smith, H.C. (1989) FEBS Letters 248, 101-104.

140 Hirling, H., Scheffner, M., Restle, T. and Stahl, H. (1989) Nature 339, 562-564.

141 Iggo, R.D. and Lane, D.P. (1989) EMBO J. 8, 1827-1831.

142 Arenas, J. and Hurwitz, J. (1987) J. Biol. Chem. 262, 4274-4279.

143 Sittler, A., Gallimaro, H., Kister, L. and Jacob, M. (1987) J. Mol. Biol. 197, 737-741.

144 Krainer, A.R., Maniatis, T., Ruskin, B. and Green, M.R. (1984) Cell 36, 993-1005.

145 Edery, I. and Sonenberg, N. (1985) Proc. Nat. Acad. Sci. U.S.A. 82, 7590-7594.

146 Patzelt, E., Thalmann, E., Hartmuth, K., Blaas, D. and Küchler, E. (1987) Nucl. Acids Res. 15, 1387-1399.

147 Rozen, F. and Sonenberg, N. (1987) Nucl. Acids Res. 15, 6489-6500.

148 Baker, B.S. (1989) Nature 340, 521-524.

149 Green, M.R. and Zapp, M.C. (1989) Nature 338, 200-201.

150 Shapiro, M.N. and Senapathy, P. (1988) Nucl. Acids Res. 15, 7155-7174.

151 Ruskin, B. and Green, M.R. (1985) Science 229, 135-140.

152 Zhuang, Y., Goldstein, A.M. and Weiner, A.M. (1989) Proc. Nat. Acad. Sci. U.S.A. 86, 2752-2756.

THE GENES ENCODING WHEAT STORAGE PROTEINS:
TOWARDS A MOLECULAR UNDERSTANDING OF BREAD-MAKING QUALITY AND ITS GENETIC MANIPULATION

Vincent Colot

Department of Molecular Genetics
IPSR, Cambridge Laboratory
Maris Lane, Trumpington, Cambridge CB2 2JB, UK

INTRODUCTION

Wheat, a hexaploid species, has always occupied a prominent position amongst crops for human consumption. Wheat provides the staple food of many nations, including, paradoxically, a number of non-wheat-producing nations of Africa. The uses to which wheat flour is put vary around the world, the most popular food-form being bread, leavened or unleavened. Other uses include the manufacturing of pasta and couscous from Durum (tetraploid) wheat, which is not suitable for bread-making. In most cases, it is the visco-elastic properties of the dough produced by mixing flour with water that determine the type of food product that can be made. These properties, in other words the balance between extensibility and elasticity, are in turn determined primarily by the protein composition of the dough and by its protein content. While the latter essentially depends on the growing conditions, protein composition is a heritable characteristic which may differ greatly depending on the variety of wheat. Consequently, considerable effort has been put into the genetic analysis of varietal differences in protein composition, in an attempt to establish correlations with bread- and pasta-making qualities. Results of these analyses have already been extensively reviewed and have shown that qualitative differences in bread- or pasta-making between wheat varieties are predominantly associated with the major storage proteins of the grain.

This chapter focuses on recent advances made in the molecular characterization of genes and gene products that correspond to the major storage proteins of the wheat grain. The impact which this new information is likely to have on our

Genetic Engineering, Vol. 12
Edited by J.K. Setlow
Plenum Press, New York, 1990

understanding of bread-making quality and on its genetic manipulation is briefly evaluated.

THE STORAGE PROTEINS OF THE WHEAT ENDOSPERM AND THEIR GENETICS

The endosperm, both by volume and dry weight, is the major
organ of the wheat grain and constitutes the bulk of the flour
produced during the milling process. It consists predominantly
of starch but also contains over 70% of the proteins of the
grain, which serve as a store of nitrogenous and sulfurous
compounds for the germinating seedling. While some of the
proteins of the endosperm fulfill metabolic or structural functions during endosperm development, most (over 80%) are storage
proteins. They are synthesized exclusively in the developing
endosperm where they are deposited as an insoluble mass and in
the form of protein bodies which progressively coalesce into a
single protein matrix (6,7).

The storage proteins of wheat endosperm are usually classified into two major groups of equal abundance, the gliadins and
the glutenins, and several minor groups still only partly
characterized (2,8). Since both gliadins and glutenins are rich
in the hydrophobic amino acids proline and glutamine, they are
often collectively called prolamins (9). However, gliadins and
glutenins differ markedly in that the gliadin fraction consists
of a complex mixture of monomeric polypeptides from about 25 kD
to about 60 kD (as estimated by SDS-polyacrylamide gel electrophoresis) whereas the glutenin fraction contains very large
protein aggregates that are maintained by intermolecular disulfide bonds (1). These aggregates can be separated into two
subgroups of monomeric proteins, the high molecular weight
subunits (HMW, 60 to 100 kD), which account for about 20% of the
total glutenin fraction, and the predominant low molecular weight
subunits (LMW, 40 to 50 kD). On the basis of electrophoretic
mobility and isoelectric focusing, HMW glutenin subunits are
further subdivided into x and y types. Finally, gliadins can be
separated by gel electrophoresis at low pH into four subgroups of
decreasing mobility, called α, β, γ and ω. A single hexaploid
wheat variety contains 3 to 5 HMW glutenin subunits, from 15 to
20 α- + β-gliadins, 7 to 16 LMW glutenin subunits and 3 to 9
components of each of the other two subgroups, the γ- and ω-
gliadins (2, P. Payne, personal communication).

Bread wheat (<u>Triticum aestivum</u>) is an amphidiploid species
which is derived from the assembly, through natural hybridization, of three different but evolutionarily related diploid
genomes (called A, B and D), each consisting of seven pairs of
chromosomes (10). Thus, most genes are effectively triplicated
and these are usually located at similar positions on the
homoeologous (homologous but non-pairing) chromosomes of the
three diploid genomes. Genetical analysis of bread wheat has

revealed that the genes which encode the gliadins and the glutenins occur, to a first approximation, at a total of nine complex loci, on the homoeologous chromosomes of groups 1 and 6 (2). The Glu-1 loci (Glu-A1, Glu-B1 and Glu-D1) encode the HMW glutenin subunits and map to the long arms of the group 1 chromosomes, whilst the three Gli-2 loci encode the α and β-gliadins and map to the short arms of the group 6 chromosomes. Finally, the three Gli-1 loci, which map to the short arms of the group 1 chromosomes, are the most complex of the nine loci and encode the LMW glutenins in addition to the γ- and the ω-gliadins. More refined genetic studies have identified additional minor loci and recombination events within the nine complex loci are progressively being documented (2,11-14). For example, there is evidence that ω-type gliadins are encoded at additional loci, Gli-A3 and Gli-B3, that map on chromosomes 1A and 1B respectively, each midway between the centromere and the relevant Gli-1 locus (14). Also, a low recombination frequency has been reported between genes encoding LMW glutenin subunits and ω-gliadins located on the 1B chromosome, which indicate that the LMW glutenin subunits are encoded on that chromosome at a locus, Glu-B3, that is separate from the Gli-B1 locus (12). This and other studies would in fact suggest that each of the three complex Gli-1 loci that were originally defined contain three distinct clusters of genes encoding LMW glutenins, γ-gliadins and ω-gliadins respectively. However, there is no evidence that no intermingling of genes ever occurs between clusters. Recombination and molecular studies finally indicate that each of the Glu-1 loci consist of two tightly linked genes that encode HMW glutenin subunits of the x and y types respectively (15).

For each of the nine major loci that encode the gliadins and the glutenins, extensive allelic variation has been documented. For instance, a total of at least 20 major HMW glutenin subunits have been identified by SDS-PAGE in surveys of cultivated and old landraces of bread wheat, and additional subunits have been found in the wild diploid and tetraploid relatives of wheat, as well as in Durum wheat species (16-19). Similarly, over 70 alleles have already been identified for the three Gli-1 and Gli-2 loci (20,21). Such an extensive variation in prolamin composition provides a large reservoir of potentially useful variation in bread-making quality and this is progressively being exploited in breeding programs. However, the full impact of this approach will only be realized with the development of predictive methods to assess the effect of given alleles on bread-making quality before breeding.

THE GENES AND THEIR PRODUCTS

The cloning and sequencing of prolamin genes has offered an easy and convenient route to obtain complete amino acid sequences. In contrast, the direct amino acid sequencing of gliadins and of glutenin subunits has been almost exclusively limited to N-terminal regions, owing to the particular amino acid composition of prolamins and to the multiplicity of components within each prolamin fraction (1, and for an interesting development concerning the second point, see 22).

Numerous genes have now been isolated from wheat and from the two closely related species barley and rye. Sequence comparisons between genes and with N-terminal sequences of wheat prolamin fractions indicate that the six wheat prolamin subgroups correspond to five distinct gene families, with α- and β-gliadins being encoded by similar genes that belong to a single family (1,23-26). DNA sequence comparisons also show that the 5 prolamin gene families of wheat have evolved from a common ancestral sequence (27,28). However, while the LMW glutenin, γ-gliadin and α,β-gliadin gene families share high and comparable levels of similarity with each other (estimated at 60 to 65% sequence identity in the large non-repetitive domain of their coding sequence), they have diverged quite extensively from the HMW glutenin gene family (1,26). Even though -gliadin genes have not yet been characterized (a major omission that can partly be explained by difficulties in cloning genes whose sequence mostly consists of repeats), analysis of a recently reported sequence of an analogous C hordein gene from barley would suggest that -gliadin genes are also quite distinct from the other prolamin genes of wheat (29). Finally, estimation of family size by Southern blot analysis of wheat genomic DNA would indicate that most, if not all, of the prolamin gene families of wheat contain genes in excess of the number of proteins they encode (2,26). Thus, pseudogenes are thought to be present among the members of these families and genes have been sequenced that contain premature termination codons, or in the case of a HMW glutenin subunit gene, an insertion sequence within the coding region (30-32). Incidentally, direct and indirect evidence suggests that pseudogenes map preferentially on the A genome of tetraploid and hexaploid wheat, and it has been proposed that the low number of active prolamin genes on the A genome reflects a selective "diploidization" of tetraploid and hexaploid wheat (33). However, the presence of most pseudogenes on the A genome could simply be coincidental and more studies are clearly required in order to confirm its significance.

The characterization over the last ten years of many active or putatively active genes that encode prolamins of wheat and of other cereals has revealed that these genes lack introns and all share a common organization of their coding sequence (27,28). The amino acid sequences derived from prolamin genes show that all prolamins possess a central domain made up of repeat motifs rich in proline and glutamine residues and which account for the particular amino acid composition of this class of proteins. The flanking regions are largely non-repetitive and contain almost all of the cysteine residues and most of the other amino acids that are found in low abundance in these proteins. Prolamin gene products are also characterized by the presence at their N-terminus of a sequence which is typical of signal peptides (1,34). This is consistent with the observation that prolamins are synthetized on ribosomes bound to the endoplasmic reticulum, and that they are translocated into the lumen of the endoplasmic reticulum prior to their deposition in protein bodies (6,7).

Although similar to each other for the reasons mentioned above, wheat prolamins belonging to separate subgroups can be distinguished by several features of their primary sequence. For instance, they differ in their overall size and in the relative size of the three distinct domains they contain (Table 1). Thus, HMW glutenin subunits have a central repetitive domain of between 500 and 700 amino acids (being either of the smaller y type or of the larger x type), N-domains of about 80 to 110 amino acids and C domains of 42 amino acids. On the other hand, ω-gliadins would consist primarily of a repetitive domain of about 285 amino acids, flanked by two short domains of around 10 amino acids each, as inferred from the amino acid sequence predicted from a recently characterized C hordein gene. Finally, the LMW glutenins, α,β- and γ-gliadins, which are similar in size and organization, contain a relatively short repetitive domain ranging from about 90 (LMW glutenins and α,β-gliadins) to about 140 (γ-gliadins) amino acids, flanked by a short N-domain (5 to 15 amino acids) and a long C domain (160 to 190 amino acids). Wheat prolamins belonging to separate subgroups not only differ in size but also in sequence. Differences are found in the N-terminal domain of prolamins, as previously shown by direct amino-acid sequencing of N-termini of purified fractions, as well as in the other two domains (1,26). Thus, each domain taken separately is diagnostic of a given prolamin subgroup. For example, the sequence of the repeat motif(s) that appear(s) in the central domain is unique to each subgroup, with the interesting exception of that found in γ-gliadins, which is identical to that of C hordeins and hence probably of ω-gliadins (Table 2).

Within each of the wheat prolamin subgroups, size and sequence differences are also present and most noticeable among them are extensions/contractions of the central repetitive domain. For example, the size polymorphism of HMW glutenin subunits is almost exclusively related to the central domain, and it can be.predicted that a similar picture will emerge for the ω-gliadins (24). However, significant size differences within the other subgroups result additionally from the extensionβcontraction of the several polyglutamine stretches present in the extended C domain of the α,β-gliadins, γ-gliadins and LMW glutenin subunits. Few studies have elaborated on the evolutionary forces that could be responsible for the fixation of these and other mutational events. Balanced selective forces have been

Table 1
Typical Length* of the Three Distinct Domains of
Prolamin Sequences

	N-terminal domain	Central repetitive domain	C-terminal domain
HMW glutenin subunits			
x-type	86	666	42
y-type	104	481	42
LMW glutenin subunits	10	89	185
α,β-gliadins	5	95	174
γ-gliadins	12	137	163
C hordeins (barley) (equivalent to wheat ω-gliadins)	12	285	10

*Values are derived from individual amino acid sequences, and represent number of amino acids.

Table 2
Consensus Repeat Motif(s) of the Central Domain of
Prolamin Sequences

HMW glutenin subunits	
x-type	GYYPTS$^{P}_{L}$QQ + PGQGQQ + GQQ
y-type	G$^{H}_{T}$YP$^{A}_{T}$SLQQ + PGQGQQ
LMW glutenin subunits	QQQ$^{Q}_{P}$PF$^{S}_{P}$
α,β-gliadins	QPQQP$^{Y}_{F}$P
γ-gliadins	QQPQQPFP
C hordeins (barley) (equivalent to wheat ω-gliadins)	QQPQQPFP

invoked to explain the extensive polymorphism of prolamins within subgroups (18), but evidence from other organisms would suggest that genetic drift and DNA turnover processes are in fact the main evolutionary forces at play here (35,36). For example, size polymorphism within the central domain of prolamins is probably the result of widely tolerated unequal crossover and/or gene conversion events in the corresponding region of the genes. Similarly, size polymorphism of the polyglutamine stretches of the C domain of certain prolamins could be a consequence of frequent slippage during replication or repair of the corresponding DNA simple sequences (26,37). However, such an extended polymorphism does not necessarily mean that prolamins are under no selective constraint, and this is clearly reflected in the common organization of their primary structure. Additional evidence for selection is provided by the fact that prolamins of the different subgroups share a short sequence (CCQQL) that is also present within several other seed proteins (26,27), suggesting that this sequence must be functionally or structurally essential, although its precise role remains to be elucidated.

ENDOSPERM-SPECIFIC EXPRESSION OF WHEAT PROLAMIN GENES

Synthesis of wheat prolamins, like that of prolamins of other cereals, occurs exclusively in the developing endosperm. Analyses of steady-state mRNAs and in vitro measurements of transcriptional activity indicate that this synthesis is controlled primarily, but not exclusively, at the level of transcription (38-43).

Recently, reverse genetics techniques have been used to gain insights into the molecular mechanisms controlling the endosperm-specific transcription of wheat prolamin genes. The rationale of identifying the cis-acting regulatory sequences necessary for gene activity usually involves the deletion or mutation of candidate sequences in vitro followed by the assay of the activity of the modified gene in a suitable system. In the absence of a transformation procedure for wheat or for any other cereal which would enable the regeneration of fertile plants, sequences of wheat prolamin genes have been introduced into tobacco, a heterologous host that could readily be transformed by using the Agrobacterium tumefaciens-mediated T-DNA transfer system and disarmed Ti-derived vectors (44). An essential discovery that emerged from these experiments is that the pattern of wheat prolamin gene expression in transformed tobacco appears to resemble that observed in wheat. More precisely, when a genomic clone containing the complete coding sequence of a HMW glutenin subunit gene as well as 2.6 kb of 5' and 1.5 kb of 3' flanking DNA was introduced into tobacco, low but significant levels of mRNA and HMW glutenin subunit were exclusively detected in the transformed tobacco seeds, and specifically in the endosperm tissue (45). Similarly, when chimeric genes containing the coding sequence of a bacterial reporter gene (CAT or GUS) and sequences 5' to the start codon of either the HMW or a LMW glutenin subunit genes were introduced into tobacco, gene activity was detected only in the endosperm tissue of transformed seeds (Figure 1) (45,46). This pattern of gene expression was in sharp contrast with that observed in control experiments based on a viral promoter sequence (-800 CaMV35S promoter) that is functional in all plant organs. Since then, endosperm-specific expression has also been obtained in tobacco transformed with chimeric genes containing the regulatory sequences of a barley or of hordein gene (47) or of a maize zein gene (48). Altogether, these observations further support the notion that the endosperm-specific expression of prolamin genes is primarily controlled at the transcriptional level. These results also indicate that no essential cis-acting regulatory sequence is located 3' to the start codon (or 3' to the transcription initiation site, 46) of these genes. Finally, they imply that some of the trans-acting factors which recognize the regulatory sequences of prolamin genes and which are necessary for endosperm-specific expression are conserved between cereals and tobacco. Thus, tobacco offers a useful surrogate system in which to analyze the regulatory mechanisms associated with prolamin gene expression.

Using this system, and the two glutenin subunit genes mentioned above, several cis-acting elements that control endosperm-specific expression have now been identified. In the case of the LMW glutenin subunit gene, sequences essential for endosperm-specific expression were located on a 160 bp fragment 5' to the TATA box (positions -326 to -160 relative to the transcription initiation site) (26, V. Colot, unpublished data). Recent results also indicate that another fragment located further 5' to the TATA box (positions -938 to -371) functions similarly to the 160 bp fragment. Finally, both fragments appear to act in concert, leading to higher levels of expression than when tested separately (V. Colot, R.B. Flavell and M.W. Bevan, unpublished data). Inspection of the two fragments reveals the presence of a 26 to 30 bp sequence, called the "endosperm box", that is conserved among many cereal prolamin genes (24,26). These observations strongly suggest that the endosperm box is an important cis-acting element regulating the endosperm-specific expression of prolamin genes, and that its reiteration leads to higher levels of expression. Experiments are in progress to test directly these hypotheses by site-specific mutagenesis and to search for protein factor(s) that may bind to the endosperm box.

In the HMW glutenin gene, a 230 bp fragment located 5' to the TATA box (positions -375 to -45 relative to the transcription

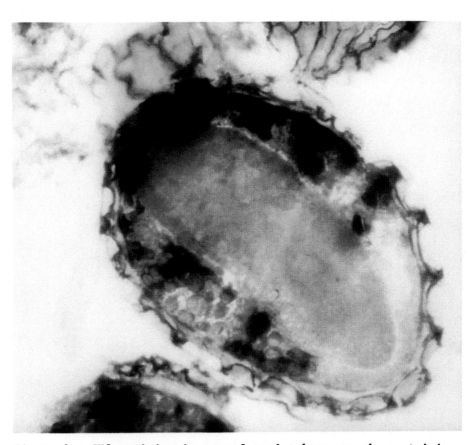

Figure 1. GUS activity in transformed tobacco seeds containing a LMW glutenin subunit-GUS gene fusion. A 1 kb sequence 5' to the start codon of LMW glutenin gene LMWG-1D1 (26) was fused to the coding sequence of the Gus gene (66). Thin sections (50 µm) of transformed tobacco seeds were incubated in X-glucuronide. Blue staining indicating GUS activity can be seen in the endosperm only.

initiation site) has been identified which when fused to a weakly expressed heterologous promoter (-90 CaMV35S promoter) enhances transcription specifically in the endosperm (M.S. Thomas and R.B. Flavell, unpublished data). This fragment possesses all the properties typical of an enhancer, and analysis of sequential deletions indicates that transcription enhancement depends on a 38 bp element located between positions -186 to -148. This sequence is common to all other HMW glutenin genes that have so far been characterized. In contrast, and except for a short motif, TGCAAAG, that resembles the TGTAAG motif contained within the endosperm box, this 38 bp fragment shows no similarity with 5' sequences of other prolamin genes of wheat. It is therefore possible that the HMW glutenin subunit genes are regulated by a set of cis-acting element(s) and trans-acting factor(s) that differs from that of the other prolamin genes of wheat. This hypothesis is supported by the observation that the lys 3a mutant of barley abolishes the transcription of all prolamin genes except those encoding D hordeins, which are closely related to HMW glutenin subunits (43). In effect, this suggests that the Lys 3a gene, and a similar gene in wheat, encode a trans-acting factor that specifically regulates the expression of genes encoding prolamins other than D hordeins and HMW glutenin subunits.

ENDOSPERM PROTEIN QUANTITY AND QUALITY FOR BREAD-MAKING

Improvement of bread-making quality is best approached through the genetic modification of the protein composition of the grain. Indeed, a high quality variety produces good bread over a fairly wide range of protein content, whereas a low quality variety will produce relatively poor quality bread even at high protein content. The HMW, together with the more numerous LMW glutenin subunits, contribute directly to the elasticity of a dough while gliadins contribute to its extensibility and stickiness.

Several models have been put forward to explain the rheological properties of a dough in relation to its protein composition (for specific references, see 3). However, they are mostly derived from the model originally proposed by Ewart which predicts that non-permanent intermolecular disulfide bonds form between glutenin subunits and that these bonds, in addition to the size distribution of the linear glutenin polymers, play a key role in determining the rheological behavior of a dough (56). According to this and other models, elasticity would result from the elastic deformation of individual HMW glutenin subunits and viscosity from slippage between the long glutenin polymers. The LMW glutenin subunits would simply contribute to the formation of

Figure 2.   Influence of the number of HMW glutenin subunits on
bread-making quality. A: Bread produced from the variety Sicco-
5 (HMW glutenin subunits 1Ax1, 1Bx7 + 1By9 and 1Dx5 + 1Dy10); B
and C: Breads produced from two near isogenic lines derived from
Sicco-5.   B possesses all the HMW glutenin subunits of Sicco-5
whereas C possesses only the subunit pair 1Bx7 + 1By9.   Only A
and B are breads of good quality. (Photograph kindly provided by
P. Payne.)

the cross-linked network while the gliadins would, through
hydrogen bonding and hydrophobic interactions, aid the slippage
of molecules within the protein network.
     Allelic differences between glutenin subunits and gliadins
have been carefully studied for their potential effects on bread-
making quality.   There is now a strong body of evidence which
suggests that the presence of certain HMW glutenin subunits is
correlated positively with good bread-making potential (3).   For
example, the pair of subunits 1Dx5 and 1Dy10 is associated with
greater elasticity than the subunit pairs 1Dx2 + 1Dy12, 1Dx3
+1Dy12 and 1Dx4 + 1Dy12 (49).   In addition, it appears that a
higher number of HMW glutenin subunits in a variety correlates
positively with better bread-making quality (Figure 2) (50,51).
This demonstrates directly that the ratio of HMW glutenin
subunits to other prolamins contributes significantly to bread-
making. quality.   Furthermore, recent studies indicate that
allelic variation of LMW glutenin subunits can also affect bread-
making quality, and that effects associated with variations in
HMW and LMW glutenin subunits are additive (52,53).   Contrary to
earlier reports, indirect biochemical evidence suggests that
allelic variation in gliadins has little effect on bread-making
quality (52).   Similarly, differences in pasta-making quality
have not been found to be correlated with differences in

gliadins, but are closely associated with quantitative (and perhaps qualitative) differences in LMW glutenin subunits (54,55).

FIRST STEPS TOWARDS A MOLECULAR UNDERSTANDING OF BREAD-MAKING QUALITY

Dough contains, in addition to the major storage proteins, many other components such as water-soluble proteins, lipids and starch. However it is clear from the above that a molecular understanding of the rheological behavior of dough, especially in relation to bread-making quality, needs to be primarily approached through an understanding of the molecular structure of its glutenin subunits and gliadins. However, this task is complicated by the multiplicity of individual components within the glutenin and gliadin fractions and more importantly by their partial insolubility in water. The implications of this latter point are that X-ray crystallography cannot easily be applied to the determination of the three-dimensional structure of wheat storage proteins and that methods of secondary structure prediction, such as those based on circular dichroism (CD) spectra, on Fourier transform infrared spectra or on amino acid sequences, are somewhat inappropriate as they are principally designed with reference to secondary structures of water-soluble, globular proteins. It should also be noted that deconvolution procedures cannot be applied to CD spectra for the accurate determination of the proportion of β-turns in a protein. This is linked to the fact that there is no unique CD spectrum associated with β-turns, each type of β-turn within a protein giving a different spectrum, and that the CD spectra of β-turns are often masked by those associated with α-helices and β-sheets. Despite these limitations, predictions based on CD spectra of purified fractions and on primary structures remain, in the absence of other methods, the best available approaches to the study of the secondary structure of glutenin subunits and gliadins to date.

Far and near UV CD spectra obtained at different temperatures indicate that the S-rich (α,β- and γ-) gliadins and the S-poor (ω-) gliadins have different secondary structures, with varying degrees of stabilization by intramolecular disulfide bonds, hydrogen bonds and hydrophobic interactions (57,58). More precisely, α,β- and γ-gliadins are predicted to contain 30 to 35% α-helices and 10 to 20% β-sheets, and the thermal stability exhibited by the putative α-helices suggests that they are most probably maintained through extensive hydrogen bonding between the many glutamine residues. In further work, LMW glutenin subunit fractions have been found to have far UV CD spectra similar to those of α,β and γ-gliadins, implying that the estimated amount of α-helix and β-sheet is roughly identical for the LMW glutenin subunits and the S-rich gliadins (59). In

contrast, analysis of CD spectra of ω-gliadins (57), of C
hordeins (60) and of HMW glutenin subunits (61,62) suggest that
these proteins are poor in α-helices and β-sheets and that they
primarily contain β-turns, stabilized by hydrophobic interactions
between aromatic residues and by hydrogen bonds involving
glutamine residues and possibly the peptide backbone.

Secondary structure predictions performed by using the
algorithm of Chou and Fasman have also been described, based on
the few known wheat and barley prolamin sequences and on the many
cognate sequences derived from cloned genes (58,60,61). These
predictions are consistent with most of the findings made with CD
spectra, and in addition show that the repeat domain that
features in all the prolamin sequences of wheat and barley is
rich in β-turns, whereas the non-repetitive N- C-domains are rich
in α-helices. However, it should be noted that the N- and C-
domains that flank the large repeat domain of HMW glutenin
subunits lack any defined structure according to CD spectra but
are predicted from their sequence to form α-helices (61,62).
This underlines the drawbacks in such predictions.

Recently, Fourier transform infrared spectroscopy has been
used in the determination of the secondary structure of α- and ω-
gliadins (63), but the full potential of such an analysis remains
to be investigated. Furthermore, the limitations mentioned above
concerning predictive approaches constructed with reference to
water-soluble, globular proteins, apply here too.

Hydrodynamic measurements have also been performed, which
indicate that the HMW glutenin subunits of wheat, like the C
hordeins of barley and hence most certainly the ω-gliadins of
wheat adopt, at least in solution, a rod-shaped conformation
(60,62). This is consistent with these proteins containing
regularly repeated β-turns, and it has been proposed that such β-
turns form a loose and elastic spiral similar to that which has
been previously described for the elastomeric mammalian protein
elastin (57,60,62). The elasticity conferred to a dough would
therefore result from the presence in HMW glutenin subunits of a
repetitive domain consisting of regularly spaced β-turns.
Further support for this model comes from the observation that
the cysteine residues of HMW glutenin subunits are located, as
indicated by sequence analysis, near the N and C termini. In
effect, such positioning would permit the formation of long
linear polymers, with either head-to-tail or head-to-head and
tail-to-tail disulfide links. On the other hand, ω-gliadins
could not form such polymers due to their lack of cysteine
residues.

As already mentioned, a tight correlation exists between
bread-making quality and the presence or absence of particular
HMW glutenin subunits. The comparison at the molecular level of
subunits of contrasting properties should therefore provide a
clear test of any model aimed at explaining, at least in part,
the molecular basis of varietal differences in bread-making

quality.   Unfortunately, HMW glutenin subunits of contrasting
properties have rarely been described as such.   Indeed, the tight
linkage that exists between the two genes present at each of the
Glu-1 loci usually means that only subunit pairs rather than
single subunits can be genetically assessed for their influence
on bread making.   This is particularly true of the subunits
encoded by the two genes at the Glu-D1 locus, since null alleles
are rarely found for either of these genes and never in commercial varieties (P. Payne, personal communication).   However, the
unusual subunit pair 1Dx5 + 1Dy12 has recently been found in an
Italian cultivar that was derived from a cross between two
varieties having the commonly occurring subunit pairs 1Dx2 +
1Dy12 and 1Dx5 + 1Dy10 respectively (11).   Since these last two
pairs are associated with contrasting qualities for bread making,
the Italian cultivar with the pair 1Dx5 +1Dy12 has been used in
different crosses to assess independently the particular contribution of each of the four subunits 1Dx2, 1Dx5, 1Dy10 and 1Dy12.
Although the results of this analysis are only partial, differences in bread-making quality appear to be mainly associated with
subunits 1Dy10 and 1Dy12, 1Dy10 being associated with better
quality attributes than 1Dy12.   Since the genes encoding both
subunits have now been cloned and sequenced, there is a unique
opportunity to compare, at the molecular level, two subunits of
contrasting properties (64).   This comparison is facilitated by
the fact that subunits 1Dy10 and 1Dy12, which are the products of
allelic genes, display only a few amino acid differences between
them.   Secondary structure predictions based on the Chou and
Fasman algorithm indicate that these differences, which
incidentally are all located within the repeat domain, lead to
subunit 1Dy10 potentially having more β-turns.   In addition, the
predicted β-turns of subunit 1Dy10 appear to be more regularly
arranged than those of subunit 1Dy12.   Thus it has been proposed
that subunit 1Dy10 produces a coil with longer regions of regular
turns, providing a basis for the higher elasticity and the better
quality of the dough produced (64).   Additional support for
subunit 1Dy10 having a different shape than subunit 1Dy12 is
provided by the observation that although subunit 1Dy10 has a
lower molecular weight than subunit 1Dy12 (as shown by the
comparison of their amino acid sequences), it migrates more
slowly than the latter on SDS-PAGE, unless 4 M urea is included
in the gel buffer.   This is consistent with subunit 1Dy10
adopting a more extended structure than subunit 1Dy12.   Furthermore, "domain swap" experiments between these two subunits have
shown that the aberrant migration of subunit 1Dy10 is caused by
one of the regions of the molecule that are responsible for at
least part of its higher β-turn potential (65).   The relationship
of these structural differences to conformation differences
between subunits is unknown at present.
     Dough elasticity is expected to depend not only on the
length and regularity of the putative β-spirals of HMW glutenin

subunits, but also on the degree of permanent and non-permanent crosslinking between subunits. In this respect, it is noteworthy that even though the cysteine residues of subunits 10 and 12 occupy identical positions, one of these residues is located within the region which imparts conformational differences between the two subunits. This residue might therefore be accessible for crosslinking to different extents in the two subunits and, as a result, could also contribute to the contrasting properties associated with subunits 1Dy10 and 1Dy12 (64).

SUMMARY AND PROSPECTS

As major differences in bread-making quality can be related to simple amino acid differences between HMW glutenin subunits as well as to modifications in the relative amount of the various prolamin types, the ability to insert directly single genes or a small number of genes into wheat is likely to have a significant impact on the improvement of dough quality. Indeed, this would first provide an attractive alternative to the lengthy process of backcrossing, which is traditionally used for the transfer of genes between wheat varieties or from the cultivated or wild diploid and tetraploid species that are related to wheat.

The likely advantages of wheat transformation are not limited to its rapidity compared to classical breeding. In effect, this technology would also provide a very efficient way of resolving problems of linkage to deleterious genes. More importantly, wheat transformation would uniquely enable the introduction of a new repertoire of genes not accessible through breeding. This repertoire would contain genes from other species and, perhaps more significantly, in vitro mutagenized or synthesized genes whose products would have specific structures and which would accumulate to specific levels within the plant. Thus, it would be possible for example to test directly the effect on bread making of HMW glutenin subunits modified to contain more regular β-turns, newly positioned cysteine residues or a modified number of such residues.

Expression of particular prolamin genes (natural or synthetic) could also be specifically altered, for example by the introduction of additional gene copies, by the modification of cis-acting regulatory sequences (or trans-acting factors), or by the synthesis of anti-sense mRNAs. The existence of post-transcriptional regulatory mechanisms acting differentially on the various prolamin gene families suggests that additional routes could also be used to modify gene expression.

The intense efforts throughout the world devoted to the design of an efficient wheat transformation system indicate that the time is close when breeders could directly tap a reservoir of new and improved prolamin genes.

Acknowledgments: I would like to thank R. Flavell, M. Bevan and P. Payne for advice in some of the experiments reported here and for critical reading of the manuscript. I am also deeply indebted to E. Heard and M. Thangavelu for their help during writing. I gratefully acknowledge support from INRA (France) and Sidney Sussex College, Cambridge. Part of the work presented here was carried out in the framework of a contract from the Commission of the European Communities.

REFERENCES

1   Kreis, M., Shewry, P.R., Forde, B.G., Forde, J. and Miflin, B.J. (1985) Oxford Surveys of Plant Molecular and Cell Biology 2, 253-317.

2   Payne, P.I. (1987) Annu. Rev. Plant Physiol. 38, 141-153.

3   Shewry, P.R., Halford, N.G. and Tatham, A.S. (1989) Oxford Surveys of Plant Molecular and Cell Biolkogy 6 (in press).

4   Feillet. P., Ait-Mouh, O., Kobrehel, K. and Autran, J.-C. (1989) Cereal Chem. 66, 26-30.

5   Autran, J.-C. and Galteirio, G.J. (1989) J. Cereal Sci. 9, 179-215.

6   Pernollet, J.-C. and Camilleri, C. (1983) Physiol. Veg. 21, 1093-1103.

7   Kim, W.T., Franceschi, V.R., Krishnan, H.B. and Okita, T.W. (1988) Planta 176, 173-182.

8   Payne, P.I., Holt, L.M., Jarvis, M.G. and Jackson, E.A. (1985) Cereal Chem. 62, 319-326.

9   Shewry, P.R., Tatham, A.S., Forde, J., Kreis, M. and Miflin, B.J. (1986) J. Cereal Sci. 4, 97-106.

10 Kerby, K. and Kuspira, J. (1987) Genome 29, 722-737.

11 Pogna, N.E., Mellini, F. and Dal Belin Peruffo, A. (1987) in Agriculture-Hard Wheat: Agronomic, Technological, Biochemical and Genetic Aspects (Borghi, B., ed.) pp. 53-69, Commission of the European Communities.

12 Singh, N.K. and Shepherd, K.W. (1988) Theor. Appl. Genet. 75, 628-641.

13 Singh, N.K. and Shepherd, K.W. (1988). Theor. Appl. Genet. 75, 642-650.

14 Payne, P.I., Holt, L.M. and Lister, P.G. (1988) in 7th Int. Wheat Genet. Symp. (Miller, T.E. and Koebner, R.M.D., eds.) pp. 999-1002, Cambridge.

15 Harberd, N.P., Bartels, D. and Thompson, R.D. (1986) Biochem. Genet. 24, 579-596.

16 Payne, P.I. and Lawrence, G.J. (1983) Cereal Res. Commun. 11, 29-35.

17 Waines, J.G. and Payne, P.I. (1987) Theor. Appl. Genet. 74, 71-76.

18 Nevo, E. and Payne, P.I. (1987) Theor. Appl. Genet. 74, 827-836.

19   Branlard, G., Autran, J.-C. and Monneveux, P. (1989) Theor.
     Appl. Genet. 78, 353-358.
20   Metakovsky, E.V., Novoselskaya, A. Yu., Kopus, M.M., Sobko,
     T.A. and Sozinov, A.A. (1984) Theor. Appl. Genet. 67, 559-
     568.
21   Gupta, R.B. and Shepherd, K.W. (1990) Theor. Appl. Genet.
     (in press).
22   Tao, H.P. and Kasarda, D.D. (1989) J. Exp. Bot. 40, 1015-
     1020.
23   Sumner-Smith, M., Rafalski, J.A., Sugiyama, T., Stoll, M.
     and Soll, D. (1985) Nucl. Acids Res. 13, 3905-3916.
24   Halford, N.G., Forde, J., Anderson, O.D., Greene, F.C. and
     Shewry, P.R. (1987) Theor. Appl. Genet. 75, 117-126.
25   Scheets, K. and Hedgcoth, C. (1988) Plant Sci. 57, 141-150.
26   Colot, V., Bartels, D., Thompson, R. and Flavell, R. (1989)
     Mol. Gen. Genet. 216, 81-90.
27   Kreis, M., Forde, B.G., Rahman, S., Miflin, B.J. and Shewry,
     P.R. (1985) J. Mol. Biol. 183, 499-503.
28   Kreis, M. and Shewry, P.R. (1989) BioEssays 10, 201-207.
29   Entwistle, J. (1988) Carlsberg Res. Commun. 53, 247-258.
30   Rafalski, J.A. (1986) Gene 43, 221-229.
31   Forde, J., Malpica, J.-M., Halford, N.G., Shewry, P.R.,
     Anderson, O.D., Greene, F.C. and Miflin, B.J. (1985) Nucl.
     Acids Res. 13, 6817-6832.
32   Harberd, N.P., Flavell, R.B. and Thompson, R.D. (1987) Mol.
     Gen. Genet. 209, 326-332.
33   Galili, G. and Feldman, M. (1985) Theor. Appl. Genet. 69,
     583-589.
34   von Heijne, G. (1985) J. Mol. Biol. 184, 99-105.
35   Dover, G.A. and Tautz, D. (1986) Phil. Trans. Roy. Soc.
     Lond. B 312, 275-289.
36   Kimura, M. (1986) Phil. Trans. Roy. Soc. Lond. B 312, 343-
     354.
37   Tautz, D. and Renz, M. (1984) Nucl. Acids Res. 12, 4127-
     4138.
38   Greene, F.C. (1983) Plant Physiol. 71, 40-46.
39   Bartels, D. and Thompson, R.D. (1986) Plant Sci. 46, 117-
     125.
40   Reeves, C.D., Drishnan, H.B. and Okita, T.W. (1986) Plant
     Physiol. 82, 34-40.
41   Tercé-Laforgue, T., Sallantin, M. and Pernollet, J.-C.
     (1987) Physiol. Plantarum 69, 105-112.
42   Boston, R.S. and Larkins, B.A. (1987) in Genetic Engineer-
     ing. Principles and Methods (Setlow, J.K., ed.), Vol. 9,
     pp. 61-74, Plenum Publ. Corp., New York, NY.
43   Sørensen, M.B., Cameron-Mills, V. and Brandt, A. (1989) Mol.
     Gen. Genet. 217, 195-201.
44   Schell, J. St. (1987) Science 237, 1176-1183.
45   Robert, L.S., Thompson, R.D. and Flavell, R.B. (1989) Plant
     Cell 1, 569-578.

46   Colot, V., Robert, L., Kavanagh, T., Bevan, M. and Thompson,
      R. (1987) EMBO J. 6, 3559-3564.
47   Marris, C., Gallois, P. and Kreis, M. (1988) Plant Mol.
      Biol. 10, 359-366.
48   Schernthaner, J.P., Matzke, M.A. and Matzke, A.J.M. (1988)
      EMBO J. 7, 1249-1255.
49   Payne, P.I., Nightingale, M.A., Krattiger, A.F. and Holt,
      L.M. (1987) J. Sci. Food Agric. 40, 51-65.
50   Payne, P.I., Holt, L.M., Harinder, K., MacCartney, D.P. and
      Lawrence, G.J. (1987) in Gluten Proteins. Proc. 3rd Inter.
      Wheat Gluten Symp. (Lásztity, R. and Békés, F., eds.) pp.
      247-253, World Scientific - Singapore.
51   Lawrence, G.J., Macritchie, F. and Wrigley, C.W. (1988) J.
      Cereal Sci. 7, 109-112.
52   Payne, P.I., Seekings, J.A., Worland, A.J. Jarvis, M.G. and
      Holt, L.M. (1987) J. Cereal Sci. 6, 103-118.
53   Gupta, R.B., Singh, N.K. and Shepherd, K.W. (1989) Theor.
      Appl. Genet. 77, 57-64.
54   Payne, P.I., Jackson, E.A. and Holt, L.M. (1984) J. Cereal
      Sci. 2, 73-81.
55   Pogna, N., Lafiandra, D., Feillet, P. and Autran, J.C.
      (1988) J. Cereal Sci. 7, 211-214.
56   Ewart, J.A.D. (1968) J. Sci. Food Agric. 19, 617-623.
57   Tatham, A.S. and Shewry, P.R. (1985) J. Cereal Sci. 3, 103-
      113.
58   Tatham, A.S., Miflin, B.J. and Shewry, P.R. (1985) Cereal
      Chem. 62, 405-412.
59   Tatham, A.S., Field, J.M., Smith, S.J. and Shewry, P.R.
      (1987) J. Cereal Sci. 5, 203-214.
60   Field, J.M., Tatham, A.S., Baker, R.M. and Shewry, P.R.
      (1986) FEBS Lett. 200, 76-80.
61   Tatham, A.S., Shewry, P.R. and Miflin, B.J. (1984) FEBS
      Lett. 177, 205-208.
62   Field, J.M., Tatham, A.S. and Shewry, P.R. (1987) Biochem.
      J. 247, 215-221.
63   Purcell, J.M., Kasarda, D.D. and Wu, C.S.C. (1988) J. Cereal
      Sci. 7, 21-32.
64   Flavell, R.B., Goldsbrough, A.P., Robert, L., Schnick, D.
      and Thompson, R.D. (1989) BioTechnology 7, 1281-1285.
65   Goldsbrough, A.P., Bulleid, N.J., Freedman, R.B. and
      Flavell, R.B. (1989) Biochem. J. 263, 837-842.
66   Jefferson, R.A. (1988) in Genetic Engineering. Principles
      and Methods (Setlow, J.K., ed.) Vol. 10, pp. 246-263, Plenum
      Publ. Corp., New York, NY.

CONTROL OF TRANSLATION INITIATION IN MAMMALIAN CELLS

Randal J. Kaufman

Genetics Institute
Cambridge, MA 02140

INTRODUCTION

Protein synthesis is an important regulatory step in gene expression. The rate of protein synthesis is controlled by the amount of mRNA available to the translational machinery and the translational efficiency of the mRNA. The former involves gene transcription, mRNA processing, mRNA transport to the cytoplasm, and the half-life of the mRNA. Translational efficiency is determined by mRNA structure, the level and activity of the translational machinery, and by factors that influence initiation and elongation rates. Most translational control is thought to be exerted at the level of initiation of protein synthesis. Initiation is controlled primarily by structure at the 5' end of the mRNA and the activity of translation initiation factors. Factors which control translation initiation may influence essentially all cellular mRNAs or may be mRNA specific.

Our knowledge of the mechanism and regulation of protein synthesis in eukaryotic cells has emerged through the study of in vitro systems reconstituted with purified cellular components (1). However, our understanding of eukaryotic protein synthesis is deficient compared to the process in prokaryotes primarily due to lack of genetic approaches in eukaryotes. One approach to study translational control in intact cells has been to perturb the cell and measure alterations in rates of either specific or general protein synthesis. Different perturbations include stress such as heat shock, modulations of cell growth such as serum deprivation, and infection by different viruses. Since viruses utilize the host cell machinery for viral mRNA and protein production and virion assembly, many viruses establish conditions within the host cell that allow them to dominate the

cellular translational machinery (2). An understanding of the
mechanisms by which viruses alter host translational machinery
has provided insights into cellular control mechanisms for
translational initiation. More recently, with the ability to
introduce and express foreign genes in mammalian cells it has
become possible to dissect control mechanisms which regulate
translation in mammalian cells through gene transfer experiments.
This approach has permitted an analysis of the importance of
sequences within the 5' end of the mRNA in translation. With the
recent isolation of many of the genes for initiation factors that
control initiation it is now possible to alter expression or
express mutant genes in cells to study the importance of a
particular factor in the control of protein synthesis initiation.
The purpose of this chapter is to summarize the recent findings
which yield insight into the control of protein synthesis in
mammalian cells.

mRNA STRUCTURAL REQUIREMENTS

Although the precise mechanism by which eukaryotic ribosomes
initiate at appropriate AUG codons within mRNA involves numerous
components and is regulated by many steps, a substantial amount
of evidence supports a scanning model for translation initiation
(3). This model proposes that a 40S ribosomal subunit binds to
the 5' end of the mRNA and migrates in the 3' direction until it
encounters the first AUG triplet which, if present in an appropriate context, can efficiently serve as the initiator codon.
Initial evidence for this hypothesis came from the realization
that eukaryotic mRNAs encode only a single polypeptide (4). The
scanning model is supported by the stimulatory effect of the m^7G
structure at the 5' end of the mRNA (5), the inability of
ribosomes to bind circular mRNAs (6,7), the ability of 40S
subunits to migrate along the mRNA (8-10), and the negative
influence of secondary structure or the insertion of additional
AUGs between the 5' end of the mRNA and the initiation codon (11-
15).

The 5' Untranslated Region and CAP Structure

The primary determinants for the efficiency by which
ribosomes bind mRNA and initiate polypeptide chain synthesis are
the structural features within the 5' noncoding region of the
mRNA. The discovery that eukaryotic mRNAs contain a unique
structure of the composition $m^7G(5')ppp(5')$ at their 5' end and
the description of proteins which can bind both to this m^7G cap
and to ribosomal subunits provided a plausible mechanism by which
the 40S ribosomal subunit could attach to the 5' end of the mRNA
(16). Indeed, early studies demonstrated that the $5'm^7G$ cap
structure of mRNAs is important in promoting ribosome binding to

mRNA (17-19).   However, the extent of cap dependence varies
between different mRNAs and translation systems (20).

In addition to the 5' cap structure, eukaryotic mRNAs
possess a 5' noncoding region of variable length preceding the
initiator methionine codon. The primary sequence and secondary
structure within this region can greatly influence translational
efficiency.   Most long 5' untranslated regions contain one or
more AUG codons followed shortly by in-frame termination codons.
Although there is little evidence that these small open reading
frames (ORFs) have general biological significance, there are
examples where small polypeptides are translated from several
viral mRNAs and where mutations within these ORFs can alter
initiation at the primary downstream AUG within the mRNA (21-23).
The best studied example demonstrating function for upstream
small ORFs is in translational control of the yeast GCN4 mRNA.
In this case the intactness of the ORFs is of critical importance
for translational control mediated by availability of charged
amino acid tRNAs (24,25).   Thus, it is expected that small ORF's
might also contribute to control of translation for other
eukaryotic mRNAs.

Experimental evidence has accumulated that increased
secondary structure within the 5' untranslated region of the mRNA
results in reduced translational efficiency (13,14,26).   Computer
modeling indicates that many of the 5' noncoding regions of mRNAs
may assume stable stem and loop structures (27).   The existence
of these structures can be experimentally tested by nuclease
mapping techniques and, in many cases, the predicted structures
are observed to occur (28,29).   Increased suppressive effects on
translation occur when the structure is positioned close to the
5' cap (30).   More stable hairpin structures are required to
inhibit downstream translation when their distance from the cap
is increased (15).   The observation that a -60 kCal structure can
cause a 40S ribosome to pause immediately upstream of the
structure is strong supportive evidence for the scanning model of
translation initiation (15).   Cellular mRNAs which have long 5'
untranslated regions which may form secondary structures and
dramatically influence translation include collagen (31), c-Abl
(32), c-Jun (33), and platelet-derived growth factor (PDGF) A
chain (34,35).   The 5' untranslated region of PDGF has a stable
stem-loop structure which acts as a potent translation inhibitor
when inserted upstream of a foreign coding region (35).   The
significance that these 5' untranslated regions play in translational control for these mRNAs is presently being studied.

The best studied example of a 5' untranslated region that
regulates translation initiation in higher cells is that of
ferritin biosynthesis controlled by iron.   Two proteins responsible for the uptake and detoxification of iron in higher eukaryotic cells are the transferrin receptor and ferritin,
respectively. The expression of both these proteins is regulated
by iron.   Whereas ferritin synthesis is inhibited in response to

iron deprivation, the synthesis of the transferrin receptor increases. The reduction in ferritin synthesis is mediated at the level of translation initiation whereas the increase in transferrin receptor synthesis is mediated through increased mRNA stability. An RNA sequence element of approximately 30 bases is necessary and sufficient for the ferritin mRNA to be translationally controlled by iron (36-38). This iron-responsive element (IRE) located in the 5' end of the ferritin mRNA can form a stable stem-loop structure which is required for iron control. The region responsible for control of mRNA stability of the transferrin receptor mRNA is located in the 3' untranslated region of the mRNA and contains 5 stem-loop structures that resemble IREs (39). A 90 kD IRE binding protein (IRE-BP) which specifically binds the IRE in the 5' end of the ferritin mRNA has been identified (40,41). IRE-BP purified by RNA affinity chromatography (42), acts as a translational repressor (43) which can exist in two conformations depending on the oxidation-reduction potential. In the absence of iron, two sulfhydryls within the protein are free and this conformation exposes a high affinity IRE binding site (20 pM). When iron is abundant, these sulfhydryls form a disulfide bond which reduces the protein's IRE binding site affinity to 3 nM and permits translation of the ferritin mRNA (44). This sulfhydryl-switch may constitute a general mechanism by which protein-nucleic acid interactions are regulated (45).

There are now several reports of mRNAs which can be translated in a cap-independent manner in mammalian cells (46-48). The best studied examples are picornavirus mRNAs which are about 7500 bases and have a unique 5' end terminated by monophosphate residues (pUpU...). Another distinct feature of the 5' end of these mRNAs is their extremely long 5' untranslated region (650 to 1300 bases) which contains a number of upstream AUG codons. Translation of these viral mRNAs in infected cells is very efficient. The mechanism for cap-independent translation of poliovirus and encephalomyocarditis virus (EMCV) mRNA has been shown to occur by internal binding of ribosomes within the 5' untranslated region of the mRNA without scanning from the 5' end (46,47). DNA transfection experiments have shown that the 5' untranslated region of poliovirus (46) or of EMC virus (47) allows efficient translation of an adjacent downstream coding region regardless of its position within the mRNA. Poliovirus infection results in inhibition of translation of capped cellular mRNAs. Thus, poliovirus infection of transfected cells completely inhibited translation of the 5' cistron which was preceded by a capped 5' beta globin untranslated region, whereas a second internal cistron (CAT) preceded by the EMC virus 5' untranslated region was efficiently translated (47). The identification of a long sequence in the 5' untranslated region of picornaviruses which participates in ribosome binding argues for the involvement of secondary and tertiary structures in this

process. Models generated with the use of computer-predicted
minimal energy folding, comparative analysis of different
picornaviruses, and RNase and chemical sensitivity suggest there
are three domains of secondary structure between residues 240 and
620 of poliovirus which play a functional role (49,50). More
recent studies have identified a 52 kD polypeptide in HeLa cells
which specifically binds a 64 nucleotide sequence within the
poliovirus 5' untranslated region (51). It is not known what
role this protein plays in translation of poliovirus mRNA or if
this protein also binds the 5' untranslated region of EMC virus.
The presence of this protein in HeLa cells suggests that the
machinery for internal ribosome binding may exist for translation
of cellular mRNAs and not just for utilization by viral mRNAs.
Cellular mRNAs that harbor long 5' untranslated regions with a
large degree of secondary structure may utilize internal ribosome
binding as a means to control their translation. Recently it was
shown that cap-independent translation of the glucose-regulated
protein of 78 kD is increased in poliovirus-infected cells (52).
GRP78 mRNA contains a potential stem-loop structure immediately
upstream from the AUG which is conserved between species and may
play a role in regulation of GRP78 translation (53). Heat shock
proteins are also more resistant than the majority of cellular
mRNAs to inhibition after poliovirus infection (54). Although
these mRNAs have a reduced requirement for eIF-4F (see below), it
has not been demonstrated that these mRNAs have internal ribosome
binding sites.

Choice of the Initiator AUG and Translation of Polycistronic mRNAs

A comparison of the sequence context of the initiation codon
within eukaryotic mRNAs has shown that the most favored initiation site has the following sequence (55): 5' - C-C-A/G-C-C-A-T-
G-G -3'. Of most importance is the purine in the -3 position and
secondarily, the G in the +4 position. Mutagenesis of bases
within this consensus sequence may reduce translation initiation
up to 10-fold (56). Although initiation at AUG codons is most
efficient, several reports have shown that initiation utilizing
leucine-charged tRNA may occur at CUG codons in the c-myc (57)
and basic fibroblast growth factor (58) mRNAs or threonine-
charged tRNA at ACG codons in several viral mRNAs (59-61).

Although ribosomes usually translate only from the 5'
proximal AUG in polycistronic mRNAs, much evidence has accumulated that many naturally occurring mRNAs can initiate at
internal AUGs (62). Translation initiation from internal AUG
codons was directly shown in experimental constructs by the
insertion or deletion of initiator or terminator codons upstream
from the translation start site of a particular open reading
frame (12,63-65). Insertion of an upstream AUG, which is out of
frame with the downstream open reading frame, can severely

suppress translation of the downstream open reading frame.
Insertion of a termination codon can reverse the suppression.
After these observations the scanning model was modified to
include potential for reinitiation at internal AUGs (65). There
are now numerous reports of experimentally constructed polycistronic mRNAs which can translate two or more proteins from
non-overlapping reading frames in the same transcription unit in
mammalian cells (66-70). The finding that efficient reinitiation
requires a minimum of 50 bases between the termination codon and
the next AUG codon has suggested that a refractory period exists
after translation termination in which scanning resumes but
reinitiation cannot occur (71).

The ability of cells to translate multiple polypeptides from
a single mRNA has provided the ability to derive expression
vectors that can efficiently express a foreign gene in the 5'
position and a selectable marker gene in the 3' position to
obtain high level expression in transfected cells (70). More
recently, expression vectors have been constructed utilizing the
ability of picornavirus internal ribosome entry sequences to
allow efficient expression of selectable and amplifiable marker
genes (such as neomycin phosphotransferase or dihydrofolate
reductase) in the 3' end of the mRNA with cloning sites for
insertion of cDNAs within the 5' end of the transcript (72).
These more efficient dicistronic expression vectors are less
likely to undergo rearrangement and deletion upon selection in
the host cell. In general the expression of the 5' cistron is
approximately 5- to 10-fold more efficient than the internal
cistron behind the EMCV sequences. These dicistronic expression
vectors will likely prove very useful to obtain high level
expression after genes have been stably introduced into a variety
of cells.

The Role of the 3' PolyA Tail in Translation Initiation

Most eukaryotic mRNAs have a polyadenylic acid [polyA] tract
at their 3' termini (73,74). These polyA tails are added post-
transcriptionally in the nucleus with an initial length of 200 to
250 adenylate residues. Following transport to the cytoplasm the
polyA is shortened to a steady state length of 50 to 70 adenylate
residues (75,76). Although it has been almost two decades since
the discovery of polyA, it is only recently that evidence has
accumulated for its function. Addition of polyA to a reticulocyte lysate translation system can enhance translation of capped
mRNAs which do not contain a polyA tail and inhibit translation
from capped polyA-containing mRNAs (77). Thus, polyA can act in
trans to activate translation of non-polyA-containing mRNAs. The
inhibition mediated by addition of excess polyA to translation
reactions containing polyadenylated mRNAs may result from
competition for a limiting polyA binding factor. PolyA may bind
and induce a conformational change in a factor required for

translation.   In the absence of polyA, mRNAs are recruited into
the 80S initiation complex at a reduced efficiency.   The identification and cloning of a polyA binding protein (PABP) yielded
insight into the mechanism by which the polyA may facilitate 80S
initiation complex formation.   PABP consists of a repeated domain
of 90 amino acids that contains a consensus RNA binding site
(78,79).   The multiple domains allow the protein to move between
different polyA tracts by an interstrand transfer mechanism (80).
Inactivation of the polyA binding protein gene in Saccharomyces
cerevisiae results in an inhibition of translation initiation and
polyA tail shortening (81).   Upon selection for reversion, seven
independent extragenic suppressor mutants were identified.   All
mutants allow translation initiation without significantly
affecting the polyA tail-shortening defect.   These suppressor
mutations result in an increase in the amount of the 60S
ribosomal subunit.   One of these suppressors encodes the
ribosomal 60S protein L46.   These experiments suggest that the
role of polyA in translation initiation is to bind a factor
[polyA binding protein] which facilitates binding of the 60S
ribosomal subunit at the mRNA 5' end.

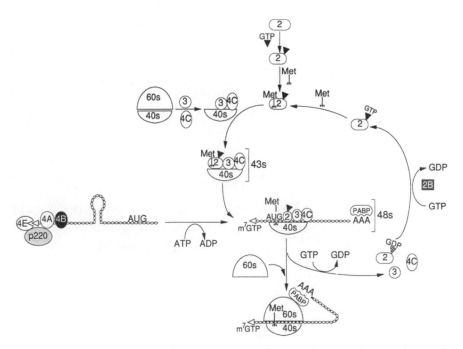

Figure 1.   Formation of ternary complex and recycling of eIF-2.

TRANSLATION INITIATION FACTORS

At least 10 different initiation factors have been purified and characterized in vitro (1,82-84). Table 1 lists the factors presently identified with their putative roles in translation initiation. Also indicated are those factors which have been cloned. Figure 1 depicts the steps that occur to initiate protein synthesis in mammalian cells. The two most well studied reactions are the eIF-2 promoted binding of met-tRNA to 40S ribosomal subunits and the recycling of eIF-2 (85) and binding of mRNA to ribosomes promoted by eIF-4A, eIF-4B and eIF-4F (16).

Table 1
Eukaryotic Initiation Factors (220)

Factor	Subunits kD		Cloned	Activities
eIF-2	α	36	+	Ternary complex formation; binding met-tRNA to 40S
	β	38	+	
	γ	55	-	
eIF-2B	α	26	-	eIF-2 recycling
	β	39	-	
	γ	58	-	
	δ	67	-	
	ε	82	-	
eIF-3	8 subunits		3/8	Binding 43S to mRNA; maintains dissociated subunits
eIF-4A	46		+	mRNA binding, ATPase, mRNA unwinding
eIF-4B	80		+	mRNA binding/stimulates 4A
eIF-4C	18		+	ribosome dissociation; stabilizes met-tRNA binding to 40S
eIF-4D	16		+	enhances 80S initiation complex reactivity with puromycin
eIF-4E	25		+	binds cap structure
eIF-4F	α	25	+	cap recognition, mRNA binding and unwinding
	β	46	+	
	γ	220	-	
eIF-5	150		-	stimulates 48S joining with 60S
eIF-6	25		-	ribosome anti-association

Formation of mRNA Ternary Complex and eIF-2 Recycling

Polypeptide chain synthesis initiates when a binary complex of eIF-2 and GTP forms and subsequently binds met-tRNA (Figure 1). This ternary complex of eIF2, GTP, and initiator met-tRNA binds the 40S ribosomal subunit to generate a 43S preinitiation complex. eIF-2 is the primary protein involved in forming the ternary complex and is comprised of three subunits: α (36 kD), β (38 kD) and γ (55 kD). This 43S preinitiation complex then binds the 5' end of the mRNA and according to the scanning model (62) this complex migrates in the 3' direction until the first appropriate AUG codon is reached and translation begins. Recent genetic studies have provided evidence that eIF-2 functions in the scanning process by selection of the initiator AUG codon. Selection in yeast S. cerevisiae for reversion of an initiator codon mutant in the His4 gene identified two suppressor mutants, sui2 and sui3, which encode the alpha and beta subunits of eIF-2, respectively (86). Mutations within the zinc finger motif in the beta subunit of eIF-2 permit initiation at a UUG codon (87). In addition, mutations in either sui2 or sui3 disrupt translational control of GCN4 by altering start site utilization at AUG codons in upstream ORFs within GCN4 mRNA (88). These data implicate a nucleic-acid-binding domain of eIF-2 as an important component of AUG site recognition for the scanning ribosome.

Although the GTP requirement for ternary complex formation can be replaced with non-hydrolyzable analogues of GTP, GTP hydrolysis is required for 60S joining to form an 80S initiation complex (89). GDP completely prevents eIF-2 from binding met-tRNA. At physiologic concentrations of magnesium, the affinity of eIF-2 for GDP is 100-fold higher than that for GTP (90,91). Thus, for eIF-2 to promote another round of initiation, GDP bound to eIF2 must be exchanged for GTP. This exchange reaction requires catalysis by the guanine nucleotide exchange factor [GEF (92), eIF-2B (93) or RF (94)]. eIF-2B is a protein complex composed of 5 subunits (82, 67, 58, 39 and 26 kD). The control of the eIF-2 recycling activity is a primary regulatory step of protein synthesis in mammalian cells.

eIF-2 recycling is controlled at one level by the phosphorylation state of the alpha subunit of eIF-2. The importance of eIF-2 phosphorylation in regulating eIF-2 activity was first demonstrated in vitro in a rabbit reticulocyte lysate translation system. Reticulocytes require heme to maintain high rates of protein synthesis. Deprivation of heme results in an inhibition of initiation of protein synthesis. The inhibition results from the generation of a repressor [heme controlled repressor (HCR) (90kD)]. HCR is a protein kinase which phosphorylates the alpha subunit of eIF-2 (94,95). Phosphorylation of the alpha subunit of eIF-2 (eIF-2α) stabilizes the eIF-2-GDP-eIF-2B complex and consequently prevents GTP exchange, eIF-2 recycling and further initiation events (85,89,96). Since small changes in the phosphorylation state of eIF-2 can dramatically affect protein synthesis and since the estimated concentration of eIF-2B in the cell is approximately an order of magnitude less than that of eIF-2, it was proposed that translation arrest results from sequestration of eIF-2B by phosphorylated eIF-2 (94-96). This eIF-2B sequestering model is supported by the observation that addition of eIF-2 or eIF-2B to the heme-deficient inactive lysate stimulates protein synthesis.

In addition to heme depletion, addition of double-stranded RNA to reticulocyte lysate results in the activation of another eIF-2α kinase, the double-stranded RNA activated inhibitor (DAI) (68 kD). Activation of this kinase appears to require a perfect RNA-RNA duplex of 50 base pairs minimum length (97), although a variety of polyanions can elicit its activation (98). Double-stranded DNA or an RNA-DNA hybrid will not activate DAI kinase (97). It is hypothesized that the double-stranded RNA promotes dimerization of the kinase which results in its autophosphorylation. The phosphorylated kinase is constitutively active and no longer requires double-stranded RNA for activity (98). In mammalian cells, induction of the double-stranded RNA activated eIF-2α kinase (DAI) is part of the cellular antiviral response induced by interferon (99). Interferon-resistant viruses such as adenovirus (100-103), vaccinia virus (103), influenza virus (104) and reovirus (105,106) have evolved mechanisms to prevent DAI kinase activation by doubled-stranded RNA.

The best studied example of viral control of host translation apparatus is that mediated upon adenovirus infection. Adenovirus encodes small RNA polymerase III gene products, VA RNA I and II, which are required for efficient initiation of mRNA translation late in adenovirus infection (107). VAI RNA can prevent activation of the interferon-induced DAI protein kinase by double-stranded RNA (100,101,108,109). It is proposed that high concentrations of VAI RNA can bind DAI kinase but not permit its dimerization and subsequent activation (108). Infection with adenovirus VAI gene deletion mutants results in activation of DAI kinase, phosphorylation of eIF-2α (104,110) and inhibition of translation initiation. The importance of VAI RNA in mediating inhibition of eIF-2α phosphorylation was directly shown by expression of a phosphorylation-resistant serine to alanine mutant of eIF-2α in human 293 cells (111). This serine to alanine mutant at residue 51 of eIF-2α was resistant to phosphorylation at residue 51 and resistant to inactivation by phosphorylation. Expression of this mutant complemented the defects in protein synthesis and infectious virus production upon infection with an adenovirus VAI RNA gene deletion mutant (111). These results demonstrate that the primary function of VAI RNA in the lytic adenovirus infection is the inhibition of eIF-2α phosphorylation by DAI kinase and that eIF-2α is the primary phosphorylation target to mediate the effect of DAI kinase activation.

In a similar manner to infection with adenovirus VAI RNA gene deletion mutants, plasmid DNA transfection into cells may result in DAI kinase activation and inefficient translation of plasmid derived mRNA (112-114). The inclusion of a VAI RNA gene in the transfection can reverse the translation defect. Alternatively, addition of a DAI kinase inhibitor, 2-aminopurine, to the culture medium also elicits a similar effect as VAI RNA (115). The defect in translation can also be corrected by expression of a mutant eIF-2α with a serine to alanine change at residue 51, the site of DAI kinase-mediated phosphorylation (116). In contrast, expression of a serine to aspartic acid mutation at residue 51, created to mimic the phosphorylated serine, inhibited translation, even in the presence of VAI RNA (116). These findings dramatically implicate the importance of eIF-2α phosphorylation in translational control. The use of site-directed mutagenesis and expression of eIF-2α mutants which cannot serve as a substrate for DAI kinase or which mimic the charge of the phosphorylated state has provided a powerful tool to dissect the role of eIF-2α phosphorylation in intact cells. Extension of this method to other proteins should provide a means to ascertain which phosphorylation reactions and substrates have physiological significance.

The translation inhibition observed in DNA transfection experiments is specific to the mRNA derived from the plasmid DNA. No detectable inhibition of global host protein synthesis was observed (115). To account for the preferential translation inhibition of plasmid-derived mRNA, it was proposed that a localized activation of DAI kinase in the vicinity of the plasmid-derived mRNA results in reduced translation. Symmetric transcription from circular plasmid DNA may generate an mRNA which contains double-stranded RNA and may directly activate the kinase. Thus, the activated kinase would bind specifically to the plasmid-derived mRNA and result in a localized phosphorylation of eIF-2. Alteratively, the activated DAI kinase may be associated with the ribosome which is bound to the mRNA having partially double-stranded RNA features. Since DAI kinase activation may alter translation of a specific mRNA, DAI activation may provide a means to control translation of specific cellular mRNAs.

Control of eIF-2α Phosphorylation by Other Viruses

Since RNA viruses and some DNA viruses produce double-stranded RNA at some point in their infectious cycle, infection results in the activation of DAI kinase and subsequent eIF-2α phosphorylation. Although it is likely that viral mRNAs still require eIF-2 for initiation, its reduced level could allow an opportunity for more efficient or abundant viral mRNAs to outcompete cellular mRNAs for the translational machinery. Some viruses may utilize eIF-2α phosphorylation to usurp translational machinery while producing a gene product which blocks DAI kinase activation later in the infectious cycle to assist late viral protein synthesis. Viruses which appear to utilize such a strategy are discussed below.

Vaccinia virus. Infection of permissive cells with vaccinia virus induces a rapid shutoff of cellular protein synthesis which occurs concomitantly with degradation of cellular mRNA and polysomes. There is an inhibitor of DAI kinase in extracts from vaccinia virus-infected cells and its generation requires protein synthesis (103). Although the mode of action of this inhibitor is unknown, it is thought to be a protein that stoichiometrically interacts with double-stranded RNA (117). A vaccinia virus-encoded gene product or induction of a cellular function which inactivates DAI kinase may be necessary late in vaccinia virus infection when double-stranded RNA is produced by symmetrical transcription of the viral genome (118).

Abortive infection by vaccinia virus is characterized by a general shutoff of cellular and host protein synthesis (119). In several cases the viral defects have been mapped to specific gene products. Vaccinia virus infection in CHO cells is abortive due to the inability to translate mRNA late in the infectious cycle. A cowpox virus 77 kD polypeptide can alleviate this block in CHO cells (120). A vaccinia virus 29 kD gene product is required for efficient replication in human cells (121). It is possible that these gene products inhibit or bypass the effects of activated DAI kinase and in their absence eIF-2α phosphorylation would result in protein synthesis shutoff and aborted infection.

Reovirus. Upon infection of cells with reovirus serotype 1 (Lang strain), there is generation of an inhibitor of the DAI kinase. Recent evidence suggests that the virus-encoded sigma 3 protein is responsible for the inhibitory activity (105). Expression of sigma 3 can stimulate translation in a transient DNA transfection assay suggesting that sigma 3 may prevent eIF-2α phosphorylation in intact cells (106). Reovirus replication is generally thought to be sensitive to interferon treatment (122). However, reovirus serotype 3 (Dearing strain) is more sensitive than reovirus serotype 1 (Bert Jacobs, personal communication). Host protein synthesis is not inhibited upon infection with serotype 1 compared to infection with serotype 3. Extracts from reovirus serotype 1-infected cells contain more kinase inhibitory activity and more sigma 3 protein than extracts from reovirus serotype 3-infected cells (123). This suggests that differences in sigma 3 expression and resulting DAI kinase inhibitory activity may be responsible for differences in interferon sensitivity and host shutoff between these serotypes.

HIV. All mRNAs of HIV-1 contain in their 5' untranslated region a sequence termed TAR that responds to trans-activation by the tat (trans-activating) protein. This RNA sequence assumes a stable stem-loop secondary structure. The TAR sequence can activate DAI kinase in vitro (124,125) and inhibit in trans the translation of other mRNAs (124). Mutations which disrupt the stem-loop structure in the TAR region of the mRNA fail to activate DAI kinase. Regeneration of the secondary structure by introduction of compensatory mutations reconstitutes the ability to activate DAI kinase and inhibit translation (124). Recently three proteins which specifically bind TAR have been identified (126). One has a molecular weight expected for DAI kinase. Further experiments will be required to determine if DAI kinase can directly bind TAR. The mechanism by which tat trans-activates HIV-1 gene expression is not clear, although evidence suggests that both transcriptional (127) and post-transcriptional components are involved (128-130). It is interesting that recently it has been shown that tat may interfere with activation of DAI kinase (221). Activated DAI kinase can induce the synthesis of B-interferon (131) resulting in an anti-viral state in the host cell infected with HIV-1. This antiviral state induced by activated DAI kinase may contribute to the latency of HIV infection.

Control of eIF-2α Phosphorylation by Stress Conditions

In addition to viral infection and plasmid DNA transfection, phosphorylation of the alpha subunit of eIF-2 correlates with an inhibition of initiation of protein synthesis in mammalian cells in response to a wide variety of different stimuli including heat shock (132-134), serum deprivation (135), glucose starvation (136), amino acid starvation (137), exposure to heavy metal ions (138) and other inducers of the stress response (139,140). Although the importance of eIF-2α phosphorylation has been proposed from correlative in vivo data and substantial in vitro results, it will now be possible, through the expression of specific phosphorylation resistant mutants of eIF-2α, to establish that phosphorylation of eIF-2α actually controls protein synthesis in cells after response to these different stimuli and to separate its effects from the many other initiation factor modifications that occur (for review see 141). In addition, it is not known what kinase is responsible for eIF-2α phosphorylation in response to these different stimuli. Regulation of eIF-2α activity may also occur through regulation of dephosphorylation. A phosphoprotein phosphatase has been reported to dephosphorylate eIF-2α although its importance in regulating eIF-2 activity is not known (142,143). An eIF-2-associated protein of 67 kD has been isolated from reticulocyte lysate which can protect eIF-2α from HCR-mediated phosphorylation (144). This unusual protein contains multiple O-linked N-acetylglucosamine residues that appear required for its ability to protect eIF-2α from phosphorylation (145). An understanding of the mechanisms that regulate phosphorylation and dephosphorylation of eIF-2α will be required to understand how control of translation through eIF-2α phosphorylation is mediated.

The mechanism by which HCR becomes activated upon heme deprivation is best understood. HCR is activated from a latent to an active form by autophosphorylation. Hemin promotes intersubunit disulfide bond formation in both latent and activated phosphorylated HCR. This thio-disulfide exchange appears to involve sulhydryl groups that are required for binding of ATP to HCR (146,147). Recent studies on the control of HCR activity have shown that the 90 kD HCR is in a complex with the cellular heat shock protein of 90 kD (hsp90) in an inactive form in hemin-supplemented lysates and dissociates from hsp90 upon activation (148). Hsp90 can also be phosphorylated (149) and its phosphorylation increases the activity of purified HCR in vitro (15). These results suggest that hsp90 may play a direct role in mediating the activation of HCR. Hsp90 may also play a role in the activation of the eIF-2α kinase which occurs in response to heat shock. Pretreatment of cells with heat, a condition which induces hsp90 expression, can abrogate the inhibitory effect of subsequent heat shock on protein synthesis (151). In this respect, it is interesting to speculate that the newly synthesized hsp90 may promote binding to an eIF-2α kinase, like HCR, and prevent heat shock induced inhibition of protein synthesis or play a role in the recovery of normal protein synthesis in heat shock treated cells.

FACTORS WHICH PROMOTE mRNA UNWINDING AND RIBOSOME BINDING TO mRNA

In contrast to initiation in prokaryotes, ATP is required for translation initiation in eukaryotes (152,153). The best understood function for the 5' m7G cap structure is its ability to promote ribosome binding to mRNA in an ATP-dependent manner. As a consequence, it was reasoned that proteins which bind to the cap structure should promote ribosome binding. The use of m7G cap-affinity columns enabled the isolation of specific proteins which bind the cap structure (16,154-158). Further studies have shown that these cap-binding proteins function to melt RNA secondary structure within the 5' end of the mRNA in an ATP-dependent manner.

Several proteins which bind the 5' m7G cap and melt secondary structure have been purified and characterized. CBPI (eIF-4E) is a 25 kD protein which binds directly to the 5' m7G cap structure in an ATP-independent manner. CBP II (eIF-4F) is a protein complex composed of three subunits: 1) eIF-4E, 2) eIF-4A, a 46 kD polypeptide which has ATPase activity, mRNA unwinding activity, and stimulates the binding of mRNA to the 40S ribosome and 3) p220. The function of p220 in the CBPII complex has not been determined. In addition, eIF-4B is absolutely required for mRNA binding to ribosomes in an ATP-dependent reaction (152,153). It is thought that eIF-4F plays a discriminatory role in selecting mRNAs for translation (159).

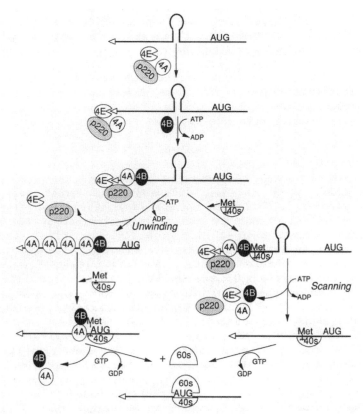

Figure 2. Role of eIF-4A, eIF-4E, p220 and eIF-4B in ribosome binding to mRNA.

eIF-4E can bind mRNA in an ATP-independent manner, although it is not known if eIF-4E is independently functional within the cell. In an ATP-dependent manner, eIF-4A binds mRNA. It is not known how p220 facilitates the binding of eIF-4A. It is not known if eIF-4E and p220 are released from mRNA before mRNA unwinding, as shown in the model. Excess free eIF-4A may maintain the mRNA free of secondary structure. Ribosomes bind after initial mRNA unwinding, however, but it is not known where the ribosome binds. Two possibilities are shown where the ribosome binds after eIF-4A-mediated unwinding and where the ribosome binds immediately after the cap-binding protein complex to initiate scanning which may unwind the mRNA.

In 1981 it was hypothesized that the cap-binding proteins facilitate ribosome binding by melting mRNA (160,161). In support of this hypothesis were the observations that eIF-4A and eIF-4B crosslink to the cap structure in an ATP-dependent manner for many mRNAs (155,162,163) but not for mRNA with unstructured 5' ends (156). Subsequently, it was shown that eIF-4A and eIF-4B can bind the cap of reovirus RNA only in the presence of eIF-4F and ATP (162). These results are consistent with the present hypothesis that eIF-4F binds to the cap via its 25 kD subunit in an ATP-independent manner and then eIF-4A and eIF-4B bind in an ATP-dependent manner (156,162,163). The roles of the cap-binding proteins in binding mRNA are depicted in Figure 2.

eIF-4E

eIF-4E contains the binding site for the cap structure and may interact with the cap independent of other polypeptides (164). However, eIF-4E also exists as a subunit of eIF-4F and, as part of eIF-4F, binds the mRNA 5' structure about 10-fold more efficiently (165). eIF-4E is probably the first factor to interact with an mRNA to initiate translation. The gene encoding eIF-4E has been isolated (166) and the deduced amino acid sequence contains eight tryptophan residues which are highly conserved between the genes of yeast, murine and human origin. Tryptophan residue 36 is in a region that is highly conserved to a region around tryptophan 360 in the influenza cap-binding protein PB2 (167), the consensus being K-X-X-X-N-X-R/K-W-A-L where X is any amino acid (166). It has been suggested that tryptophans might be involved in cap recognition (168). Gene disruption experiments in yeast have demonstrated that eIF-4E is essential for growth (169). The estimated abundance of eIF-4E in cells (approximately 0.1/ribosome) would suggest that it may be a limiting factor in promoting ribosome binding to mRNAs in a cap-dependent manner (170). With the availability of the cDNA encoding eIF-4E, it will be possible to study the effects of overexpression of wild-type and mutants of eIF-4E on translational control in mammalian cells.

The activity of eIF-4E appears to be modulated by phosphorylation. Dephosphorylation of eIF-4E correlates with inhibition of protein synthesis initiation in response to heat shock (170). Translation extracts prepared from heat shock treated cells are inefficient in protein synthesis. Addition of eIF-4F can restore translation of mRNA, whereas addition of other initiation factors had no effect (171). It is possible that the reduction of eIF-4F activity results from a dephosphorylation of the eIF-4E subunit. Dephosphorylation of eIF-4E also occurs upon entry into mitosis, a condition in which protein synthesis is reduced to 25% (172). One site of phosphorylation in eIF-4E has been identified at serine residue 53 (173). The serine at

residue 53 has been mutated to an alanine to prevent phosphorylation. The resultant protein was capable of binding a m'GTP
sepharose column indicating its ability to bind a cap structure.
However, when translation reaction mixtures were resolved on
sucrose gradients, the wild-type eIF-4E was bound to the 48S
initiation complex in the presence of GMP-PNP (164), whereas the
alanine mutant was not (222). These results suggest that
phosphorylation of eIF-4E is necessary for it to perform its role
of transferring mRNA to the 48S complex.

eIF-4A

eIF-4A can exist in a free form or as part of the eIF-4F
complex. In both forms eIF-4A has an ATPase activity (174) and
as part of the eIF-4F complex, eIF-4A can crosslink to the mRNA
cap in the presence of ATP (161-163). The primary role identified for eIF-4A is its helicase activity which likely acts to
melt secondary structure of mRNA in an ATP-dependent manner. The
helicase activity of eIF-4A is more effective as part of eIF-4F
and is stimulated by the presence of eIF-4B. The unwinding
activity of eIF-4F shows a greater dependence on the presence of
a cap structure compared to the unwinding activities of eIF-4A
and eIF-4B. The observation that eIF-4A can crosslink to ATP is
consistent with its ATPase and RNA helicase activities (175).
eIF-4A and eIF-4F can unwind RNA in a 5' to 3' direction as well
as in a 3' to 5' direction (176). Unwinding mediated by eIF-4F
in the presence of eIF-4B was stimulated by the presence of the
5' cap structure whereas unwinding in the 3' to 5' direction was
independent of the presence of the cap structure (176). Importantly, the unwinding activity of eIF-4A or eIF-4E absolutely
required the presence of eIF-4B (176).

The cloning and sequence analysis of eIF-4A indeed demonstrated that this factor is a member of a family of at least
seven proteins which exhibit ATP-dependent helicase activities
(177). This family of related ATPases contains the consensus
sequence A-X-S/T-G-S/T-G-K-T which is present in many ATP-
hydrolyzing proteins (178). The GKT tripeptide is present in all
other ATP-binding proteins. Mutation of the lysine residue to an
asparagine resulted in a loss of ATP crosslinking to eIF-4A,
suggesting the requirement for the lysine in the consensus as an
important determinant in ATP binding for eIF-4A (179). Another
consensus sequence motif is the DEAD box which is present in a
special version of the B motif of ATP-binding proteins (178).
This family of helicases also contains a third consensus
sequence, HRIGR, which is speculated to be important in polynucleotide binding. Two functional eIF-4A genes have been
identified in the mouse genome which exhibit differential tissue
expression (180). In yeast eIF-4A is encoded by two duplicated
genes, TIF1 and TIF2, which are essential for growth (181).

eIF-4B

eIF-4B is a single polypeptide of 80 kD that may function as a dimer of identical subunits (182). eIF-4B can crosslink to the 5' 7-mG cap in the presence of eIF-4F and ATP (162) and can stimulate the mRNA binding and ATPase activity of eIF-4A and eIF-4F (151). eIF-4B can stimulate the release and rebinding (recycling) of eIF-4F bound to the mRNA cap (183). Since eIF-4B exhibits a preferential binding to polynucleotides containing an AUG (184,185), it is speculated that eIF-4B plays a role in ribosome recognition of the appropriate initiator ATG. The cloning of the cDNA encoding eIF-4B identified eIF-4B as a 611 amino acid protein which is a member of a class of RNA binding proteins, of which the polyA binding protein is a member (79, 186). The most distinguishing feature of this class is that they contain one or more repeats in their aminoterminal region of an RNA binding site consensus sequence composed of a conserved octamer and a pentapeptide separated by 30 residues. The presence of this domain implicates a role for eIF-4B in direct RNA binding as supported by experimental evidence (154,187).

eIF-4B exhibits a high variable degree of serine phosphorylation which correlates with initiation activity. eIF-4B is dephosphorylated in response to heat shock (133) and becomes phosphorylated after addition of serum (135,188). Protein kinase C, cAMP-dependent protein kinase, S6 protein kinase, protease-activated kinase I and casein kinase II can phosphorylate eIF-4B in vitro (189). Extracts from serum-starved neuroblastoma cells show a defect in initiation of protein synthesis that can be reversed by addition of exogenous eIF-4B (190). The amino acid sequence of eIF-4B identified a clustering of serine residues within the carboxyterminal portion of the protein. Further studies are required to identify the role(s) that the different eIF-4B phosphorylations play in regulating protein synthesis.

The Role of p220 in Translation Initiation in Intact Cells

The role for p220 in translation initiation is not known. The importance of p220 has been studied from monitoring its role in poliovirus infection. Poliovirus encodes a protease whose expression results in the cleavage of the large subunit of eIF-4F (p220) (191,192). This cleavage is probably indirectly induced (191,193) by the poliovirus protease p2A (191,194). Although they do not contain a 5' cap structure, the 5' untranslated regions of picornavirus mRNAs can effectively compete with cellular mRNA for ribosome binding (195). Thus, cleavage of p220 is thought to preferentially inhibit host cap-dependent protein synthesis. This is supported by the observations that inhibition of host protein synthesis in poliovirus-infected cells roughly coincides with cleavage of p220 (191), and addition of purified eIF-4F can restore the translation defect in lysates prepared from poliovirus-infected cells (196, see 197 for review). However, several observations suggest a more complex mechanism. First, p220 cleavage occurs about 30 min after significant inhibition of host protein synthesis occurs (191). Second, poliovirus infection in the presence of an inhibitor of RNA replication results in complete cleavage of p220 but only 55 to 77% inhibition of host protein synthesis, suggesting that complete inhibition of host protein synthesis requires another event (198). Third, expression of the poliovirus p2A gene in transfected COS-1 cells results in p220 cleavage, with little effect on cellular protein synthesis. However, it was observed that expression was reduced due to a reduction in mRNA transcription in cells which have cleaved p220 (199). Finally, although infection with a p2A defective poliovirus mutant does not result in cleavage of p220 and does not inhibit host protein synthesis early in infection, late in infection inhibition of both host and viral protein synthesis occurs (200). In this latter case, the indiscriminate shutoff may result from phosphorylation of eIF-2α (201).

Increased phosphorylation of eIF-2α is observed in poliovirus-infected cells (202). It has been suggested that there is an inhibitor generated in poliovirus-infected cells which limits the extent of eIF-2α phosphorylation (203) and other studies have shown that activated DAI kinase is degraded in poliovirus infection (202). These results suggest the shutoff of protein synthesis in poliovirus infection is a consequence of synergistic actions of p220 cleavage and limited eIF-2α phosphorylation. Two mechanisms have been proposed. One mechanism proposes that after poliovirus infection, p220 is cleaved and thus prevents mRNA unwinding of secondary structure. The secondary structure activates DAI kinase and results in a generalized inhibition of host mRNA translation. This type of mechanism has been proposed for the role of the TAR region in HIV-1 mRNA (124). Another mechanism is that the pronounced requirement for p220 may result from the poorer initiation frequency which occurs upon eIF-2α phosphorylation. As depicted in Figure 3, p220 may be required for the initial ribosome to bind mRNA but not for further reinitiation events. When initiation frequency is high, the mRNA is coated with ribosomes which prevent mRNA secondary structures from forming. When the initiation is infrequent, such as for in vitro translation systems or when eIF-2α becomes phosphorylated in vivo, increased secondary structure within the mRNA may form and result in increased requirement for eIF-4F in order to unwind the mRNA to permit ribosome binding. At high rates of translation initiation the unwinding function of eIF-4F is essentially replaced by ribosomes. This model predicts that the primary ribosome binding event requires eIF-4F but that subsequent ribosome binding may occur independent of eIF-4F when the initiation frequency is high since the translocation of the ribosome will unwind mRNA secondary structure. This model is

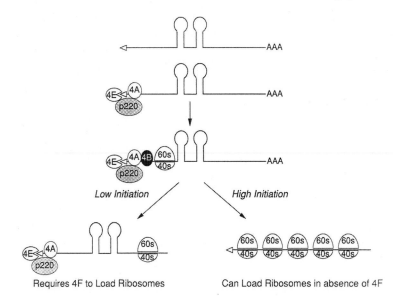

Figure 3. Model for synergistic actions of eIF-2α phosphorylation and p220 cleavage in poliovirus infection.

consistent with the observation that inhibition of host protein synthesis occurs early after poliovirus infection, before significant cleavage of p220 occurs (191).

Is Translation Coupled to mRNA Transport and/or Stability?

An important aspect concerning the role of translation in the regulation of gene expression is its effect on mRNA transport and/or stability. Increasing data suggest that translation may be coupled to mRNA degradation for specific mRNAs (204). Most evidence stems from the use of an inhibitor of polypeptide chain elongation, cycloheximide. In the presence of cycloheximide, many unstable mRNAs become stable. Although some investigators have attributed this effect to the loss of an unstable nuclease involved in cleavage of these mRNA, more recent experiments suggest that in some instances decay may be associated with a ribosome-associated nuclease which recognizes the 3' end of the mRNA. The best studied example is that of beta tubulin mRNA decay coupled to translation in response to tubulin pool size (205,206). The destabilizing element in tubulin mRNA was localized to the four aminoterminal amino acids of the tubulin subunit. Decay of tubulin mRNA also requires translation into 90 amino acids beyond this signal and suggests that the nascent polypeptide must be exposed from the ribosome in order to mediate decay. A reasonable hypothesis is that the four terminal amino acids must complex with a growing microtubule to stabilize the mRNA. In the presence of excess tubulin monomer, the nascent polypeptide would bind a free tubulin monomer and be rapidly degraded. In a similar manner, decay of histone mRNA in response to cessation of DNA synthesis requires the ribosome to translocate across 300 bases of a terminal stem-loop structure (207). Specific sequences within the 3' end of the mRNA which may be targets for nucleases have been identified. Many short-lived mRNAs such as the lymphokine GM-CSF (208) and oncogenes such as c-fos (209) contain one or more copies of a short sequence motif, AUUUA, in the 3' untranslated region of the mRNA. This sequence contributes to selective mRNA degradation and in all cases, cycloheximide can prevent mRNA degradation. A correlation has been observed between the stability of the polyA tail length and the absence of the AU sequence motif (210) and suggests the AU sequence acts by influencing interactions between polyA and specific binding proteins. An additional component for mRNA decay in c-fos has been localized in the translated region encoding the zinc finger and leucine zipper of the protein (209). It is interesting to speculate that the nucleic acid binding properties of the nascent polypeptide in this instance may influence mRNA decay directly.

Data have accumulated suggesting that protein synthesis may be coupled to mRNA transport from the nucleus. Nonsense mutations within coding regions can elicit a decrease in mRNA levels at the mouse immunoglobulin locus (211), the human β-globin locus (212), the triose phosphate isomerase genes (213) and the DHFR gene (214). A translation-like mechanism has been implicated in the transport of tRNA molecules from nucleus to cytoplasm (214). Ribosome-like particles have been observed on the cytoplasmic face of nuclear pores (216) and might serve as exit ports for RNA molecules being translated (217,218). In addition, inhibition of protein synthesis by amino acid starvation or cycloheximide treatment delays the nuclear processing time for five different RNA molecules but has little effect on transcription or mRNA stability (219).

CONCLUSION

Control of protein synthesis initiation is a critical regulatory step in gene expression. There are potentially many factors that affect initiation rates in mammalian cells which include the number of ribosomes, the status of the cytoskeletal network, mRNA primary sequence and secondary structure, and multiple initiation factors. Dissection of the events in translation has been difficult due to the inability to study non-covalent protein-protein and protein-RNA interactions as well as the lack of genetic approaches for mammalian cells. To date most studies have focused on mRNA structural requirements and the role of specific initiation factors responsible for promoting ribosome

binding to mRNA. The most well studied factors which regulate
translation are eIF-2 and the factors eIF-4A, eIF-4B and eIF-4F
which are required for ribosome binding to mRNA. In general, the
activities of these factors within cells appear to be primarily
controlled by post-translational modifications and not by
alterations in their rates of synthesis and degradation. Of all
post-translational modifications of translation initiation
factors which influence translation, phosphorylation of eIF-2 has
been proven to be a major single control step. Although many of
the other factors are also phosphorylated, it is not known what
important consequences these other phosphorylation reactions may
have. The ability to express cloned genes which harbor specific
mutations to inhibit a particular phosphorylation or to mimic a
phosphorylation allows an approach to dissect the importance of
the individual phosphorylation reactions in translational
control. An understanding of the protein kinases and phosphatases which mediate the phosphorylation reactions is required
to understand what regulates the activity of these initiation
factors.

Much has been learned about the mechanisms by which a
variety of viruses usurp the translational apparatus to direct
the synthesis of viral-specific polypeptides. For example, many
viruses likely utilize eIF-2α phosphorylation to facilitate a
host to virus shift in mRNA utilization. Studies with picornaviruses have provided key insights to our knowledge about how
ribosomes bind mRNA. We now know that ribosomes are capable of
binding directly within the mRNA rather than to the 5' cap
structure. The precise mechanism by which cap-dependent and
independent translation occur requires further studies as well as
an investigation of which and how many cellular genes are
regulated by ribosomes binding to specific sites within mRNAs.
This mechanism may account for much of the specificity observed
in translation control for cellular genes. Finally, more studies
are required to determine how translation may be linked to
processes which regulate mRNA levels such as by affecting mRNA
transport and/or stability. Our understanding of the mechanisms
by which specific mRNAs are chosen for translation will provide
insights to how protein synthesis is coupled with other cellular
processes such as organelle assembly, cell division and
differentiation.

As we obtain a greater understanding of what factors and
mRNA sequences are required for the selective utilization of mRNA
for translation, it will be possible to engineer cells to produce
high levels of specific polypeptides at a designated time. These
processes are crucial to the development of optimal expression
systems for mammalian cells. The identification of specific
ribosome binding sites has permitted the engineering of functional polycistronic mRNAs which have the capacity to code for
multiple polypeptides. The study of how specific mRNAs are
recruited into polysomes in response to specific signals, such as

ferritin mRNA in response to iron levels, should provide the
ability to turn on the utilization of an mRNA at any particular
time in the growth cycle of a cell.

REFERENCES

1   Benne, R., Brown-Leudi, M. and Hershey, J.W.B. (1979)
    Methods Enzymol. 60, 15-35.
2   Schneider, R.J. and Shenk, T. (1987) Annu. Rev. Biochem. 56,
    317-332.
3   Kozak, M. (1980) Cell 22, 7-8.
4   Jacobson, M.F. and Baltimore, D. (1968) Proc. Nat. Acad.
    Sci. U.S.A. 61, 77-84.
5   Shatkin, A.J. (1976) Cell 9, 645-653.
6   Kozak, M. (1979) Nature 280, 82-85.
7   Konarska, M., Filipowicz, W., Domday, H. and Gross, H.J.
    (1981) Eur. J. Biochem. 114, 221-227.
8   Kozak, M. and Shatkin, A.J. (1978) J. Biol. Chem. 253, 6568-
    6577.
9   Kozak, M. (1979) J. Biol. Chem. 254, 4731-4738.
10 Kozak, M. (1980) Cell 22, 459-468.
11 Kozak, M. (1983) Cell 34, 971-978.
12 Liu, C-C., Simonsen, C.C. and Levinson, A.D. (1984) Nature
    309, 82-85.
13 Pelletier, J. and Sonenberg, N. (1985) Cell 40, 515-526.
14 Kozak, M. (1986) Proc. Nat. Acad. Sci. U.S.A. 83, 2850-2854.
15 Kozak, M. (1989) Mol. Cell. Biol. 9, 5134-5142.
16 Sonenberg, N. (1988) Prog. Nucl. Acids Res. Mol. Biol. 35,
    173-207.
17 Lockard, R.E. and Lane, C. (1978) Nucl. Acids Res. 5, 2850-
    2854.
18 Muthukrishnan, S., Both, G.W., Furuichi, Y. and Shatkin,
    A.J. (1975) Nature 255, 33-37.
19 Rose, J.K. and Lodish, H.F. (1976) Nature 262, 32-37.
20 Banerjee, A.K. (1980) Microbiol. Rev. 44, 175-205.
21 Hennessy, K., Wang, F., Woodland Bushman, E. and Kieff, E.
    (1986) Proc. Nat. Acad. Sci. U.S.A. 83, 5693-5697.
22 Khalili, K., Brady, J. and Khoury, G. (1987) Cell 48, 639-
    645.
23 Hackett, P.B., Petersen, R.B., Hensel, C.H., Albericio, A.,
    Gunderson, S.I., Palmenberg, A.C. and Barany, G. (1986) J.
    Mol. Biol. 190, 45-57.
24 Thireos, B., Driscoll Penn, M. and Greer, H. (1984) Proc.
    Nat. Acad. Sci. U.S.A. 81, 5096-5100.
25 Mueller, P.P. and Hinnenbusch, A.B. (1986) Cell 45, 201-207.
26 Pelletier, J. and Sonenberg, N. (1987) Biochem. Cell. Biol.
    65, 576-581.
27 Jaeger, J.A., Turner, D.H. and Zuker, M. (1989) Proc. Nat.
    Acad. Sci. U.S.A. 86, 7706-7710.

28 Pavlakis, G.N., Lockard, R.E., Vamvakopoulos, N., Rieser, L., RajBhandary, U.L. and Vournakis, J.N. (1980) Cell 19, 91-102.

29 Muesing, M.A., Smith, D.H. and Capon, D.J. (1987) Cell 48, 691-701.

30 Pelletier, J. and Sonenberg, N. (1985) Mol. Cell. Biol. 5, 3222-3330.

31 Bornstein, P., McKay, J., Devarayalu, S. and Cook S.C. (1988) Nucl. Acids Res. 16, 9721-9736.

32 Bernards, A., Rubin, C.M., Westbrook, C.A., Paskind, M. and Baltimore, D. (1987) Mol. Cell Biol. 7, 3231-3236.

33 Ryder, K. and Nathans, D. (1988) Proc. Nat. Acad. Sci. U.S.A. 85, 8464-8467.

34 Betsholtz, C.A., Johnsson, A., Heldin, C-H., Westermark, B., Lind, P., Urdea, M.S., Eddy, R., Shows, T.B., Philpott, K., Mellor, A.L., Knott, T.J. and Scott, J. (1986) Nature 320, 695-699.

35 Rao, C.D., Pech, M., Tobbins, K.D. and Aaronson, S.A. (1988) Mol. Cell. Biol. 8, 6017-6936.

36 Ariz, N. and Munro, H.N. (1987) Proc. Nat. Acad. Sci. U.S.A. 89, 8478-8482.

37 Hentze, M.W., Caughman, S.W., Rouault, T.A., Barriocanal, J.G., Dancis, A., Harford, J.B. and Klausner, R.D. (1987) Science 238, 1570-1573.

38 Hentze, M.W., Rouault, T.A., Caughman, S.W., Dancis, A., Harford, J.B. and Klausner, R.D. (1987) Proc. Nat. Acad. Sci. U.S.A. 84, 6730-6734.

39 Casey, J.L., Hentze, M., Koeller, D.M., Caughman, S.W., Rouault, T.A., Klausner, R.D. and Harford, J.B. (1988) Science 240, 924-928.

40 Leibold, E.A. and Munro, H.N. (1988) Proc. Nat. Acad. Sci. U.S.A. 85, 2171-2175.

41 Rouault, T.A., Hentze, M.W., Caughman, S.W., Harford, J.B. and Klausner, R.D. (1988) Science 241, 1207-1210.

42 Rouault, T.A., Hentze, M.W., Haile, D.J., Harford, J.B. and Klausner, R.D. (1989) Proc. Nat. Acad. Sci. U.S.A. 86, 5768-5772.

43 Walden, W.E., Daniels-McQueen, S., Brown, P.H., Gaffield, L., Russell, D.A., Bielser, D., Bailey, L.C. and Thach, R.E. (1988) Proc. Nat. Acad. Sci. U.S.A. 85, 9503-9507.

44 Haile, D.J., Hentze, M.W., Rouault, T.A., Harford, J.B. and Klausner, R.D. (1989) Mol. Cell. Biol. 9, 5055-5061.

45 Hentze, M.W., Rouault, T.A., Harford, J.B. and Klausner, R.D. (1989) Science 244, 357-359.

46 Pelletier, J. and Sonenberg, N. (1988) Nature 334, 320-325.

47 Jang, S.K., Davies, M.V., Kaufman, R.J. and Wimmer, E. (1989) J. Virol. 63, 1651-1660.

48 Pilipenko, E.M., Blinov, V.M., Romanova, L.I., Sinyakov, A.N., Maslova, S.V. and Agol, V.I. (1989) Virology 168, 201-209.

49 Skinner, M.A., Racaniello, V.R., Dunn, G., Cooper, J., Minor, P.D. and Almond, J.W. (1989) J. Mol. Biol. 207, 379-392.

50 Jean-Jean, O., Weimer, T., Recondo, A-M., Will, H. and Rossignol, J-M. (1989) J. Virol. 63, 5451-5454.

51 Meerovitch, K., Pelletier, J. and Sonenberg, N. (1989) Genes Dev. (in press).

52 Sarnow, P. (1989) Proc. Nat. Acad. Sci. U.S.A. 86, 5795-5799.

53 Ting, J., Wooden, S.R., Kriz, R., Kelleher, K., Kaufman, R.J. and Lee, A.S. (1987) Gene 55, 147-152.

54 Munoz, A., Alonso, M.A. and Carrasco, L. (1984) Virology 137, 150-159.

55 Kozak, M. (1984) Nucl. Acids Res. 12, 857-872.

56 Kozak, M. (1986) Cell 44, 283-292.

57 Hann, S.R., King, M.W., Bentley, D.L., Anderson, C.W. and Eisenman, R.N. (1988) Cell 52, 185-195.

58 Prats, H., Kaghad, M., Prats, A.C., Klagsburn, M., Lelias, J., Liauzun, P., Chalon, P., Tauber, J.P., Amalric, F., Smith, J.A. and Caput, D. (1989) Proc. Nat. Acad. Sci. U.S.A. 86, 1836-1840.

59 Becerra, S.P., Rose, J.A., Hardy, M., Baroudy, B.M. and Anderson, C.W. (1985) Proc. Nat. Acad. Sci. U.S.A. 82, 7919-7923.

60 Curran, J. and Kolalofsky, D. (1988) EMBO J. 7, 245-251.

61 Peabody, D.S. (1987) J. Biol. Chem. 262, 11847-11851.

62 Kozak, M. (1989) J. Cell. Biol. 108, 229-241.

63 Dixon, L.K. and Hohn, T. (1984) EMBO J. 3, 2731-2736.

64 Hughes, S., Mellstrom, K., Koslik, E., Tamanoi, R. and Brugge, J. (1984) Mol. Cell. Biol. 4, 1738-1746.

65 Kozak, M. (1986) Cell 47, 481-483.

66 Mertz, J.E., Murphy, A. and Barkan, A. (1983) J. Virol. 45, 36-46.

67 Herman, R.C. (1986) J. Virol. 58, 797-804.

68 Peabody, D.S. and Berg, P. (1986) Mol. Cell. Biol. 6, 2695-2703.

69 Peabody, D.S., Subramani, S. and Berg, P. (1986) Mol. Cell. Biol. 6, 2704-2711.

70 Kaufman, R.J., Murtha, P. and Davies, M.V. (1987) EMBO J. 6, 187-193.

71 Kozak, M. (1987) Mol. Cell. Biol. 7, 3438-3445.

72 Wood, C.R., Davies, M.V., Wasley, L.C. and Kaufman, R.J. (1990) (submitted).

73 Kates, J. (1970) Cold Spring Harbor Symp. Quant. Biol. 35, 743-752.

74 Darnell, J.E., Wall, R. and Tushinski, R.J. (1971) Proc. Nat. Acad. Sci. U.S.A. 68, 1321-1325.

75 Sheiness, D. and Darnell, J.E. (1973) Nature New Biol. 241, 265-268.

76 Jeffery, W.R. and Brawerman, G. (1974) Biochemistry 13, 4633-4637.
77 Munroe, D. and Jacobson, A. (1990) Genes and Dev. (in press).
78 Sachs, A.B., Bond, M.W. and Kornberg, R.D. (1986) Cell 45, 827-835.
79 Mattaj, I.W. (1989) Cell 57, 1-3.
80 Sachs, A.B., Davis, R.W. and Kornberg, R.D. (1987) Mol. Cell. Biol. 7, 3268-3276.
81 Sachs, A.B. and Davis, R.W. (1989) Cell 58, 857-867.
82 Merrick, W.C. (1979) Methods Enzymol. 60, 101-108.
83 Schreier, M.H., Erni, B. and Staehelin, T. (1977) J. Mol. Biol. 116, 727-754.
84 Voorma, H.O., Thomas, A., Goumans, H., Amesz, H. and Vvan der Mast, C. (1979) Methods Enzymol. 60, 124-135.
85 Pain, V. (1986) Biochem. J. 235, 625-637.
86 Cigan, A.M., Pabich, E.K., Feng, L. and Donahue, T.F. (1989) Proc. Nat. Acad. Sci. U.S.A. 86, 2784-2788.
87 Donahue, T.F., Cigan, A.M., Pabich, E.K. and Valavicius, B.C. (1988) Cell 54, 621-632.
88 Williams, N.P., Hinnebusch, A.G. and Donahue, T.F. (1989) Proc. Nat. Acad. Sci. U.S.A. 86, 7515-7519.
89 Proud, C.G. (1986) Trends Biochem Sci. 11, 73-77.
90 Walton, G.M. and Gill, G.N. (1975) Biochim. Biophys. Acta 390, 231-245.
91 Rowlands, A.G., Panniers, R. and Henshaw, E.C. (1988) J. Biol. Chem. 263, 5526-5533.
92 Panniers, R. and Henshaw, E.C. (1983) J. Biol. Chem. 258, 7928-7934.
93 Konieczny, A. and Safer, B. (1983) J. Biol. Chem. 258, 3402-3408.
94 Matts, R.L., Levin, D.H. and London, I.M. (1983) Proc. Nat. Acad. Sci. U.S.A. 80, 2559-2563.
95 Siekierka, J., Manne, V. and Ochoa, S.(1984) Proc. Nat. Acad. Sci. U.S.A. 81, 352-356.
96 Ochoa, S. (1983) Arch. Biochem. Biophys. 223, 325-349.
97 Minks, M.A., West, D.K., Benvin, S. and Baglioni, C. (1979) J. Biol. Chem. 254, 10180-10183.
98 Katze, M.G., DeCorato, D., Safer, B., Galabru, J. and Hovanessian, A.G. (1987) EMBO J. 6, 689-697.
99 Pestka, S., Langer, J.A., Zoon, K.C. and Samuel, C.E. (1987) Annu. Rev. Biochem. 56, 727-778.
100 O'Malley, R.P., Mariano, T.M., Siekierka, J. and Mathews, M.B. (1986) Cell 44, 391-400.
101 Kitajewski, J., Schneider, R.J., Safer, B., Munemitsu, S.M., Samuel, C.E., Thimmappaya, B. and Shenk, T. (1986) Cell 45, 195-200.
102 Reichel, P.A., Merrick, W.C., Siekierka, J. and Mathews, M. (1985) Nature 313, 196-200.
103 Rice, A.P. and Kerr, I.M. (1984) J. Virol. 50, 229-236.

104 Katze, M.G., Tomita, J., Black, R., Krug, R.M., Safer, B. and Hovanessian, A.J. (1988) J. Virol. 62, 3710-3717.
105 Imani, F. and Jacobs, B.L. (1988) Proc. Nat. Acad. Sci. U.S.A. 85, 7887-7891.
106 Giantini, M. and Shatkin, A.J. (1989) J. Virol. 63, 2415-2421.
107 Thimmappaya, B., Weinberger, C., Schneider, R.J. and Shenk, T. (1982) Cell 31, 543-551.
108 Kostura, M. and Mathews, M.B. (1989) Mol. Cell. Biol. 9, 1576-1586.
109 Schneider, R.J., Safer, B., Munemitsu, S.M., Samuel, C.E. and Shenk, T. (1985) Proc. Nat. Acad. Sci. U.S.A. 82, 4321-4325.
110 O'Malley, R.P., Duncan, R.F., Hershey, J.W.B. and Mathews, M.B. (1989) Virology 169, 112-118.
111 Davies, M.V., Furtado, M., Hershey, J.W.B., Thimmappaya, B. and Kaufman, R.J. (1989) Proc. Nat. Acad. Sci. U.S.A. 86, 9163-9167.
112 Akusjarvi, G., Svensson, C. and Nygard, O. (1987) Mol. Cell. Biol. 7, 549-551.
113 Svensson, C. and Akusjarvi, G. (1984) Mol. Cell. Biol. 4, 736-742.
114 Kaufman, R.J. (1985) Proc. Nat. Acad. Sci. U.S.A. 82, 689-693.
115 Kaufman, R.J. and Murtha, P. (1987) Mol. Cell. Biol. 7, 1568-1571.
116 Kaufman, R.J., Davies, M.V., Pathak, V.K. and Hershey, J.W.B. (1989) Mol. Cell. Biol. 9, 946-958.
117 Whitaker-Dowling, P. and Younger, J.S. (1984) Virology 137, 171-181.
118 Colby, C. and Duesberg, P.H. (1969) Nature 222, 940-944.
119 Condit, R.A. and Motyczka, G. (1981) Virology 113, 224-241.
120 Spehner, D., Gillard, S., Drillien, R. and Kirn, A. (1988) J. Virol. 62, 1297-1304.
121 Gillard, S., Spehner, S., Drillien, R. and Kirn, A. (1986) Proc. Nat. Acad. Sci. U.S.A. 83, 5573-5577.
122 Wiebe, M.E. and Joklik, W.K. (1975) Virology 122, 381-391.
123 Munemitsu, S.M. and Samuel, C.E. (1984) Virology 136, 133-143.
124 Edery, I., Petryshyn, R. and Sonenberg, N. (1989) Cell 56, 303-312.
125 SenGupta, D.N. and Silverman, R.H. (1989) Nucl. Acids Res. 17, 969-978.
126 Gatignal, A., Kumar, A., Rabson, A. and Jeang, K-T. (1989) Proc. Nat. Acad. Sci. U.S.A. 86, 7828-7832.
127 Sharp, P.A. and Marciniak, R.A. (1989) Cell 59, 229-230.
128 Cullen, B.N. (1986) Cell 46, 973-982.
129 Rosen, C.A., Sodroski, J.G., Goh, W.C., Dayton, A.F., Lippke, J. and Haseltine, W.A. (1986) Nature 319, 355-359.

130 Braddock, M., Chambers, A., Wilson, W., Esnouf, M.P., Adams, S.E., Kingsman, A.J. and Kingsman, S.M. (1989) Cell 58, 269-279.

131 Zinn, K., Keller, A., Whittemore, L.A. and Maniatis, T. (1988) Science 240, 210-213.

132 De Beneddetti, A. and Baglioni, C. (1986) J. Biol. Chem. 261, 338-342.

133 Duncan, R.F. and Hershey, J.W.B. (1984) J. Biol. Chem. 259, 11882-11889.

134 Duncan, R.F. and Hershey, J.W.B. (1989) J. Cell. Biol. 109, 1467-1481.

135 Duncan, R.F. and Hershey, J.W.B. (1985) J. Biol. Chem. 260, 5493-5497.

136 Scorsone, K.A., Panniers, R., Rowlands, A.G. and Henshaw, E.C. (1987) J. Biol. Chem. 262, 14538-14543.

137 Clemens, M.J., Galpine, A., Austin, S.A., Panniers, R., Henshaw, E.C., Duncan, R., Hershey, J.W.B. and Pollard, J. (1987) J. Biol. Chem. 262, 767-771.

138 Hurst, R., Schatz, J.R. and Matts, R.L. (1987) J. Biol. Chem. 262, 15939-15945.

139 Duncan, R.F. and Hershey, J.W.B. (1987) Mol. Cell. Biol. 7, 1293-1295.

140 Duncan, R.F. and Hershey, J.W.B. (1987) Arch. Biochem. Biophys. 256, 651-661.

141 Hershey, J.W.B. (1989) J. Biol. Chem. 264, 20823-20826.

142 Crouch, D. and Safer, B. (1980) J. Biol. Chem. 255, 7918-7924.

143 Pato, M., Alstom, R.S., Crouch, D., Safer, B., Ingebritsen, R.S. and Cohen, P. (1983) Eur. J. Biochem. 132, 283-287.

144 Datta, B., Chakrabarti, D., Roy, A.L. and Gupta, N.K. (1988) Proc. Nat. Acad. Sci. U.S.A. 85, 3324-3328.

145 Datta, B., Ray, M.R., Chakrabarti, D., Roy, A.L. and Gupta, N.K. (1989) J. Biol. Chem. 264, 20620-20624.

146 London, I.M., Levin, D.H., Matts, R.L., Thomas, N.S.B., Petryshyn, R. and Chen, J-J. (1987) in The Enzymes (Boyer, P.D. and Krebs, E.G., eds.) 3rd Ed., Vol. 17, pp. 359-380, Academic Press, New York, NY.

147 Chen, J-J., Yang, J.M., Petryshyn, R., Kosower, N. and London, I.M. (1989) J. Biol. Chem. 264, 9559-9564.

148 Matts, R.L. and Hurst, R. (1989) J. Biol. Chem. 264, 15542-15547.

149 Kelley, P.M. and Schlesinger, M.J. (1982) Mol. Cell. Biol. 2, 267-274.

150 Szyszka, R., Kramer, G. and Hardesty, B. (1989) Biochemistry 28, 1435-1438.

151 Lindquist, S. (1986) Annu. Rev. Biochem. 55, 1151-1191.

152 Benne, R. and Hershey, J.W.B. (1978) J. Biol. Chem. 253, 3078-3087.

153 Trachsel, H., Erni, B., Schreier, M.H. and Staehelin, T. (1977) Mol. Cell. Biol. 116, 755-767.

154 Abramson, R.D., Dever, T.E., Lawson, T.G., Ray, K.B., Thach, R.E. and Merrick, W.C. (1987) J. Biol. Chem. 262, 3826-3832.

155 Grifo, J.A., Tahara, S.M., Leis, J.P., Morgan, M.A., Shatkin, A.J. and Merrick, W.C. (1982) J. Biol. Chem. 257, 5246-5252.

156 Lee, K.A.W., Guertin, D. and Sonenberg, N. (1983) J. Biol. Chem. 258, 707-710.

157 Ray, B.K., Lawson, T.G., Kramer, J.C., Cladaras, M.H., Grifo, J.A., Abramson, R.D., Merrick, W.C. and Thach, R.E. (1985) J. Biol. Chem. 260, 7651-7658.

158 Sonenberg, N. and Shatkin, A.J. (1977) Proc. Nat. Acad. Sci. U.S.A. 74, 4288-4292.

159 Ray, B.K., Brendler, T.G., Adya, S., Daniels-McQueen, S., Miller, J.K., Hershey, J.W.B., Grifo, J.A., Merrick, W.C. and Thach, R.E. (1983) Proc. Nat. Acad. Sci. U.S.A. 80, 663-667.

160 Sonenberg, N., Guertin, D., Cleveland, D. and Trachsel, H. (1981) Cell 27, 563-572.

161 Sonenberg, N. (1981) Nucl. Acids Res. 9, 1643-1656.

162 Edery, I., Humbelin, M., Darveau, A., Lee, K.A.W., Milburn, S., Hershey, J.W.B., Trachsel, H. and Sonenberg, N. (1983) J. Biol. Chem. 258, 11398-11403.

163 Grifo, J.A., Tahara, S.M., Morgan, M.A., Shatkin, A.J. and Merrick, W.C. (1983) J. Biol. Chem. 258, 5804-5810.

164 Hiremath, L.S., Hiremath, S.T., Rychlik, W., Joshi, S., Domier, L.L. and Rhoads, R.E. (1989) J. Biol. Chem. 264, 1132-1138.

165 Lee, K.A.W., Edery, I. and Sonenberg, N. (1983) J. Virol. 54, 515-524.

166 Rychlik, W., Domier, L.L., Gardner, P.R., Hellman, G.M. and Rhoads, R.E. (1987) Proc. Nat. Acad. Sci. U.S.A. 84, 945-949.

167 Fields, S. and Winter, G. (1982) Cell 28, 303-313.

168 Ishida, T., Katsuta, M., Inoue, M., Yamagata, Y. and Tomita, K. (1983) Biochem. Biophys. Res. Commun. 115, 849-854.

169 Altman, M., Handschin, C. and Trachsel, H. (1987) Mol. Cell. Biol. 7, 998-1003.

170 Duncan, R.F., Milburn, S.C. and Hershey, J.W.B. (1987) J. Biol. Chem. 262, 380-388.

171 Panniers, R., Steward, E.B., Merrick, W.C. and Henshaw, E.C. (1985) J. Biol. Chem. 260, 9648-9653.

172 Bonneau, A-M. and Sonenberg, N. (1987) J. Biol. Chem. 262, 11134-11139.

173 Rychlik, W., Russ, M.A. and Rhoads, R.E. (1987) J. Biol. Chem. 262, 10434-10437.

174 Grifo, J.A., Abramson, R.D., Salter, C.A. and Merrick, W.C. (1984) J. Biol. Chem. 259, 8648-8654.

175 Sarkar, G., Edery, I. and Sonenberg, N. (1985) J. Biol. Chem. 260, 13831-13837.

176 Rozen, F., Edery, I., Meerovitch, K., Dever, T.E., Merrick, W.C. and Sonenberg, N. (1990) Mol. Cell. Biol. (in press).

177 Linder, P., Lasko, P.P., Ashburner, M., Leroy, P., Nielsen, P.J., Nishi, K., Schnier, J. and Slonimski, P.P. (1989) Nature 337, 121-122.

178 Walker, J.E., Saraste, M., Runswick, M.J. and Gay, N.J. (1982) EMBO J. 1, 945-951.

179 Rozen, F., Pelletier, J., Trachsel, H. and Sonenberg, N. (1989) Mol. Cell. Biol. 9, 4061-4063.

180 Nielsen, P.J. and Trachsel, H. (1988) EMBO J. 7, 2097-2105.

181 Linder, P. and Slonimski, P.P. (1989) Proc. Nat. Acad. Sci. U.S.A. 86, 2286-2290.

182 Abramson, R.D., Dever, T.E. and Merrick, W.C. (1988) J. Biol. Chem. 263, 6016-6019.

183 Ray, B.K., Lawson, T.G., Abramson, R.D., Merrick, W.C. and Thach, R.E. (1986) J. Biol. Chem. 261, 11466-11470.

184 Butler, J.S. and Clark, J.M. (1984) Biochemistry 23, 809-815.

185 Goss, D.J., Woodley, C.L. and Wahba, A.J. (1987) Biochemistry 26, 1151-1156.

186 Milburn, S.C., Hershey, J.W.B., Kelleher, K. and Kaufman, R.J. (1990) (submitted).

187 Pelletier, J. and Sonenberg, N. (1985) Mol. Cell. Biol. 5, 3222-3230.

188 Morley, S.J. and Traugh, J.A. (1989) J. Biol. Chem. 264, 2401-2404.

189 Tuazon, P., Merrick, W.C. and Traugh, J.A. (1988) J. Biol. Chem. 264, 2773-2777.

190 Salimans, M.M., Van Heugten, H.A.A., Van Steeg, H. and Voorma, H.O. (1985) Biochim. Biophys. Acta 824, 16-26.

191 Etchison, D., Milburn, S.C., Edery, I., Sonenberg, N. and Hershey, J.W.B. (1982) J. Biol. Chem. 257, 14806-14810.

192 Etchison, D. and Fout, S. (1985) J. Virol. 54, 634-638.

193 Lloyd, R.E., Toyoda, H., Etchison, D., Wimmer, E. and Ehrenfeld, E. (1986) Virology 150, 299-303.

194 Toyoda, H., Nicklin, M.J.H., Murray, M.G., Anderson, C.W., Dunn, J.J., Studier, F.W. and Wimmer, E. (1986) Cell 45, 761-770.

195 Wimmer, E. (1972) J. Mol. Biol. 68, 536-542.

196 Tahara, S.M., Morgan, M.A. and Shatkin, A.J. (1981) J. Biol. Chem. 256, 7691-7694.

197 Sonenberg, N. (1987) Adv. Virus Res. 33, 175-204.

198 Bonneau, A-M. and Sonenberg, N. (1987) J. Virol. 61, 986-991.

199 Davies, M.V., Sonenberg, N. and Kaufman, R.J. (1990) (submitted).

200 Bernstein, H., Sonenberg, N. and Baltimore, D. (1985) Mol. Cell. Biol. 5, 2913-2923.

201 O'Neill, R.E. and Racaniello, V.R. (1989) J. Virol. 63, 5069-5075.

202 Black, T.L., Safer, B., Hovenessian, A. and Katze, M.G. (1989) J. Virol. 63, 2244-2251.

203 Ransone, L.J. and Dasgupta, A. (1988) J. Virol. 62, 3551-3557.

205 Yen, T.J., Machlin, P.S. and Cleveland, D.W. (1988) Nature 305, 580-585.

206 Gay, D.A., Sisodia, S.S. and Cleveland, D.W. (1989) Proc. Nat. Acad. Sci. U.S.A. 86, 5763-5767.

207 Graves, R.A., Pandley, N.B., Chodchoy, N. and Marzluff, W.F. (1987) Cell 48, 615-626.

208 Shaw, G. and Kamen, R. (1986) Cell 46, 659-667.

209 Shyu, A.B., Greenberg, M.E. and Belasco, J.G. (1989) Genes Dev. 3, 60-72.

210 Wilson, R. and Triesman, R. (1988) Nature 336, 396-399.

211 Bauman, B., Potash, M.J. and Kohler, G. (1985) EMBO J. 4, 2056-2060.

212 Baserga, S.J. and Benz, Jr., E.J. (1988) Proc. Nat. Acad. Sci. U.S.A. 85, 557-561.

213 Daar, I.O. and Maquat, L.E. (1988) Mol. Cell. Biol. 8, 802-813.

214 Urlaub, G., Mitchell, P.J., Ciudad, C.J. and Chasin, L.A. (1989) Mol. Cell. Biol. 9, 2868-2880.

215 Zasloff, M. (1983) Proc. Nat. Acad. Sci. U.S.A. 80, 6436-6440.

216 Unwin, P.N.T. and Milligan, R.A. (1982) J. Cell. Biol. 93, 63-75.

217 Blobel, G. (1985) Proc. Nat. Acad. Sci. U.S.A. 82, 8527-8531.

218 Dworetzky, S.J. and Feldherr, C.M. (1988) J. Cell Biol. 106, 575-584.

219 Muralidhar, M.G. and Johnson, L.F. (1988) J. Cell Physiol. 135, 115-121.

220 Safer, B. (1989). Eur. J. Biochem. 186, 1-3.

221 Roy, S., Katze, M.G., Edery, I., Hovanessian, A.G. and Sonenberg, N. (1990) Science (in press).

222 Josh-Barue, S., Rychlik, W. and Rhoads, R.E. (1990) J. Biol. Chem. 265 (in press).

ELECTROPORATION OF BACTERIA:

A GENERAL APPROACH TO GENETIC TRANSFORMATION

William J. Dower

Department of Molecular Biology
Affymax Research Institute
Palo Alto, CA 94304

INTRODUCTION

Studies on the molecular genetics of many prokaryotic species have been limited by the lack of convenient and reliable methods of gene transfer. Until recently, for those working with many types of bacteria, the only route to genetic transformation of their favorite species was via conjugation or protoplast transformation, methods which are often tedious, irreproducible and usually applicable only to certain strains. In the last few years, a general method for transforming intact bacteria has become available. This method is known as electroporation or electro-transformation, and numerous reports have shown it to be effective for a wide variety of prokaryotic species. When a mixture of cells and transforming DNA is subjected to a brief electrical pulse of rather high voltage, a substantial portion of the cells is usually killed; but from among the survivors, transformants are recovered at a frequency often much higher than that obtained with the same species transformed by other methods (1-10).

Our early work relied on E. coli and Campylobacter jejuni as models for developing general methods of bacterial electroporation (2,4,11), and the information gathered from these model species and certain gram positive species (1,12) has been successfully applied to a large number of other prokaryotes. (Chassy, et al., provide an interesting account of the history of these developments in Ref. 13.)

An unexpected finding of the work with E. coli was the observation of unprecedentedly high levels of transformation obtained by electroporation. Efficiencies exceeding 10^{10} transformants per μg of plasmid DNA, frequencies of up to 80% of the

surviving cells and DNA capacities of nearly 10 μg of transforming DNA per ml are possible with very simple electro-
transformation procedures.

In this chapter, I discuss some of the theoretical aspects
of electroporation that especially pertain to bacterial cells,
the effects of certain physical and biological variables and the
practical use of electroporation to transform bacteria.

ELECTROPORATION THEORY

When cells suspended in an electrolyte are exposed to an
electric field, a potential forms across the regions of the
membrane facing the electrodes. The magnitude of this potential
depends on the voltage drop across the length of the cell; this,
in turn, depends on the size of the cell and the strength of the
field. When a certain critical potential is reached, the
membrane is sufficiently perturbed to become permeable to a
variety of compounds normally excluded from the cell. These
compounds include charged species and macromolecules (14,15). As
the electric field and the transmembrane potential increase, more
and more cells experience irreversible damage and cannot be
recovered as colony-forming units. For animal cells or plant
protoplasts this simplistic description of membrane permeabilization leading to the entry of macromolecules is fairly easy to
accept, and the mechanisms not hard to imagine. However, with
bacterial cells possessing complex enveloping structures, it is
more difficult to envision how reversible membrane permeabilization leads to the uptake of macromolecules. Enlightening data on
this process are sparse, and we are far from understanding how it
actually occurs.

THE ELECTRICAL PULSE

Shape of the Pulse

Pulses of several shapes, or waveforms, have been used to
transform bacteria. Square (16), truncated sine (17), and
exponential decay (1-12,16-25) pulses are effective. Because of
the commercial availability of relatively inexpensive capacitor
discharge devices, the exponential decay waveform is, by far, the
most commonly used pulse.

A diagram of an exponential decay pulse is shown in Figure
1. When the charge stored in a capacitor is directed to electrodes flanking a sample of suspended cells, the potential across
the sample rises rapidly to a peak voltage, V_0. The voltage of
the pulse is taken as this peak voltage. As the capacitor
discharges through the conductive sample, the voltage declines
over time as a function of the size of the capacitor and the

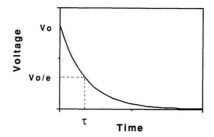

Figure 1. Exponential decay waveform.

resistance in the circuit. The length of the pulse is described
by τ, the RC time constant. τ is the time over which the voltage
declines from V_0 to V_0/e (= 0.37 V_0), and may be calculated as τ
= RC (where τ is time in seconds, R is resistance in ohms, and C
is capacitance in farads).

THE ELECTRIC FIELD

Strength of the Field

The most important electrical effectors in the permeabilization of a cell are the amplitude and duration of the electric
field. The field may be thought of as a voltage gradient between
the electrodes. That is, the field, E, equals the voltage
difference across the electrodes divided by the distance between
the electrodes (E = V/d). The amplitude of the field may be
adjusted by varying the imposed voltage or the inter-electrode
distance. Fields up to 1 kV/cm are often used to electroporate
cultured mammalian cells, conditions that might be obtained by
applying 400 V across a sample placed between electrodes 0.4 cm
apart. Because bacterial cells are much smaller than mammalian
cells they experience a smaller voltage drop in a given field and
should require greater fields to reach the critical transmembrane
potential to become permeabilized. In practice, this is the
case: intense fields of 5 to 20 kV/cm are usually required for
bacterial electroporation.

Shape of the Field and Electrode Configuration

The usual electrode arrangement consists of two parallel,
conductive plates a fixed distance apart. This parallel plate
geometry produces fields that are roughly uniform; that is, most
cells will experience the same field regardless of their location
in the sample. For most purposes of electro-transformation,
uniform fields are preferred.

Electrodes with gaps of 0.05 to 1.0 cm are available
commercially. Narrow gap cuvettes are capable of producing very

high fields with pulse generators of limited voltage output;
however, gaps of less than 0.1 cm are difficult to work with and
limit the sample volume. We prefer electrodes with 0.1 to 0.2 cm
gaps for bacterial electroporation. Several materials have been
used for electrode construction. These include aluminum, gold,
platinum and stainless steel. Of these, I am aware of no
evidence of differences in transformation or viability attributable to the electrode material.

THE PULSE GENERATOR

There are many pulse generators designed specifically for
electroporation available from commercial sources (reviewed by
Potter (26)), but the choice of commercial exponential-pulse
generators capable of the high voltages necessary for transforming most bacteria is limited. Most reports to date describe the
use of the Bio-Rad Gene Pulser, a device capable of up to 25
kV/cm with 0.1 cm electrodes. A detailed description of the Bio-
Rad device, as set up specifically for electro-transformation, is
found in the appendix to Ref. 4. Another commercial device
useful for bacterial electroporation is the BTX Transfector-100,
capable of producing fields of 18 kV/cm with 0.05 cm electrodes.
A sophisticated homemade device specifically designed for
bacterial electroporation using a sine wave pulse is described in
Ref. 17. A more detailed description of the relevant electrical
theory and equipment for electroporation of bacteria is found in
Refs. 4 and 27.

EXPERIMENTAL CONDITIONS FOR ELECTROPORATION

Physical Variables

The most important electrical variables are field strength
and pulse length. This was clearly demonstrated with
Campylobacter jejuni (2) and E. coli (4,17,22). With pulses of a
few milliseconds, the field was raised from 1 kV/cm to greater
than 10 kV/cm and the recovery of transformants increased
greatly. With longer pulses, this dependence on the field
strength is also seen, but the peak of maximum transformation is
shifted to a lower field. Thus, a longer pulse may be used to
compensate for a lower field. In our hands this range of
compensation is limited. We have not obtained high efficiency
transformation of E. coli with fields less than 3 kV/cm even with
very long pulses of almost 1 second. Jessee (23) has reported
rather high transformation efficiencies ~10^9/μg) by low voltage
electroporation of E. coli that have been treated with PEG.
    Recently we have extended our studies of voltage dependence
to fields up to 25 kV/cm, and the results are shown in Figure 2.

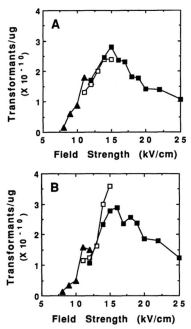

Figure 2. Effect of field strength on the electro-transformation
of cells grown at 37° and 44°. MC1061 cells were grown in L-
broth at (A) 37°C or (B) 44°C and transformed in cuvettes with
inter-electrode distances of 0.2 cm (▲) or 0.1 cm (□, ■) with
pulses of 4.5 to 5.0 msec.

In this experiment, the recovery of transformants continues to
increase to a maximum at a field of 15 kV/cm. At this field
strength, the transformation efficiency is 3 to 4 x 10^{10} trans-
formants/µg of pUC18 DNA. In contrast to the effect we show
here, our previous report described a plateau and decline in
transformation at fields from 10 to 15 kV/cm (4). The higher
field strengths in the earlier report were produced with a spark
plug-style mini-electrode. The results shown in Figure 2 were
obtained with a 0.1 cm cuvette-style electrode. The geometries
of these two electrodes are quite different and this could
explain the different effects we have observed. For example, the
cuvettes have a much greater mass of chilled metal that would
more rapidly remove heat generated by the pulse. An effect like
this would be more noticeable at higher voltages where the
greater currents cause more heating of the sample.

Electroporation Medium

The overriding consideration in the choice of medium for
bacterial electroporation is its conductivity, a property closely
related to the ionic strength. At the high voltages necessary

for permeabilizing bacteria, medium that is too conductive will cause explosive arcing in the cuvette with the annoying result of loss of the sample. Many species may tolerate washing and suspension in a medium of very low ionic strength such as water. The practice of electroporation at low temperature (recommended for other reasons--see below) further reduces the incidence of electrical volatility. For cells requiring the presence of some ionic species, the empirical approach of finding the lowest ionic strength compatible with the well-being of the cells, and making some compromise in the amplitude of the electrical field is necessary. Glycerol (10%) may be used as a cryoprotectant to preserve frozen aliquots of "electro-competent" cells of many types. E. coli stored in this way have nearly the same competence as freshly prepared cells and may be frozen and thawed several times with little loss of activity.

For our earliest attempts at bacterial electroporation, we employed a buffered sucrose solution for no better reason than that it was the low conductivity, iso-osmotic medium we used for electroporation of cultured mammalian cells (11). We soon found that sucrose is not necessary for electro-transforming E. coli. We also found that increasing the osmolarity of the medium with various agents (sucrose, mannitol, sorbitol) has no effect on E. coli electro-transformation until, at high concentrations (30 to 50%) the viability of the cells is reduced and recovery of transformants declines commensurately. These experiments rendered unlikely one of our proposed mechanisms of electroporation, the bulk flow model, in which a substantial volume of the medium enters the permeabilized cells driven by an osmolarity difference.

Several studies have described effects of divalent cations on electroporation. Transformation of Campylobacter jejuni is significantly inhibited by 1 mM of Ca^{+2}, Mn^{+2}, or Mg^{+2} (2); E. coli transformation is inhibited by $MgCl_2$ from 1 mM (50%) to 8 mM (>90%) (24); EDTA however, does not seem to be stimulatory (Dower, unpublished results).

Electroporation Temperature

Besides providing a beneficial effect on electrical volatility of the sample, low temperature greatly improves the efficiency of electro-transformation of many, perhaps most, bacteria. Figure 3 shows the effect of temperatures from -7 to +25° on the electroporation of E. coli. We observed a plateau of high efficiency from -4 to +4° with markedly decreasing transformation as the temperature was raised above +4°. It is likely that this inflection point is related to the state of the membrane and will vary with different species or strains. We have only examined one strain of E. coli (WM1100) at this resolution, but for many bacteria electro-transformation is much more efficient at 0° than at room temperature.

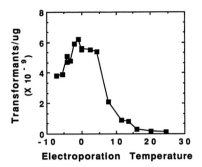

Figure 3. Effect of electroporation temperature on the recovery
of transformants. Electroporation cuvettes and samples of WM1100
cells (in 10% glycerol) were equilibrated to temperatures of-
7.2°C to +24.5°C and pulsed immediately at 12.5 kV/cm, 4.5 to 5.0
msec.

                                    DNA

        The concentration of transforming DNA present during
electroporation is directly related to the proportion of cells
that become transformed. With E. coli, this relationship holds
over at least 6 logs, and at high DNA concentrations nearly 80%
of the surviving cells are transformed (4). This is very
different from the transformation of chemically-treated competent
cells where saturation occurs at DNA concentrations about 100-
fold lower, and where a much smaller fraction of the cells are
competent to become transformed. The difference in DNA capacity
may derive from the difference the role DNA binding plays in the
two types of transformation. Chemically-mediated transformation
follows binding of the DNA to the cells (28), and non-transforming DNA competes for the limited number of binding sites.
Binding to a limited number of sites seems not to occur in
electroporation (4,17), although non-specific interaction at a
very large number of sites may occur and serve to increase the
effective concentration of DNA in the vicinity of each cell.
        The extremely high efficiencies and frequencies of electro-
transformation observed with a few species may be unusual. A
number of genera such as lactobacilli and streptococci show a
somewhat lower DNA capacity and a much smaller sub-population of
electro-competent cells (1 to 10%) (12,13).
        The effect of size and conformation of the plasmid DNA has
been tested in several species. Powell et al. have measured the
activity of plasmids from 4.4 to 26 kb in electro-transformation
of lactic streptococci and reported no relation of size to
transforming activity (12). McIntytre and Harlander electro-
transformed Lactococcus lactis with plasmids of 9 to 30 kb at
similar efficiencies (21). By far the largest molecule reported

Table 1
Effect of Plasmid Size and Conformation on
Electro-Transforming Activity

DNA	Size (kbp)	Form	Transformation Efficiency Tfs/µg	Tfs/pMol
pUC18	2.7	SC	6.1×10^9	1.1×10^{10}
		OC	6.5×10^9	1.2×10^{10}
pBR329	4.2	SC	3.5×10^9	9.7×10^9
		OC	3.1×10^9	8.6×10^9
M13mp18RF	7.3	SC	2.1×10^9	1.0×10^{10}
pLAFR3	21	SC	6.0×10^8	8.3×10^9
		OC	5.8×10^9	8.1×10^9

SC, supercoiled; OC, open circles.

to be introduced into bacteria by electroporation is a 200 kb plasmid transformed into Bacillus cereus. This huge plasmid transformed this species to about the same frequency as did a much smaller 3 kb plasmid (19). We have measured the activity of supercoiled plasmids of 2.7 to 21 kb in electro-transforming E. coli and found all to produce the same molar efficiency. Furthermore, converting the plasmids to relaxed circular form did not affect their transforming activity (Table 1). Although there are some differences besides size distinguishing these plasmids, these results suggest that the probability that a plasmid will transform a cell during electroporation is unrelated to its size or degree of supercoiling up to at least 20 kb and possibly much more. In contrast, Hanahan has shown chemically-mediated transforming activity to be inversely related to plasmid size (29).

Cell Concentration

Because the proportion of cells that become transformed is dependent on the DNA concentration, we attempted to obtain more transformants by increasing the number of cells present. We found that our recovery of transformants rose with increasing cell concentration up to 3 to 5×10^{10} cells/ml. With the low DNA concentrations typical of most cloning applications the yield of transformants plateaus when the cell concentration reaches about 5×10^{10} cells/ml (4, and unpublished data). It is clear from these data that the transformation efficiency may be adjusted by changing the cell concentration, the very highest efficiencies requiring 3 to 5×10^{10} cells/ml and lower efficiencies for subcloning obtained by diluting the cells for more economical use.

The tactic of increasing the cell concentration to obtain higher efficiencies is applicable to other species as well, but there are cases when higher cell concentration may be detrimental. Some cells secrete nucleases and increasing the number of cells may create problems with vector degradation. In preparations containing many lysed cells there may also be nucleases present that can cause problems at higher cell concentrations.

BIOLOGICAL VARIABLES

Growth of Electro-Competent Cells

A most important feature of the preparation of E. coli for electroporation is the timing of the harvest. Figure 4 displays the effect of harvesting the cells at various times during growth from early exponential through stationary phase. The electrocompetence of the cells declines substantially as they pass into late exponential and the growth of the population slows. In other species too, cells in early to mid exponential growth are more efficiently transformed than those in late exponential and stationary phase (2). This effect, while probably common to many species, may not be universal. McIntyre and Harlander have reported that preparations of Lactococcus lactis are best transformed when taken from stationary cultures (21). It may be that for some species, gram positives in particular, the changes in the cell wall that occur during stationary phase may improve the electrically-induced uptake of DNA.

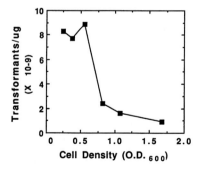

Figure 4. Electro-transformation of cells harvested at various stages of growth. WM1100 cells were grown in L-broth and harvested at OD_{600} of 0.25 to 1.7. After preparation as described (4), the samples were adjusted to the same cell concentration and pulsed with pUC18 DNA at 12.5 kV/cm, 4.5 to 5.0 msec.

We have prepared electro-competent E. coli from a variety of media that would be described as rich or very rich, and we have seen little difference in the transformability of the cells. Cells grown in a very rich medium such as super broth (3.2% tryptone, 2.0% yeast extract, 0.5% NaCl) remain in exponential growth at a higher density and produce a better yield of cells of very high electro-competence.

The temperature of growth has been reported by Taketo to affect the competence of E. coli. Cells grown at 44° were electro-transformed to efficiencies 5- to 10-fold higher than those grown at 37° (24). The electro-transformations in this report were done at a single, relatively low field strength of 6.25 kV/cm. We have repeated these experiments transforming at several field strengths in the higher range and found that the higher growth temperature may shift the voltage dependence curve slightly, but does not significantly improve transformation at the optimal field for our tester strain, MC1061 (see Figure 2).

Strains of Cells

The efficiencies of electro-transformation vary considerably among different strains of E. coli. A notable difference is that between Rec$^+$ and recA strains. In Table 2 a comparison of MC1061 and WM1100 (an otherwise isogenic, Rec$^+$/recA pair) shows transformation of the recA strain to be several times less efficient than the parental Rec$^+$ strain. The frequencies of transformation (calculated as transformants/survivor) of the two strains are

Table 2
Electroporation of E. coli Rec$^+$ and recA

Strain	DNA	Efficiency Tfs/µg	Frequency Tfs/survivor
MC1061 (Rec$^+$)			
Batch 1	pUC18	2.5×10^{10}	4.2×10^{-4}
Batch 2	pUC18	2.9×10^{10}	3.1×10^{-4}
WM1100 (recA)			
Batch 1	pUC18	7.5×10^9	3.5×10^{-4}
Batch 2	pUC18	9.0×10^9	3.0×10^{-4}

WM1100 is a recA derivative of MC1061.
Each 40 µl sample of cells was transformed with 4 pg of plasmid DNA by pulsing 5 msec at 12.5 kV/cm. Transformation efficiency was calculated as transformants per µg of plasmid DNA, and transformation frequency was calculated as transformants per surviving cell.

nearly identical, and the difference in transformation efficiency is entirely attributable to the reduced viability (2- to 3-fold) of the recA strain.

With some other species, differences in the electro-competence between strains has been reported to be several orders of magnitude (3,8,11). The reason for this is not known, but may have to do with plasmid establishment, maintenance, or expression rather than the particular electroporation conditions.

Cell Walls and Capsules

The present accumulation of evidence leaves one with the strong impression that gram negative cells are generally more receptive to the electrically induced uptake of DNA. The thicker, denser cell wall of gram positive organisms may be largely responsible for this difference. Consistent with this possibility is the observation by Powell et al. that a very gentle disruption of the cell wall of lactic streptococci increases by up to 100-fold the recovery of electro-transformants (12).

Other components of the cell envelope might be expected to affect the transit of macromolecules from the exterior to the interior of a cell. Differences in the structure of the LPS coat of Salmonella have a pronounced effect on the competence of the cells for chemically-mediated transformation (29), but these differences seem not to influence the electro-transformation of the cells (K. Sanderson, unpublished). Klebsiella pneumoniae with an extensive mucopolysaccharide capsule were electro-transformed at a reasonably high efficiency (~10^6/μg pBR322). Removal of the capsule by modified growth conditions did not improve this efficiency (V. Piantinida and F. Bayliss, unpublished). While it seems likely that some types of capsules will impede the entry of DNA, the presence of extensive envelope structures does not necessarily inhibit electro-transformation.

State of the Cells Following the Pulse

E. coli cells are quite unstable immediately after the pulse, and their viability and transformability are greatly enhanced by the rapid addition of growth media (4). The cells are not permeable to DNA during this time as DNA added within a few seconds following a pulse has little or no transforming activity (W.J. Dower, unpublished). This period of vulnerability has not been reported for other species, but until known otherwise for a given species, the rapid addition of outgrowth medium following the pulse is prudent. As is typical for conventional transformation systems, a period of incubation allowing expression of the marker products is usually required before positive selection is applied. This expression, or outgrowth, period is

0.5 to 1 h for E. coli and 1 to 2 doubling periods for other species.

An interesting effect observed by Calvin and Hanawalt was the formation of micro-colonies of non-septated cells upon plating electroporated E. coli. They found the addition of pantoyl lactone, a compound known to reverse irradiation-induced inhibition of septation, increased the recovery of electro-transformants in some strains (17). This phenomenon is apparently strain specific as we have not found pantoyl lactone to improve the transformation of the strains we have tested.

PRACTICAL ASPECTS OF THE
ELECTRO-TRANSFORMATION OF E. COLI

Since the detailed procedure for high efficiency electro-transformation of E. coli appeared (4), we have obtained additional information in our laboratory and from many colleagues on the practical use of the technique. Some improvements to the original procedure are described below. The most critical requirements for obtaining the highest transformation efficiencies with a particular strain of E. coli are 1) proper growth and preparation of the cells, 2) a high concentration of cells, 3) well chilled samples and electrodes, 4) proper pulse amplitude and duration and 5) the rapid addition of outgrowth medium to the cells immediately following the pulse.

Preparation of the Cells

The preferred method of growing and harvesting the cells is essentially that described in Ref. 4. The importance of harvesting the culture when it is still growing exponentially deserves emphasis. Electro-competence of E. coli declines as the growth of the culture slows, and for some strains this decline is striking (see Figure 4). Because electro-transformations are best done at very high cell concentrations, the recovery of large numbers of cells from each culture is desirable. Cultures grown in very rich medium such as super broth remain in exponential growth at a higher density and provide excellent yields of cells with very high electro-competence. Harvesting at an OD_{600} of 0.4 to 0.5 yields cells of the highest competence. If one attempts to improve the yield of cells by growing to higher density, the culture must be monitored carefully and harvested at the first sign that growth is slowing. This is usually between OD_{600} of 0.6 to 1.2 depending on the strain and the medium. The cells should be chilled, pelleted and washed several times in cold water. Cells to be stored are finally washed in 10% glycerol and resuspended in 0.2 to 0.5% of the original culture volume in 10% glycerol.

The manner of freezing the cells may affect the quality of a preparation. Quick freezing, in liquid nitrogen for example, seems deleterious. Transferring aliquots of 40 to 250 µl of concentrated cells (in 10% glycerol) to ice cold microcentrifuge tubes and placing these tubes in dry ice pellets until the cells are frozen, consistently gives preparations of the highest electro-competence. The cells should be stored at -70° and thawed on ice immediately before use.

Electro-transformation of the Cells

An optional modification to the original protocol is the use of higher field strengths obtained with the 0.1 cm gap cuvette-style electrodes. A 5 msec pulse at 14 to 16 kV/cm produces efficiencies ~ 2-fold higher than previously reported (see Figure 2). If arcing is often encountered with particular preparations of cells or DNA, a return to the wider electrode gap and lower field strength (0.2 cm, 12.5 kV/cm) recommended in the original protocol may be necessary.

Problems with arcing using either of these conditions is cause by excessive conductivity which can be traced to the cells, the DNA, or the temperature. Preparations of cells that are too conductive may not have been washed properly, or may contain an unusually high number of cells that have lysed, dumping their ionic contents into the medium. Washing with solutions of low ionic strength (distilled water is recommended) and resuspending the cells well with each wash should produce preparations of adequate resistivity. Preparations of certain strains tending to lyse may be improved by the use of a non-ionic, osmotic stabilizing agent such as sucrose (5 to 10%) as the suspending medium.

The ionic strength of the DNA sample must be rather low. DNA resuspended in TE (10 mM Tris, pH 8, 1 mM EDTA) and added in a volume less than 20% that of the cells should work well. Conductivity problems arise, however, if the DNA solution contains residual salt from a CsCl gradient or from an alcohol precipitation (precipitation from ammonium acetate can be a problem, especially if the 70% ethanol wash is omitted), or has been taken directly from an enzyme reaction containing substantial concentrations of mono- and divalent cations. The usual cure in these cases is dilution or alcohol precipitation (or reprecipitation) and washing. The temperature of the sample and the cuvettes has a significant effect on the ionic activity and conductivity. A sample that behaves at 0°, may arc spectacularly if the electrodes are not well chilled.

By attention to the principles described here and the use of the electrical circuit described in reference 4, one can nearly eliminate problems with electrical arcing. When arcing does occasionally occur, it is usually not difficult to find and correct the problem.

Transformations with DNA are often done with DNA taken directly from enzyme reactions, particularly ligation reactions. One concern, as described above, is the contribution of reaction buffer salts to the ionic strength of the electro-transformation medium, and steps to keep the final concentration to less than 2 or 3 mEq are necessary. In the special case of ligation reactions, arcing caused by the carryover of salt is usually not a problem (though some ligation mixtures containing PEG can be troublesome), but typical ligation mixtures do cause some inhibition of electro-transformation. One μl of our standard ligation mix (50 mM Tris-Cl pH 7.8, 10 mM $MgCl_2$, 1 mM DTT, 1 mM ATP) added to a 40 μl sample causes about 50% decrease in recovery of transformants, 3 μl causes a 90% decrease. We find that ethanol precipitation and resuspension in TE relieves this inhibition. Dilution is an even simpler expedient reported to eliminate arcing and reduce inhibition of electro-transformation (31).

Inert nucleic acids may be used as carriers for the recovery of small amounts of transforming DNA without interfering with electro-transformation. Up to 1 μg of carrier DNA or tRNA may be added to a 40 μl sample with no effect on the efficiency of transformation of plasmid DNA.

POTENTIAL USES

cDNA Cloning

The remarkably high transformation efficiencies available with electroporation are quite useful when a limited amount of DNA is available. A good example is constructing cDNA libraries from small amounts of tissue and screening for rare sequences. Most high efficiency cDNA cloning strategies utilize lambda phage vectors and rely on the very efficient gene transfer provided by in vitro packaging of the recombinant DNA. The efficiencies obtained with electro-transformation have revived interest in cDNA cloning strategies based on plasmid vectors by providing efficiencies comparable to that of the lambda systems. Good packaging extracts (10^9 pfu/μg) produce about 3 x 10^{10} pfu per pMol of correctly ligated insert DNA; by comparison a good recA E. coli strain will electro-transform at about 2 x 10^{10} cfu per pmol of correctly ligated insert DNA (and Rec+ strains at 2- to 3-fold higher). Thus, advantage can be taken of the much greater array of useful plasmid vectors in library construction. Using standard cDNA synthesis protocols and the electro-transformation procedure described in (4), Boettger obtained 10^6 recombinants per μg of RNA. With a simplified cDNA synthesis procedure, ligating a blunt-ended cDNA to blunt-ended plasmid and electro-transformation into DH5, >10^6 transformants per μg of RNA were obtained, as high as that often obtained by linker addition and cloning into phage (K. Reed, unpublished).

Genomic Libraries

The construction of genomic libraries in plasmid vectors has been limited by the rather low capacity of competent cells for DNA. The much higher DNA capacity available with electroporation allows production in a single electro-transformation of libraries of up to 10^7 recombinants. Such libraries would require dozens, perhaps hundreds, of chemically-mediated transformations even with the best available competent cells. In addition, the molar efficiency of electro-transformation does not decline with increasing plasmid size (up to at least 20 kb), a property of some importance when cloning genomic fragments.

Plasmid Rescue

The characteristic of high DNA capacity may be exploited in other ways. The rescue of plasmids (and adjacent sequences) from a complex genome is an example. In this case, a small quantity of transforming DNA, and a very large amount of non-transforming DNA is present. With chemical transformations, the non-transforming DNA competes effectively with the plasmid DNA, and very few transformants are recovered. A single electro-transformation has a capacity of >2 μg of total DNA from which thousands of transformants may be isolated in a typical plasmid rescue experiment.

DNA Capacity and Multiply-Transformed Cells

There is a caveat concerning the use of electroporation with a large mass of transforming DNA. We have shown that increasing the DNA concentration increases the probability (P) that any cell will become transformed with at least one plasmid (4). We assume, though we have not shown, that the probability that a cell will become transformed with n plasmids is P^n. The transformation of cells with multiple plasmids is usually undesirable, and could cause some confusion in the subsequent analysis of a library. Potential problems are avoided by adjusting the concentration of transforming DNA to a level calculated to give an acceptably small fraction of doubly transformed cells. To illustrate, the data from Figure 3 of reference 4 show that in this particular experiment cells were transformed with a probability of ~ 3×10^{-3} when the concentration of transforming DNA was 10 ng/ml. We might therefore expect double transformants at the rate of $(3 \times 10^{-3})^2$, or about 1 in 10^5, an acceptable frequency for most applications. If 1 to 10% of the plasmids in a ligation have taken up an insert and circularized, then 2 ng of active DNA (in 20 to 200 ng of total DNA) may be used to electro-transform 200 µl of cells to yield 10^6 to 10^7 total transformants, only 10 to 100 of which may be double transformants.

PRACTICAL APPLICATION OF ELECTRO-TRANSFORMATION
TO CELLS OTHER THAN E. COLI

General Recommendations

It is now evident that many, and perhaps most, types of bacteria may be transformed by electroporation. It is also evident that the transformation efficiencies vary greatly among species (3,8). Several reviews have appeared listing an impressive variety of species that have been electro-transformed to reasonable efficiencies (13,27), and it is not my purpose here to update the list. Rather, I will suggest some general principles to follow in applying the technique to as yet untested bacterial species, and I'll discuss some of the problems occasionally encountered.

The most general and obvious suggestion for initial attempts to transform a particular species is to try the conditions reported useful for a similar species. If this information is lacking, begin with the conditions described for E. coli (4) for gram negative cells, or for Lactobacilli and lactic streptococci (1,12,21) for gram positive cells. Much is known about the practice of electro-transforming these species and this is a useful starting point for cells of average size (see below for discussion of cell size). Once some activity is detected, further refinements of conditions to suit the particular target species is usually straightforward.

Preparation of the Cells

Because the growth phase of greatest electro-competence seems to be mid-exponential for many species, it is prudent also to use the cells of untested species in this state (although this is not always the case, see Ref. 21). There are other aspects of the growth of some cells that can greatly affect their competence. Campylobacter, for example, was initially found to electro-transform much better when harvested from plates rather than liquid culture (2). When initial attempts to electro-transform a species fail, alternative growth conditions may be required. In general, conditions allowing cells to grow rapidly are preferred.

Harvesting the cells is usually done by chilling an exponential culture, pelleting and washing the cells extensively in a cold solution of as low an ionic strength as the cells tolerate. For many species this solution is water. Washing in the cold may not be necessary, and some species of cold-labile bacteria have been prepared and transformed at room temperature with reasonably good results. Cell types that are not easily pelleted or are damaged by centrifugation may be washed by filtration.

With E. coli and several other species, the concentration of cells in the electroporation sample has a profound effect on the yield of transformants. This phenomenon may hold for many species and if possible, electroporation should be done at high cell density ($>10^{10}$/ml) to improve transformation efficiencies. For those types of cells that secrete nucleases, increasing the cell concentration may not be feasible. Problems with nucleases may be lessened by adding the DNA to the cells immediately before pulsing, adding large amounts of carrier DNA to the samples, or electroporating at very low temperature.

The availability of frozen aliquots of competent cells is a great convenience. Many species can be resuspended in 10% glycerol and stored at -70° with little or no loss of activity. This is not always the case however, and initial attempts to electroporate a species (or strain) should be done with fresh cells.

Electro-transformation of the Cells

Electroporation of bacteria is often most effective with the sample at low temperature. If the cells can tolerate cold, the sample, the electrodes, and the electrode holder should be chilled in ice before the pulse. Electroporation in the cold has the additional benefit of reducing the likelihood of electrical arcing during the pulse.

DNA added to the cells should be low in ionic strength and free of constituents such as SDS or phenol that might be deleterious should they enter the cells during permeabilization. The presence of restriction barriers is a substantial obstacle to the transformation of many species, and initial optimization studies should be done with modified DNA isolated from the target strain. If this is not possible, large amounts of DNA (>1 μg per sample) should be used in the hopes of detecting the 10^2- to 10^4-fold lower efficiencies to be expected with unmodified DNA transported into a restriction[+] background. A general method of circumventing restriction barriers is not yet available.

The strength of the field required for effective permeabilization is probably related to the size of the bacterium. Bacteria are typically 0.5 to 2 micron across and require for permeabilization fields of 5 to 20 kV/cm with pulses a few milliseconds in duration. Larger species of 3 to 5 micron require lower field strengths of about 2 to 5 kV/cm, and very small species of 0.1 to 0.3 micron may require intense fields of greater than 20 kV/cm. Some experimentation is required to find the appropriate electrical conditions. This is accomplished by selecting a pulse length of 5 msec and a field strength in the middle of the range for bacteria of similar size, and varying this field by increments of about 10%. For example, a useful field strength profile for cells 1 micron in diameter would be generated with fields of 10 to 20 kV/cm in 1 or 2 kV/cm increments. The transformation efficiency will rise to a maximum as the field is increased and then decline at higher fields as cell death becomes excessive.

Many of the reports referred to above are based on data collected when fields of only about 6 kV/cm were available. Subsequent experiments have shown that some species are more effectively transformed with fields of 10 to 20 kV/cm (21). Consequently, conditions reported successful for many species may yet be improved with higher fields.

Outgrowth

The rapid addition of outgrowth medium to E. coli immediately following their exposure to the pulse greatly enhances the recovery of electro-transformants. The reason for this seems to be increased survival of the cells, and the effect may extend to other species; it is therefore recommended for trials with untested species. Before positive selection is applied, an outgrowth period equivalent to 1 to 2 doubling periods, as with conventional transformation methods, is customary.

Refractory Species

Many species have been successfully transformed in initial trials of electroporation using "standard" conditions, but some species are not so compliant. When the standard conditions fail, there are several areas where adjustments might be made. Restriction barriers are a common cause of failure. Besides the general approaches of using modified DNA or large amounts of unmodified DNA, there are several tactics that may be effective with certain species. Heating the cells prior to transformation, presumably inactivating the restriction systems, is occasionally helpful. Exploiting the double-stranded specificity of many restriction nucleases and transforming with singe-stranded circular DNA may work for some types of bacteria, although there is a wide variation in receptivity to single-stranded DNA among different types of cells. Another possibility is the modification of the transforming DNA with a collection of commercially available modification methylases that might, by chance, methylate the appropriate bases and provide protection from the host nucleases. Yet another approach is the selection of a restriction⁻ strain by electroporation with unmodified marker DNA.

Identifying the appropriate growth conditions for the cells is an area where a certain amount of experimentation might be required. Changing media, growing on agar or in liquid, or harvesting at different stages of growth may improve the transformation of some difficult species. In some cases, growth

conditions (or treatments) that compromise the cell well or
capsule may be effective.

The general recommendations of field strength based on cell
size may not hold for all species, and trials of a greater range
of field strength and pulse length can lead to success. An
important indicator of the effect of the electrical conditions on
the cells is their survival. A substantial portion of the cells
is usually killed by pulses effective for transformation.

Adjustments to the electroporation medium may be helpful,
but there are an inconveniently large number of variations that
might be tried. Changes in the osmotic strength, the ionic
strength, pH, divalent cations and so on might improve
recoveries; but, as discussed earlier, addition of charged
compounds will tend to limit the voltage that may be applied.

For an especially resistant species, finding the key to
successful electro-transformation may come more from a knowledge
of the idiosyncrasies of that species than from the numerous
rules of thumb that I have offered here.

PROSPECTS

Many species of prokaryotes have now been transformed by
electroporation, some to very high frequencies and efficiencies.
The method may be applicable to nearly all types of bacteria, and
should provide a significant advance in the study of the molecular genetics of heretofore untransformable species.

Prokaryotic electroporation is still a fairly new technique
and there is much to learn about both the theoretical and
practical aspects of the phenomenon. Although many species have
been electro-transformed in initial trials, there are others that
have resisted substantial efforts to obtain reasonable (or any)
transformation. These failures may be due to some intrinsic
problem the cells have in establishing or maintaining plasmid
DNA, or to some structural characteristic that renders them
resistant to the electrical pulses as presently applied. As we
learn more about the mechanisms involved, protocols for over-
coming these obstacles will likely emerge, providing the means to
transform these organisms. A very common problem is the need to
introduce unmodified DNA into restriction+ cells. Although a
trial-and-error approach to this may be fruitful in specific
cases, a good general strategy to circumvent restriction barriers
is sorely needed.

The availability of simple, reliable protocols for the
transformation of E. coli to extremely high efficiencies and
frequencies and at high DNA capacity, is useful addition to the
repertoire of recombinant DNA techniques. Furthermore, the
remarkable levels seen with E. coli may portend greatly improved
transformation of many other interesting prokaryotes.

Acknowledgments: I want to thank my many colleagues who have shared their findings and their insights on the practical use of electro-transformation.   I especially thank Greg Heinkel, Connie Langer and Paul Zoller of Bio-Rad Laboratories for providing comments, data and specialized equipment.

REFRENCES

1   Chassy, B.M. and Flickinger, J.L. (1987) FEMS Microbiology Lett. 44, 173-177.

2   Miller, J.F., Dower, W.J. and Tompkins, L.S. (1988) Proc. Nat. Acad. Sci. U.S.A. 85, 856-860.

3   Luchansky, J.B., Muriana, P.M. and Klaenhammer, T.R. (1988) Molec. Microbiol. 2, 637-646.

4   Dower, W.J., Miller, J.F. and Ragsdale, C.R. (1988) Nucl. Acids Res. 16, 6127-6145.

5   Allen, S.P. and Blaschek, H.P. (1988) Appl. and Environ. Microbiol. 54, 2322-2324.

6   Snapper, S.B., Lugosi, L., Jekkel, A., Melton, R.E., Kieser, T., Bloom, B.R. and Jacobs, W.R. (1988) Proc. Nat. Acad. Sci. U.S.A. 85, 6987-6991.

7   Somkuti, G.A. and Steinberg, D.H. (1988) Biochimie 70, 579-585.

8   Wirth, R., Friesenegger, A. and Fiedler, S. (1989) Mol. Gen. Genet. 216, 175-177.

9   Ito, K., Nishida, T. and Izaki, K. (1988) Agric. Biol. Chem 52, 293-294.

10 Lalonde, G., Miller, J.F., Falkow, S. and O'Hanley, P. (1989) Amer. J. Vet. Res. (in press).

11 Dower, W.J. (1987) Molecular Biology Reports (Bio-Rad Laboratories) 1, 5.

12 Powell, I.B., Achen, M.G., Hillier, A.J. and Davidson, B.E. (1988) Appl. Environ. Microbiol. 54, 655-660.

13 Chassy, B.M., Mercenier, A. and Flickinger, J. (1988) Trends in Biotechnology 6, 303-309.

14 Neumann, E., Schaefer-Ridder, M., Wang, Y. and Hofschneider, P.H. (1982) EMBO J. 1, 841-845.

15 Knight, D.E. and Scrutton, M.C. (1986) Biochem. J. 234, 497-506.

16 Harlander, S.K. (1987) in Streptococcal Genetics (Ferretti, J.J. and Curtiss, R.C., eds.) pp. 229-233, ASM Publications, Washington, DC.

17 Calvin, N.M. and Hanawalt, P.C. (1988) J. Bacteriol. 170, 2796-2801.

18 Fiedler, S. and Wirth, R. (1988) Anal. Biochem. 170, 38-44.

19 Belliveau, B.H. and Trevors, J.T. (1989) Appl. Environ. Microbiol. 55, 1649-1652.

20 David, S., Simons, G. and DeVos, W.M. (1989) Appl. Environ. Microbiol. 55, 1483-1489.

21   McIntyre, D.A. and Harlander, S.K. (1989) Appl. Environ.
     Microbiol. 55, 604-610.
22   Taketo, A. (1988) Biochim. Biophys. Acta 949, 318-324.
23   Jessee, J. (1988) Abstracts 88th Annual Meeting ASM, ASM
     Publications, Washington, DC.
24   Taketo, A. (1989) Biochim. Biophys. Acta 1007, 127-129.
25   Van Der Lelie, D., Van Der Vossen, J.M. and Venema, G.
     (1988) Appl. Environ. Microbiol. 54, 865-871.
26   Potter, H. (1988) Anal. Biochem. 174, 361-373.
27   Shigekawa K. and Dower, W.J. (1988) BioTechniques 6, 742-
     751.
28   Weston, A., Brown, M.G.M., Perkins, H.R., Saunders, J.R. and
     Humphreys, G.O. (1981) J. Bacteriol. 145, 780-787.
29   Hanahan, D. (1983) J. Mol. Biol. 166, 557-580
30   MacLachlan, P.R. and Sanderson, K.E. (1985) J. Bacteriol.
     161, 442-445.
31   Wilson, T.A. and Gough, N.M. (1988) Nucl. Acids Res. 16,
     11820.
32   Bottger, E.C. (1988) BioTechniques 6, 878-880.

THE ISOLATION AND IDENTIFICATION OF cDNA GENES BY THEIR HETEROLOGOUS EXPRESSION AND FUNCTION

Gordon G. Wong

Genetics Institute, Inc.
87 Cambridge Park Drive
Cambridge, MA 02140-2387

INTRODUCTION

The most common and direct strategy for the isolation of eukaryotic genes encoding polypeptides has been based on their protein amino acid sequence structure. However, frequently the purification of sufficient amounts of homogeneous protein for amino acid sequence determination is technically and logistically not feasible. In these situations, if the gene's phenotype has been well characterized, an alternative gene isolation strategy may be utilized.

The strategy is based on highly efficient heterologous gene expression systems and is essentially an identity test for cDNA genes by their structure and function. The experimental method is to convert an mRNA source encoding the polypeptide or biological activity into a complex cDNA gene library. The cDNA library is contained within a heterologous gene expression plasmid. Each cDNA gene, of the library, is analyzed for its ability to instruct a heterologous cell specifically to express the particular phenotype, polypeptide or biological activity.

In general a mammalian host cell and expression vector system are utilized because of their prolific and faithful transcription of heterologous genes into functional protein. By this method, cDNA genes for polypeptide growth factor (1-7), cell surface antigens (8-14), growth factor receptors (15-19) and gene transcription factors (20) such as DNA binding proteins have been isolated. The strategy for this gene isolation method is diagrammed in Figure 1.

There are in the heterologous expression-function gene isolation strategy some possible theoretical flaws. The isolated gene may not directly encode the relevant protein or biological

activity. The putative gene may actually activate the endogenous
gene or encode a necessary cofactor or effect a post-
translational modification to render an inactive polypeptide
active. In addition even with the relevant cDNA gene there may
be a need for unique post-translational modifications. This is
typically the situation for Factor IX (21), which requires gamma
carboxylation for biological activity. Irrespective of these
reservations, if the biological activity can be shown to be
encoded simply by a single mRNA then the isolation of its cDNA
form may be approachable by the heterologous expression-function
strategy.

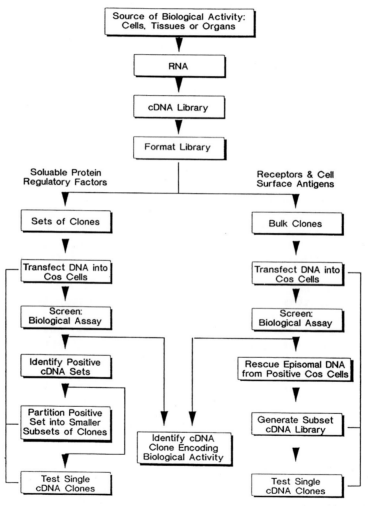

Figure 1. Heterologous expression - function gene isolation
strategy.

There are a number of technologies essential for this cDNA gene isolation strategy. Two of the technologies, the COS cell based transient gene expression vector system (22-24) and the construction of highly complex cDNA libraries, are well described and established and need no further discussion here (23,25). However the theory and methodologies of preparing the cDNA library for its transfection into COS cells and for screening the cDNA gene products have been inadequately explained and described. in this chapter we will discuss these issues first and then review a number of examples of cDNA gene isolation for cell surface associated proteins, growth factors and DNA binding proteins to illustrate the theory and methods.

ORGANIZATION OF THE cDNA GENE LIBRARY FOR TRANSFECTION INTO COS CELLS

Within the relevant plasmid cDNA library we have presumed that there will be cDNA genes encoding the specific biological activity of interest. We have chosen to identify these particular cDNA genes by the heterologous expression-function screening strategy. If we had limitless resources, we could analyze each plasmid cDNA clone individually by transfection of its DNA into COS cells and then test the resulting cDNA gene's expression product. Alternatively, we can prepare the cDNA library for analysis to take advantage of the COS cell's prolific gene expression capabilities. The format of the cDNA library for transfection into COS cells is a consequence of three parameters: the replication properties of SV40 replication origin-containing expression vectors in COS cells (26), the nature of the biological activity being sought and the type of assay utilized.

COS cells are African green monkey kidney cells, CV-1, which have been transformed with a replication origin defective DNA tumor virus, SV40 (22). Though these cells express high levels of T antigen there are no endogenously generated SV40 virus. However the T antigen is functional and can drive the replication of the SV40 replication origin containing exogenously introduced DNAs to high copy numbers.

About 10 to 15% of an exponentially growing population of COS cells will take up exogenous DNAs (24). if these DNAs contain a SV40 origin there will be T antigen-driven replication of the exogenous DNAs to copy numbers as high as 10^5 per COS cell (26). When the transfecting DNAs are a heterogeneous population of plasmids, each transfected COS cell will amplify a unique subset of plasmids from the population. This subset consists of about 10 to 50 different plasmids. Therefore if the DNA being transfected is a complex population of plasmids, of about 10^2 or greater, then we would expect every receptive COS cell to contain a unique set of amplified heterologous plasmid DNAs. If this complex population of plasmids is a cDNA library contained within a SV40 origin-based expression vector then we would expect that each DNA-receptive COS cell could be expressing a unique set of heterologous genes. This particular property of COS cells forms the basis of the expression-function cDNA gene cloning system.

If the biological activity of the cDNA gene in question can be detected at the single cell level in the form of a cellular or morphological change or by the expression of an intracellular enzyme or cell surface protein then it may be possible to treat a complex cDNA library as a single unit. If the cDNA gene of interest encodes a secreted growth factor, then the cumulative expression of the growth factor from a number of transfected COS cells will be necessary for detection. In this scenario the cDNA library will need to be partitioned into subsets of cDNA clones for transfection and analysis.

In practice the plasmid cDNA libraries can be plated at densities of about 5,000 bacterial colonies per 15 cm petri plate and then replica plated to generate a master and a series of replicas. The replicas are used for hybridization analysis of particular genes and for organizing the clones into sets for transfection. If the library is to be used just once the colonies could be plated and directly collected into sets.

If the biological activity is a ligand binding protein, receptor or intracellular enzyme that is detectable at or from a single transfected COS cell, then the cDNA library can be analyzed more simply. The expression cDNA library can be plated and then pooled together directly without generating a master or replica copies. Alternatively the cDNA library may be grown as a liquid culture in suspension for routine large scale plasmid DNA preparation. With the latter method there is the concern that complex populations of plasmids may undergo selective amplification. However this method has been successfully and routinely used in isolating membrane-associated proteins and receptors (8-14).

The method of plasmid DNA preparation can affect its heterologous expression. The large scale plasmid DNA purification methods with a cesium chloride differential centrifugation step consistently yield plasmid with high levels of expression. For the experiments where the entire cDNA library is treated as one set of plasmids such DNA purification methods are practical. However, for experiments where the cDNA library is partitioned into sets of clones, typically with about 200 sets of 1,000 clones each, it is impractical to include a cesium chloride differential centrifugation step. In such situations generally a "mini" preparation procedure is used.

The quality of plasmid DNA from such "mini" preparation procedures is sufficient for restriction enzyme analysis and cloning manipulations but usually variable for transfection-expression experiments. From a comparison of some of the standard "mini" preparation procedures, we find that the alkaline

lysis method described by Sambrook et al. (23) is simple and
reliable for preparing large numbers of sets.

THE TRANSFECTION OF DNA INTO COS CELLS

Though there are numerous methods for transfecting mammalian
cells with heterologous genes, there has not been a systematic
study of their efficiency for COS cells. We have selected the
DEAE-Dextran method (28) for its simplicity, reliability and
reproduciblity in its delivery of very small amounts of DNA into
a high frequency of receptive cells.

For the expression isolation of cDNA genes for cytokines,
growth factors or cell surface proteins where the cDNA library is
partitioned into sets of plasmids, there can be a sizeable number
of manipulations. Approximately one hundred sets of plasmids,
and therefore 100 10-cm petri plates of COS cells, can be easily
handled for transfection. When the culture plates are about 75%
confluent the COS cells are transfected with a serum-free DEAE-
Dextran DNA mixture. After 12 to 16 hr, the transfection mixture
is removed and cells are treated with chloroquine for 3 hr.
During chloroquine treatment the COS cells become vacuolized.
Twelve to eighteen hr post-chloroquine treatment, the medium is
removed and replaced with about 4 to 5 ml of medium for
accumulation of secreted cDNA expression products. In general
high levels of secreted heterologous proteins can be found in the
COS cell medium from about 44 to 72 hr post-transfection, though
there is usually continued expression from 72 to 96 hr. For the
expression of receptors and intracellular proteins, expression
levels at about 60 hr post-transfection are sufficient for
detection in a screen for functional activity.

For experiments where the screen for functional biological
activity occurs at the single cell level, as in receptors, the
entire cDNA library is treated as a single set of plasmids. For
a cDNA library of about 10^6 complexity, only about 5×10^5 COS
cells need to be transfected to ensure that each plasmid is
tested at least once.

SOME GENERAL COMMENTS ON cDNA LIBRARY CONSTRUCTION

For this cDNA gene isolation strategy we have presumed that
the biological activity is encoded by an mRNA. It may be prudent
to verify this assumption directly by translating isolated mRNA
believed to encode the biological activity. mRNA can be
heterologously translated in either an _in vitro_ translation
system or _in vivo_ in _Xenopus_ oocytes.

The presence of biological activity in the translation
products would show that the biological activity is encoded by a
translatable mRNA and does not require unique post-translational

modifications. However frequently the mRNA of interest is present at concentrations too low to yield detectable translation products or may be refractory to heterologous in vitro translation. In this situation it may be advisable to use biochemical techniques to demonstrate that the biological activity arises from a polypeptide and to determine some of its biochemical properties, e.g., molecular weight, post-translation modifications, glycosylation and relative abundance. These latter data are generally helpful in designing the appropriate identity test of the cDNA gene expression products.

As a general assumption we can presume that for biological activities such as secreted growth factors and cytoplasmic enzymes the concentration of the biologically active polypeptide will be directly proportional to the concentration of its mRNA. The concentration of the relevant mRNA will determine the size of the cDNA library that should be screened to ensure some likelihood of detection.

The relative abundance and molecular weight of the biological activity will provide some limits as to the complexity of cDNA libraries to be screened. As you would expect, the mRNA for a growth factor of 20 to 30 kD, present in nanomolar concentrations in the conditioned medium, should be relatively abundant in the cell's mRNA population, typically 0.1 to 01%. If the relevant mRNA is present at about 0.01%, and there is about a 50% efficiency in generating full-length cDNA clones for a 2 kb mRNA and about 50% of the cDNA clones ligated into the expression vector are compatible for gene expression, then a cDNA library of about 40,000 to 100,000 clones needs to be screened.

In contrast, if the biological activity is about 100 kD in mass, it may be encoded by a long mRNA which could be problematic for conversion into complete cDNA. With the present cDNA reaction methods, it appears that the likelihood of an mRNA being converted to a complete or "full-length" cDNA is dependent on specific RNA structure rather than just on its length. Finally, large cDNA fragments may have a competitive disadvantage for ligation into plasmid vectors and consequently be under-represented (27). These observations would suggest that for high molecular weight proteins a proportionally more complex cDNA library may need to be analyzed.

Therefore, efforts directed to surveying sources of "high" biological activity are well spent as this would reduce the complexity of the cDNA library required for screening. For biological activities such as those involved with cellular morphological change or receptors it is difficult to analyze what would be the most abundant source of relevant mRNA. In these situations it may be necessary to accept the premise that the relevant mRNA is infrequent and to devise biological screens which can analyze a large number of cDNA genes.

With the present technology, the complexity of the cDNA libraries is not limited by the availability or quality of mRNA but rather is constrained by the biological assay. Assays which can detect the relevant gene product at the single cell level can be used to screen libraries of 10^6 to 10^7 clones. For low abundance polypeptides such as the interleukin receptors this level of complexity is required. However biological assays which require large amounts of test material or require many manipulations are likely to limit the size of the cDNA library that can be analyzed.

The methods and techniques for the isolation of mRNAs and their conversion into cDNAs and highly complex cDNA plasmid libraries are well described in manuals by Sambrook et al. (23) and Ausubel et al. (25). In addition most of the molecular biology techniques have become standard and routine for most fields of biological investigation and thus expertise is usually available within a department. Nevertheless there still abounds a somewhat unusual mystique surrounding the construction of complex cDNA libraries in bacterial plasmids. We recommend that the novice molecular biologists take advantage of the commercially available kits for mRNA isolation and cDNA reactions, and if necessary the services of a contract molecular biology company.

DETECTION OF TRANSLATION PRODUCTS FROM COS CELLS TRANSFECTED WITH EXPRESSION cDNA LIBRARIES

Single Cell Phenotype Detection Assays

The most critical step in the expression-function isolation of cDNA genes is the detection system. If the phenotype is a morphological change or expression of a cell surface marker or cytoplasmic protein or enzyme that is detectable on a single cell level then there are two experimental routes.

The ideal experimental route (Figure 2) is to devise a phenotypic screen such that the relevant COS cells are identified and isolated for "rescue" of their expression plasmid DNA. The rescued plasmid DNAs will constitute a heterogeneous population of different cDNA genes due to the nature of the heterogeneous DNAs transfected into COS cells. This population of episomal DNAs will generate a library that is a subset of the original complex cDNA library. Plasmid DNA is purified from the subset library and reiterated through the transfection expression-function screening process.

By this reiteration, the subset cDNA library is further reduced in complexity to yield a second subset cDNA library enriched for the relevant gene sequences. This enrichment should be reflected in an increase in the frequency of positive COS cells in the assay. Because the transfected COS cell does contain a population of plasmids, eventually individual single

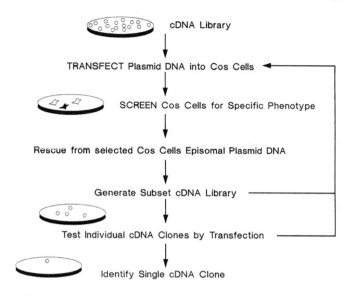

Figure 2.   Single cell expression screen with rescue of episomal plasmid DNAs.

cDNA clones will need to be transfected to identify the cDNA gene.

An alternate experimental route is followed when the expression plasmid cannot be rescued from the COS cells. Typical examples of this route are an _in situ_ enzymatic staining detection system or a labelled ligand used in direct detection of a receptor. With these methods it is necessary to organize the cDNA library into subsets of clones before the transfection-expression screen. This can be done in a pyramid fashion as shown in Figure 3. The cDNA library is plated and a master and replica copy are generated. Subsets of colonies are derived from the replicas. If the complete cDNA library contains the relevant cDNA gene, then a number of the transfected COS cells will express the phenotype. This number should be directly related to the frequency of the relevant gene. A second set of COS cell plates can be transfected with subsets of cDNA clones from the complete library. A subset containing the relevant cDNA will have a greater number of positive COS cells per plate than in the starting complete library.

The number of clones in each subset will determine the number of subsets and thereby the number of COS cell plates which will have cells that score in the assay. If the number of clones in a subset is determined optimally then only a small percentage of the subsets will lead to positive COS cell plates. The number of clones in a subset will also determine the complexity of the library that can be screened and the number of reiterations that

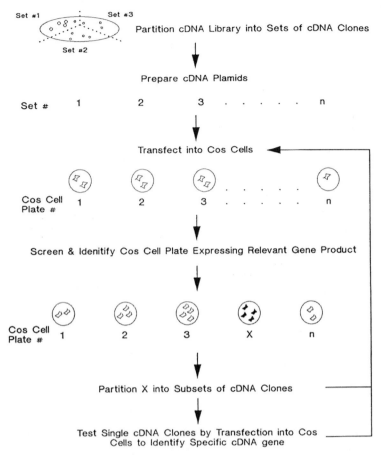

Figure 3.   Partitioned cDNA expression library, transfection, expression and identification of cDNA clones.

will be required before a sufficiently enriched subset library can be analyzed by single cDNA clone-COS cell plate transfections.

The best examples of expression-function identification and isolation of genes by single cell phenotype detection strategies are the isolation of genes for cell membrane-associated proteins and receptors.   Seed et al. (8-11,13,14) have developed an efficient expression-function isolation method for these genes. The method is to construct a relevant cDNA library, transfect the library into COS cells, select the COS cells expressing the relevant cell surface protein, and "rescue" cDNA plasmids from these cells.   The selective process of DNA transfection into COS cells results in only a small fraction of the cell population expressing the cell surface protein.   The rescued plasmids are used to generate a secondary cDNA library which is now enriched for sequences which may encode the specific membrane protein. The process of transfection, screening and plasmid rescue is reiterated. Eventually the library is reduced to a complexity where individual single cDNA clones can be analyzed. There are two critical points to consider.

The first point concerns the nature of the proteins and their biological assays. Unlike secreted proteins, cell surface proteins should accumulate on the cell surface to a plateau number that is likely to be dependent on the cell membrane surface area and other mechanisms regulating the number of cell surface proteins rather than just on the transcriptional activity of the heterologous promoter. With an active expression plasmid such as pXM (2) or πH3M (9), full length cell surface protein or receptor cDNA genes generally produce about 10^5 proteins per cell (15-17,19). Cells expressing this number of cell surface proteins can be easily recovered by "panning" with antibodies attached to plastic dishes (29) or to magnetic beads (30), or by fractionation on a cell sorter with a fluoresceinated ligand (31).

The second point of the strategy centers on the ability to rescue functional plasmid from the transfected COS cells which exhibit the specific phenotype. Expression plasmids with SV40 origins will, in COS cells, replicate up to about 50,000 to 100,000 episomal copies per cell. This DNA can be recovered by the Hirt procedure (32) and transformed into bacteria by one of the high efficiency transformation protocols (33) or by electro- poration (34).

It is clear that this strategy must rely on either the ability of COS cells to amplify only a small number of distinct plasmids or on the experimental ability to introduce only a small number of plasmids into a COS cell. In the early reports cDNA libraries were transfected into COS cells by spheroplast fusion on the assumption that this method would deliver only a small number of plasmids per COS cell and thereby each COS cell would test different cDNA genes. In later work, Allen and Seed (14), D'Andrea et al. (17) and others (18) realized that the standard DEAE-Dextran transfection procedure still only resulted in only a small number, (about 10 to 50) of unique cDNA plasmids, from a population, being amplified and expressed by a single receptive COS cell. Transfected COS cells expressing the relevant cell surface antigen are recovered and isolated from the population of transfected COS cells. This effectively enriches for COS cells harboring gene sequences encoding the cell surface antigen.

The degree of enrichment with each cycle of the transfection-expression selection is reduced as the complexity of the library is reduced. In the end, generally after about 3 to 4 cycles, single individual clones are picked for analysis. Alternatively, with a small set of clones, the library can be arranged into overlapping sets of clones for sib selection and isolation of the relevant plasmid. By this strategy, large cDNA libraries of high complexity on the order of 10^5 to 10^6 clones can be easily screened.

Seed and Aruffo (8) were the first to apply this strategy with antibody panning screens to isolate and identify a number of T cell surface antigens, e.g., CD2, CD7 (9) and CD28 (10) and Stengelin et al. (13) isolated the two FcRII receptor genes.

Often, however, there are no available antibodies to a receptor. A more direct experimental approach would be to use the ligand itself in a screen. The use of an indirect ligand-mediated selection was first reported by Allen and Seed (14) in isolating human FcRI. These investigators treated transfected COS cells with mouse Ig antibody, a ligand for FcRI, which were then panned by plastic dishes coated with goat antibody to mouse Ig.

Staunton et al. (12) was the first to demonstrate that a ligand, LFA-1, can be used directly to recover transfected COS cells expressing the corresponding receptor, ICAM-1. A second ligand, ICAM-2, was isolated by screening COS cells transfected with an endothelial cell cDNA library, in the presence of antibody which blocked ICAM-1 action.

The strategy of panning with ligand has not been successful for growth-factor receptor cDNA genes. It may be that soluble growth factors, unlike ligands which are cell surface-associated proteins, are not compatible with presentation on a two-dimensional surface. However, such receptors have been cloned by a functional assay of ligand binding.

If the soluble ligand, e.g., growth factor or cytokine, can be fluoresceinated with a biotin-avidin conjugate then it may be used to screen transfected COS cells by fluorescence-activated cell sorting (FACS). Recently Yamasaki et al. (16) screened transfected COS cells with biotinylated IL-6 conjugated with fluoresceinated avidin. The cells which bound IL-6 were recovered on the basis of their high fluorescence. Plasmids were recovered, amplified and then retransfected and FACS sorted. The process is reiterated until a subset cDNA library enriched for the relevant cDNA genes is generated. Single plasmids from this library were individually transfected and analyzed to identify the IL-6 receptor cDNA gene.

A technically simpler alternative to fluorescence labels is to conjugate a radiolabel to a ligand, though the biological activity of most cytokines can be compromised by such procedures. However some cytokines such as IL-1 and gamma interferon can be radiolabeled to high specific activity and used directly to screen transfected COS cells by ligand binding for receptor expression. This approach was taken by Sims et al. (15) to isolate the IL-1 receptor. A highly complex cDNA library was partitioned into subsets of clones, transfected into COS cells and screened for high affinity binding of IL-1. The binding of IL-1 was measured by autoradiography. A subset of cDNA clones was found to confer specific binding of iodinated IL-1 by COS cells. From this subset, a single cDNA clone encoding the IL-1 receptor was isolated.

For some cytokines that can be labelled to very high specific activity or with a high energy emitter it may be possible directly to visualize a single COS cell expressing about 100,000 to 1,000,000 receptor proteins on its cell surface. This direct approach was used by Munro and Maniatis (18) to isolate the murine gamma interferon receptor. The murine gamma interferon receptor is present at low concentrations in most cell types. Murine gamma interferon which could be phosphorylated to high specific activity was used to detect COS cells expressing gamma interferon-binding proteins. cDNA libraries of greater than 100,000 clone complexity were transfected onto a single 10 cm plate of about 10^6 COS cells, and then screened for expression of functional gamma interferon-binding proteins. In this protocol, Munro and Maniatis (18) did not fix the transfected COS cells after a binding reaction with phosphorylated gamma interferon but rather treated the cells with a glycerol mixture. This allowed the COS cells to remain viable during autoradiography at -80 C and permitted recovery of episomal DNA.

Murine gamma interferon is biologically inactive on simian cells and presumably will not complex to the endogenous COS cell interferon receptors. COS cells expressing the murine gamma interferon binding protein would bind phosphorylated gamma interferon and be detected as spots on the autoradiograph. COS cells corresponding to such spots were isolated and processed for their episomal plasmid DNA. These plasmids were amplified and further analyzed by transfection and binding assay. In the secondary transfection and binding experiment, the number of autoradiograph spots and therefore the number of COS cells binding the ligand increased as the relevant cDNA gene was enriched.

In some situations, the ligand can only be radiolabelled to a modest level of specific activity and thereby a single cell detection strategy is precluded. An alternative approach is to partition the cDNA library into sets or pools of cDNA clones. Each set or pool can be independently transfected and tested for ligand binding. This strategy was demonstrated in the isolation of the murine erythropoietin receptor.

Murine erythroleukemia (MEL) cells have about 500 to 600 low-affinity binding sites for murine or human erythropoietin. This low number of sites made purification of the putative receptor intractable. In addition, readily available human erythropoietin, which is active on murine hematopoietic cells, is sensitive to most iodination procedures and can be radiolabelled to a specific activity of only about 1,000 dpm per femptomole. If we presume that most heterologously expressed receptor proteins would number about 10^5 copies per COS cell, this low level of specific activity would make any single cell phenotype detection scheme impractical. An alternative scheme was to consider the cumulative erythropoietin binding ability of a number of COS cells transfected with the same cDNA gene. A 10 cm dish of about 10^6 to 10^7 COS cells was found to bind nonspecifically about 1,000 dpm of iodinated erythropoietin. To generate a specific signal of about 1,000 dpm, a signal detectable above the nonspecifically bound erythropoietin, about 10^4 COS cells expressing a putative erythropoietin binding protein would be necessary. However each 10 cm plate can contain about 5×10^6 COS cells. If each COS cell has about a 10% chance of taking up transfected DNA, then there are about 5×10^5 receptive COS cells. 10^4 COS cells are required to bind a detectable level of erythropoietin and therefore a single 10 cm dish of COS cells can test [5×10^5 COS cells/10^4 cells per signal] x 50 cDNA plasmids per cell = 2,500 different cDNA plasmids.

The experimental strategy was to partition a MEL cDNA library into sets of clones, transfect each set into a 10 cm petri dish of COS cells and simply assay for erythropoietin binding. D'Andrea et al. (17) screened approximately 350,000 cDNA clones of MEL cDNA library which had been partitioned into 200 sets. Two sets of clones contained COS cells which bound iodinated erythropoietin. From these two sets, two independent cDNA clones were isolated, each of which encoded a 507 amino acid polypeptide with an aminoterminal secretory sequence and a hydrophobic membrane-spanning region. Both cDNAs could direct COS cells to express high-affinity binding sites for erythropoietin and could generate a mature polypeptide consistent with the putative erythropoietin receptor's molecular mass (17).

cDNA genes isolated by the expression-function cloning strategies are not necessarily full-length. Both for the IL-6 receptor (16) and murine gamma interferon receptor (18) the clones identified and isolated had the requisite biological activity, i.e., the ability to bind ligand specifically; however, both clones were incomplete for 3' coding sequence. With such partial clones it was straightforward to isolate the entire gene by standard procedures (23,25).

For both the IL-6 and gamma interferon receptor proteins, their deleted forms were still membrane associated. We could envision a scenario where the medium of the transfected COS cells should be assayed for ligand binding activity. The presumption is that a partial cDNA gene encoding just the ligand binding domain might lead to the expression of a secreted truncated receptor.

Detection Systems for Soluble Growth Factors and Cytokines

Presently, there are no rules which can predict which polypeptide cDNA genes will be effectively transcribed and translated in heterologous cells. Some secreted proteins like Factor VIII appear to require further major research and experimentation to achieve even minimal levels of expression in

mammalian cells (35), while other proteins like tissue
plasminogen activator (36) and GM-CSF (1) are very efficiently
expressed. In general, cytokines, interleukins and other soluble
regulatory proteins appear to be heterologously expressed very
efficiently and are biologically active at picomolar
concentrations.

The assays for some secreted regulatory proteins such as GM-
CSF are responsive to biologically active protein in the 1 to 10
picomolar range (37). Under optimal conditions, one dish of COS
cells, about 5×10^6 cells, when transfected with an expression
plasmid encoding a heterologous gene such as GM-CSF, will
generate about 0.1 to 0.5 µg of heterologous protein per ml by 72
hr post-transfection. For a protein of approximately 10 kD, 1
picomolar is 10 pg per ml. Therefore this COS conditioned medium
containing such a heterologous cytokine could be active in a
standard hematopoietic colony-stimulating activity bioassay to a
1 in 10,000-fold dilution.

The expression level of a particular heterologous protein by
COS cells is directly proportional to the concentration of
protein-encoding plasmid in the DNA transfection mixture.
Maximum protein expression levels are generally reached when the
plasmid DNA concentration is about 1 µg/ml of DEAE-Dextran
transfection mixture. In a COS cell, when the concentration of
transfecting DNA is not limiting, the concentration of a
particular heterologous mRNA and hence translation product is
directly proportional to the concentration of DNA template. For
example if a hypothetical cytokine cDNA gene is transfected as a
single saturating cDNA, the resulting conditioned medium would be
active to a 1 to 10,000-fold dilution; however, if the
hypothetical cDNA is present at only 1% of the transfecting DNA
(10 ng of cDNA and 990 ng of carrier DNA) then the conditioned
medium would be active to only a 1 to 100-fold dilution.
Correspondingly if the hypothetical cDNA clone constitutes only
0.1% of the transfecting DNA, the conditioned medium would be
active to 1 to 10-fold dilution in the assay. For hematopoietic
growth factors, with a colony forming assay, biological activity
can be detected in medium conditioned by COS cells transfected
with pXM expression plasmid encoding GM-CSF, IL-3, etc., out to a
1 to 50,000 dilution. With factor-dependent cell lines such COS
conditioned medium may be active to a final 1 to 200,000-fold
dilution. Therefore under optimal COS cell transfection-
expression conditions with highly sensitive assays, cDNA
libraries partitioned into subsets of about 1,000 clones can be
easily screened by bioassay for a functional cytokine cDNA gene.

Isolation of Colony Stimulating Factors

The molecular cloning of the cDNA gene for human
Granulocyte-Macrophage Colony Stimulating Factor (GM-CSF) was the
first successful demonstration of the expression-function cDNA

gene isolation strategy (1). GM-CSF is a member of the group of Hematopoietic Colony Stimulating Factors (37), protein factors which regulate hematopoiesis in vitro and in vivo. GM-CSF directs hematopoietic progenitor cells to proliferate and differentiate along the granulocyte and macrophage cell lineages. Until recently the existence of CSFs was speculative and required extensive biochemistry and eventual molecular cloning to demonstrate that such effects were due to protein entities.

The medium conditioned by the HTLV I transformed spleen cell line Mo (38) was known to be a relatively abundant source of human GM colony stimulating activity. However it proved difficult to purify sufficient quantities of protein for amino acid sequence determination. The difficulties arose from the low levels of the putative CSF, the apparent heterogeneity in its carbohydrate structure making purification problematic and logistic difficulties of the hematopoietic CSF assay when human bone marrow cells were used as targets. The alternative strategy of gene isolation by expression-function was attempted. The Mo cell line was assumed to express human GM-CSF protein and thereby must express human GM-CSF mRNA. The mRNA from Mo cells could be translated by Xenopus oocytes to yield just detectable levels of colony stimulating activity (39). However, this result was inconsistent and therefore prohibited the use of the oocyte translation method as a cDNA gene isolation strategy. The heterologous expression-function cDNA gene isolation strategy presumed that if the full-length cDNA for the putative CSF could be linked to an appropriate promoter it would be transcribed and translated in a mammalian host cell.

mRNA was isolated from the Mo cell line and converted into cDNA and ligated into the expression plasmid p91023B (1). A cDNA library was plated and then replica plated to generate a series of master plates and copies for plasmid DNA purification. This cDNA library was partitioned into sets of about 500 to 600 colonies per set or pool. From each pool or set of clones, plasmid DNA was purified and transfected into COS cells. The medium conditioned by these transfected COS cells was harvested and assayed for GM-CSF activity. From about 100 sets of plasmid DNAs, one set led to conditioned medium which exhibited GM colony-stimulating activity. The positive set contained about 500 to 600 different cDNA clones. Presumably one of these clones encoded a cDNA gene which when transfected into COS cells led to the expression of GM colony-stimulating activity. The specific cDNA clone could be identified by a number of different steps.

The positive set of about 500 to 600 clones could be subdivided into smaller pools of 50 and then retested by transfection-expression and bioassay. From the positive subset individual clones could be tested. As with the single cell phenotype assays described in the previous section, we would expect the conditioned medium from the subset to be more biologically active then the original positive pool. And when

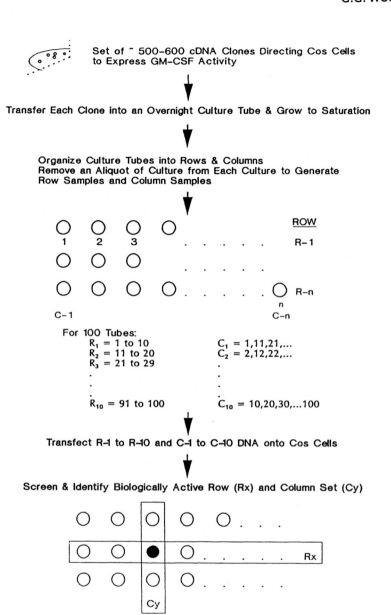

Set of ~ 500–600 cDNA Clones Directing Cos Cells to Express GM–CSF Activity

Transfer Each Clone into an Overnight Culture Tube & Grow to Saturation

Organize Culture Tubes into Rows & Columns
Remove an Aliquot of Culture from Each Culture to Generate
Row Samples and Column Samples

For 100 Tubes:

$R_1 = 1$ to 10 $C_1 = 1,11,21,...$
$R_2 = 11$ to 20 $C_2 = 2,12,22,...$
$R_3 = 21$ to 29

$R_{10} = 91$ to 100 $C_{10} = 10,20,30,...100$

Transfect R–1 to R–10 and C–1 to C–10 DNA onto Cos Cells

Screen & Identify Biologically Active Row (Rx) and Column Set (Cy)

Test cDNA Clone from Intersection of R-x and C-y

Figure 4.   Partitioned cDNA expression library, transfection, expression and identification of cDNA clones.

single clones are tested the relevant cDNA clone would lead to highly active COS conditioned medium. The increase in biological activity as the positive pool is subdivided is a good test that the original positive pool was not an artifact. For the identification of the GM-CSF cDNA gene an alternative route was taken.

From the master plate individual colonies, which constituted the positive set of plasmids, were clonally transferred into overnight culture tubes and grown to saturation. The culture tubes were arranged into a two-dimensional matrix (Figure 4). Aliquots were taken from each overnight culture and organized into sets representing each row and column of the matrix. By this process each cDNA clone was essentially tested twice, once as part of a row set and once as part of a column set. Plasmid DNA was purified from each row and column set and tested for GM colony stimulating activity. Two positive sets arose, one from each of the row and column sets. The intersection of the row and column sets identified the cDNA clone which was present in both row and column sets. By this method a cDNA encoding human GM-CSF was isolated.

The basic expression-function strategy applicable for cytokine and growth factor genes has been successfully used to isolate the cDNA genes for human IL-3 (2), human IL-6 (4), mouse IL-4 (3), mouse IL-7 (5), human IL-9 (7) and human HILDA/DIA (6). In some situations this direct and rapid strategy may be the only feasible method when purification is impractical. This is a similar situation for transcription regulatory proteins and DNA binding proteins which may be present in only a few hundred copies per cell. The ability of some of these proteins to tightly bind specific DNA motifs constitutes a highly specific biological assay which can be used in an expression-function cDNA gene isolation strategy.

The Isolation of cDNA Genes for DNA Binding Proteins

Recently Tsai et al. (20) used the heterologous expression-function approach to isolate the erythroid specific DNA binding protein GF-1. The strategy consisted of constructing a cDNA library from MEL cells which express GF-1 binding protein. The library was partitioned into pools of 1,000 to 2,000 clones and transfected into COS cells. The COS cells were harvested two days post-transfection, lysed and cytoplasmic proteins were extracted. The extracts were then tested for specific binding to the GF-1 recognition sequence in a gel retardation assay. By this strategy, Tsai et al. (20) identified two positive pools from about 300,000 clones whose extracts would retard a DNA fragment containing the GF-1 binding site. From each pool, two similar cDNA clones were isolated. One of the clones generated a cytoplasmic extract which behaved identically to a MEL cell preparation of GF-1, while the second clone generated an extract which only partially retarded the DNA fragment as compared to the control. Both cDNA clones on sequencing were found to encode the same polypeptide. The second clone which led to only a partial effect in the gel retardation assay was truncated in the 3' coding region of the polypeptide.

SUMMARY

Functional cDNA genes for cell surface proteins, receptors, growth factors and nuclear and cytoplasmic proteins can be isolated by a heterologous expression-function gene isolation strategy. The method of this strategy is dependent on highly efficient mammalian expression vectors, the properties of COS cells as hosts for SV40-origin-containing mammalian expression vectors, the well-established technology of constructing highly complex plasmid cDNA libraries and on the sensitivity of biological assays.

These technologies are combined to identify a specific gene sequence by its biological phenotype. In theory this can be accomplished by individually testing each single gene of the cDNA library by expression in COS cells. In practice complex cDNA libraries are formed so that multiple cDNA genes can be tested and analyzed simultaneously. When the biological assay can discriminate phenotypic changes at the single cell level as typified by cellular morphology, cell membrane proteins or receptors or intracellular enzymes, then a reiterative expression screening method can be used. This was demonstrated by the isolation of cDNA genes encoding cell membrane proteins (7-10,12).

For the isolation of cDNA genes for growth factors and cytoplasmic enzymes by the expression-function cDNA gene isolation strategy, an alternative method is followed. The cDNA library is partitioned into sets of cDNA clones for transfection and analysis. The number of clones in a set is dependent on the efficiency of the expression vector used and on the sensitivity of the biological assay. For secreted cytokines and growth factors which can be simply detected by factor dependent cell lines, about 1,000 cDNA clones can be tested as a single set. By this strategy the genes for numerous hematopoietic growth factors, cytokines and interleukins have been isolated.

With such numerous successes we can expect that the heterologous expression-function cDNA gene isolation strategy will become an important molecular biology method.

Acknowledgments: I wish to thank Drs. A. Dorner and P. Crosier for their critical reading of the manuscript and helpful comments on organization and structure. In addition I wish to thank Mr. T. Maher, Ms. A. DiCiccio and Ms. J. Geissler for their help in preparing the manuscript and figures.

REFERENCES

1   Wong, G.G., Witek, J.S., Temple, P.A., Wilkens, K.M., Leary,
     A.C., Luxenberg, D.P., Jones, S.S., Brown, E.L., Kay, R.M.,
     Orr, E.C., Shoemaker, C., Golde, D.W., Kaufman, R.J.,
     Hewick, R.M., Wang, E.A. and Clark, S.C. (1985) Science 228,
     810-815.

2   Yang, Y.-C., Ciarletta, A.B., Temple, P.A., Chung, M.P.,
     Kovacic, S., Witek-Giannotti, J.S., Leary, A.C., Kriz, R.,
     Donahue, R.E., Wong, G.G. and Clark, S.C. (1986) Cell 47, 3-
     10.

3   Lee, F., Yokota, T., Otsuka, T., Meyerson, P., Villaret, D.,
     Coffman, R., Mosmann, T., Rennick, D., Roehm, N., Smith, C.,
     Zlotnik, A. and Arai, K.-I. (1986) Proc. Nat. Acad. Sci.
     U.S.A. 83, 2061-2065.

4   Wong, G.G., Witek-Giannotti, J.S., Temple, P.A., Kriz, R.,
     Ferenz, C., Hewick, R.M., Clark,S.C., Ikebuchi, K. and
     Ogawa, M. (1988) J. Immunol. 140, 3040-3044.

5   Namen, A.E. Lupton, S., Hjerrild, K., Wignall, J.,
     Mochizuki, D.Y., Schmierer, A., Mosley, B., March, C.J.,
     Urdal, D., Gillis, S., Cosman, D. and Goodwin, R.G. (1988)
     Nature 333, 571-573.

6   Moreau, J.-F., Donaldson, D.D., Bennett, F., Witek-
     Giannotti, J., Clark, S.C. and Wong, G.G. (1988) Nature 336,
     690-692.

7   Yang, Y.-C., Ricciardi, S., Ciarletta, A., Calvetti, J.,
     Kelleher, K. and Clark, S.C. (1989) Blood 74, 1880-1884.

8   Seed, B. and Aruffo, A. (1987) Proc. Nat. Acad. Sci. U.S.A.
     84, 3365-3369.

9   Stengelin, S., Stamenkovic, I. and Seed, B. (1987) EMBO J.
     7, 1053-1059.

10   Aruffo, A. and Seed, B. (1987) Proc. Nat. Acad. Sci. U.S.A.
     84, 8573-8577.

11   Simmons, D., Makgoba, M.W. and Seed, B. (1988) Nature 331,
     624-627.

12   Staunton, D.E., Dustin, M. and Springer, T.A. (1989) Nature
     339, 61-64.

13   Stengelin, S., Stamenkovic, I. and Seed, B. (1988) EMBO J.
     7, 1053-1059.

14   Allen, J.M. and Seed, B. (1989) Science 243, 378-381.

15   Sims, J.E., March, C.J., Cosman, D., Widmer, M.B.,
     MacDonald, H.R., McMahan, C.J., Grubin, C.E., Wignall, J.M.,
     Jackson, J.L., Call, S.M., Friend, D., Alpert, A.R., Gillis,
     S., Urdal, D.L. and Dower, S.K. (1988) Science 241, 585-589.

16   Yamasaki K., Taga, T., Hirata, Y., Yawata, H., Kawwanishi,
     Y., Seed, B., Taniguchi, T., Hirano, T. and Kishimoto, T.
     (1988) Science 241, 825-828.

17   D'Andrea, A.D., Lodish, H.F. and Wong, G.G. (1988) Cell 57,
     277-285.

18   Munro, S. and Maniatis, T. (1989) Proc. Nat. Acad. Sci.
     U.S.A. 86, 9248-9252.

19   Gearing, D.P., King, J.A., Gough, N.M. and Nicola, N.A.
     (1989) EMBO J. 8, 3667-3676.

20   Tsai, S.-F., Martin, D.I.K., Zon, L.I., D'Andrea, A.D.,
     Wong, G.G. and Orkin, S.H. (1989) Nature 339, 446-451.

21   Kaufman, R.J., Wasley, L.C., Furie, B.C., Furie, B. and
     Shoemaker, C.B. (1986) J. Biol. Chem. 261, 9622-9628.

22   Gluzman, Y. (1981) Cell 23, 175-182.

23   Sambrook, J., Fritsch, E.F. and Maniatis, T. (1989)
     Molecular Cloning. A Laboratory Manual (2nd Ed.) Cold
     Spring Harbor Laboratory Press, Cold Spring Harbor, NY.

24   Kaufman, R.J. (1987) in Genetic Engineering, Principles and
     Methods (Setlow, J.K. ed.) Vol. 9, pp. 155-189, Plenum
     Press, New York, NY.

25   Current Protocols in Molecular Biology (1987) (Ausubel,
     F.M., Brent, R., Kingston, R.E., Moore, D.D., Seidman, J.G.,
     Smith, J.A. and Struhl, K., eds.) Wiley Interscience,
     Boston, MA.

26   Mellon, P., Parker, V., Gluzman, Y. and Maniatis, T. (1981)
     Cell 27, 279-288.

27   Dugaiczyk, A., Boyer, H.W. and Goodman, H.M. (1975) J. Mol.
     Biol. 96, 171-184.

28   McCutchan, J.H. and Pagano, J.S. (1968) J. Nat. Can. Inst.
     41, 351-357.

29   Wysocki, L.J. and Sato, V.L. (1987) Proc. Nat. Acad. Sci.
     U.S.A. 75, 2844-2848.

30   Vartdal, F., Kvazheim, G., Lea, T.E., Bosnes, V.,
     Gauderneck, G., Ugelstad, J., Albrechtsen, D. (1987)
     Transplantation 43, 366-371.

31   Parks, D.R., Lanier, L.L. and Herzenberg, L.A. (1986) in The
     Handbook of Experimental Immunology (Weir, D.M., Herzenberg,
     L.A. and Blackwell, C.C., eds.) 4th edition, Oxford.

32   Hirt, B. (1967) J. Mol. Biol. 26, 365-369.

33   Hanahan, D. (1983) J. Mol. Biol. 166, 557-580.

34   Dower, W.J., Miller, J.F. and Ragsdale, C.W. (1988) Nucl.
     Acids Res. 16, 6127-6145.

35   Kaufman, R.J., Pittman, D.D., Wasley, L.C., Wang, J.H.,
     Israel, D.I., Giles, A.R. and Dorner, A.J. (1985) in Pure
     VIII and the Promise of Biotechnology (Roberts, H.R., ed.)
     pp. 119-145, Baxter Health Care, Brussels, Belgium.

36   Kaufman, R.J., Wasley, L.C., Spiliotes, A.J., Gossels, S.D.,
     Latt, S.A., Larsen, G.R. and Kay, R.M. (1985) Mol. Cell.
     Biol. 5, 1750-1759.

37   Metcalf, D. (1984) The Hematopoietic Colony Stimulating
     Factors, Elsevier Press, Amsterdam.

38   Golde, D.W., Quan, S.G. and Cline, M.J. (1978) Blood 52,
     1068-1072.

39   Luis, A.J., Golde, D.W., Quon, D.H. and Lasky, L.A. (1982)
     Nature 298, 75-76.

MOLECULAR CLONING OF GENES ENCODING TRANSCRIPTION FACTORS WITH THE USE OF RECOGNITION SITE PROBES

Harinder Singh

Howard Hughes Medical Institute
Department of Molecular Genetics and Cell Biology
University of Chicago
Chicago, IL 60637

INTRODUCTION

In genetically tractable prokaryotes and eukaryotes, most DNA binding transcription factors have been identified as the products of trans-acting regulatory loci. In many complex eukaryotic organisms a similar approach to their identification has not been possible. Instead the recent application of sensitive DNA binding assays, in particular DNase I footprinting (1) and gel electrophoresis of protein-DNA complexes (2,3) has led to the detection and characterization of numerous transcription factors. Each of these proteins binds selectively to a distinctive transcriptional control element and is thereby implicated in regulating the activity of target genes containing such an element (4). The isolation of molecular clones encoding transcription factors facilitates a genetic cum biochemical analysis of their structural and functional properties. Prior to the application of the cloning strategy described below, genes encoding transcription factors could be isolated only by screening recombinant DNA libraries with antibody (5-7) or oligonucleotide probes (6,8,9). The latter are generated from partial amino acid sequences of the relevant proteins. Both screening strategies are dependent on the availability of substantial amounts of the purified transcription factor. Even though the purification of transcription factors has been greatly facilitated by the development of improved DNA-affinity matrices (10-12), the requirement for very large amounts of starting material (tissue or cells) makes purification on a preparative scale difficult. The strategy described in this chapter obviates purification of a transcription factor for the purpose of isolating its gene. It

simply requires an appropriate recombinant DNA library constructed for expression in E. coli and a DNA recognition site
probe.   Therefore, this strategy is ideally suited for isolating
clones encoding rare transcription factors.

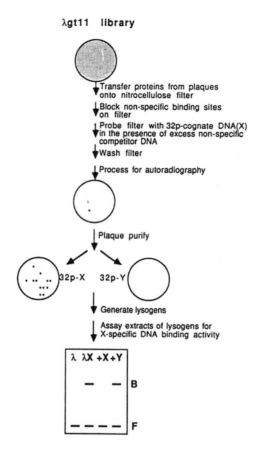

Figure 1. Outline of the strategy for the molecular cloning of
transcription factors with the use of the expression vector
λgt11. X is a recognition site DNA probe, whereas Y is a control
probe that lacks the given recognition site or contains a mutant
version of it. The initial phase involves the identification of
λgt11 recombinants that are specifically defected with DNA probe
X (λX). After plaque purification, the gel electrophoresis DNA
binding assay is used to analyze extracts of λX and λgt11 (λ)
lysogens. Radiolabeled X-DNA is used as a probe in the binding
reactions. F and B refer to free and bound X-DNA respectively.
Reactions in lane +X and +Y are carried out with the λX extracts
and contain an excess of either unlabeled X-DNA or unlabeled Y-
DNA as competitors.

CLONING STRATEGY

The cloning strategy depends on the functional expression in E. coli of high levels of the DNA binding domain of a transcription factor and a strong interaction between this domain and its recognition site. If these conditions are fulfilled, a recombinant clone encoding a transcription factor can be detected by probing protein replica filters of an expression library with radiolabeled recognition site DNA. An outline of the steps involved in identifying and analyzing such a clone, with the use of a recombinant library constructed in the expression vector λgt11, is depicted in Figure 1. The initial phase involves the identification of a recombinant clone that is specifically detected with the binding site DNA probe (X) but not with DNA probes that lack the given binding site or contain a mutant version of it (Y). Such a clone is then shown to encode a β-galactosidase fusion protein of the expected DNA binding specificity. This strategy is derived from that developed for the isolation of genes with the use of antibodies to screen recombinant expression libraries (13-15).

With a ^{32}P-labeled recognition site DNA probe having a specific activity of 10^6 cpm/pmol (ca. 10^8 cpm/µg), it is possible to detect 10^{-2} fmol of active protein in a plaque (assuming a 1:1 stoichiometry for the protein-DNA complex). This detection limit represents 1 pg of a β-galactosidase fusion protein (ca. MW 170,000), which is an amount that is well below the expected level of expression for such a protein in a plaque of the desired λgt11 phage. In fact, overexpression of the lacZ fusion gene should result in the accumulation of ca. 100 pg of the fusion protein in a phage plaque, assuming that there are 10^5 infected cells/plaque and that the β-galactosidase fusion protein represents 1% of the total protein mass (0.1 pg) of an infected cell. The sensitivity of detection achieved with a ^{32}P-labeled recognition site probe (see above) is comparable to that attained with an ^{125}I-labeled primary antibody (16) or a detection system based on a secondary antibody conjugated with alkaline phosphatase (17). A comparison of the signals generated by a DNA binding site probe and an antibody directed against the corresponding protein is illustrated in Figure 2. The λgt11 recombinant (λEB) encodes a β-galactosidase fusion protein that contains the DNA binding domain of the Epstein Barr virus nuclear antigen EBNA-1 (18,19). A protein replica filter prepared from a mixed plating of λEB and control λgt11 recombinant phage was screened initially with a recognition site DNA probe (oriP) that contains two high affinity binding sites for EBNA-1 and subsequently with antibodies directed against EBNA-1 (20). In this case, the higher signal obtained with the DNA binding site probe was attributed to a less sensitive secondary antibody conjugate containing horseradish peroxidase (17) used in immunoscreening. Note that the patterns of plaques detected by the two types of

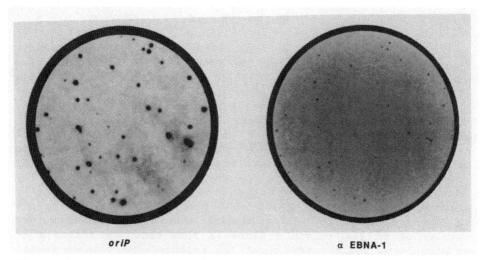

oriP   α EBNA-1

Figure 2. A comparison of fidelity of detection achieved with DNA binding site and antibody probes. A filter prepared from a mixed plating of λEB (ca. 50 pfu) and control λgt11 recombinants (ca. 5 x 10^3 pfu) was screened initially with oriP DNA (2 x 10^6 cpm/ml) and subsequently with α-EBNA1 antibodies (19). In the latter screen, the secondary antibody was a conjugate of goat anti-rabbit IgG with horseradish peroxidase.

probes are superimposable. Therefore, a DNA binding site probe can be used to detect a suitable recombinant phage with the same fidelity as an antibody.

CONSTRUCTION OF EXPRESSION LIBRARY

cDNA Synthesis and Cloning

Successful screening is critically dependent on the frequency with which functional recombinants (in-frame fusions of the DNA binding domain with a bacterial protein segment) are represented in a given cDNA expression library. The cDNA library should be made from mRNA isolated from a cell or tissue source with the highest levels of the desired transcription factor. First-strand cDNA synthesis should be carried out with the use of random primers rather than oligo(dT), since the DNA binding domain may be encoded in the aminoterminal part of the desired protein (5' end of the corresponding mRNA). Adaptors rather than linkers are preferred for ligating the cDNA inserts to the vector, since they avoid digestion of the cDNA with a restriction enzyme (21,22).

Expression Vectors

The phage λgtll appears most suitable for expression screening. It offers the advantages of high cloning efficiency, the expression of relatively stable β-galactosidase fusion proteins and a simple means of preparing protein replica filters. An alternate bacteriophage λ expression vector (λZAP) could be used. This vector obviates subcloning of cDNA inserts into plasmid vectors for their analysis (23). The presence of multiple cloning sites makes possible the use of "forced cloning" strategies for expression of cDNA inserts from its lac promoter. Unlike λgtll, λZAP expresses fusion proteins containing a small aminoterminal segment of β-galactosidase. Therefore, the stability of λZAP encoded fusion proteins may be different from their counterparts encoded in λgtll.

Preparation of Protein Replica Filters

Protein replica filters suitable for screening with DNA recognition site probes are most easily prepared by a series of steps derived from the immunoscreening protocol (19,20,24, see below). This simple procedure has permitted the isolation of many clones encoding different transcription factors, e.g., MBP-1, Oct-2, E12, XBP, YB-1, IRF-1, CREB (see Table 1). Vinson et al. (25) have shown that processing dried nitrocellulose replica filters through a denaturation/renaturation cycle, using 6 M guanidine hydrochloride, significantly enhances the signal from a λgtll recombinant encoding C/EBP (see below). However, it is not possible from this report to compare the sensitivity of the two protocols directly in detecting the C/EBP phage, since with the former the replica filters are not dried. The denaturation/renaturation cycle may in some cases increase the detection signal by facilitating the correct folding of a larger fraction of the E. coli-expressed protein. Alternatively, it may help to dissociate insoluble aggregates that form as a consequence of overexpression. This modified procedure has been successfully used to isolate clones encoding Oct-1, Pit-1, PRDI-BF and RF-X (see Table 1). Since both protocols have been successfully applied, each is detailed below.

1. Plate λgtll library on host strain E. coli Y1090 (3 to 5 x 10^4 pfu/150 mm plate).
2. Incubate the LB-plates at 42°C until tiny plaques are visible (ca. 3 hr).
3. In the meantime soak 132 mm nitrocellulose filters in 10 mM IPTG for 30 min and then air dry them.
4. Overlay each LB-plate with an IPTG-impregnated filter. Avoid trapping air bubbles between the filter and the top agar.
5. Incubate the LB-plates at 37°C for 6 hr.

Table 1
List of Clones Encoding Transcriptional Regulatory Proteins,
Isolated by Screening λgt11 cDNA Libraries with
Recoginition Site DNA Probes

Clone	Binding Site	Genes/Genomes That Contain The Binding Site	References
MBP-1	GGGGATTCCCC	MHC I, β2, Ig$\varkappa$, SV40, HIV, IL-2R, β-IFN	19
Oct-2	ATGCAAAT	IgH, Ig$\varkappa$, H2B, U1, U2, U6, SV40	26,27
Oct-1	ATGCAAAT	IgH, Ig$\varkappa$, H2B, U1, U2, U6, SV40	28
E12	GGCAGGTGG	Ig$\varkappa$	29
XBP	ND	MHC II (Aα)	30
RF-X	CCCCCTAGCAACAG	MHC II (α-DR)	31
YB-1	GACTAACCGGTTT	MHC II (α-DR)	32
IRF-1	AAGTGA	β-IFN	33
PRDI-BF	GAGAAGTGAAAGTG	β-IFN	T. Maniatis, personal communication
Pit-1	GATTACATGAATATTCATGA	Prolactin, Growth hormone	34
CREB	TGACGTC	Somatostatin, enkephalin	35
CNBP	GTGCGGTG	HMG-CoA reductase, LDL receptor	47
HBP-1	ACGTCA	Wheat histone H3 gene	48

Sequences in the binding site column generally represent one
member of a set of related motifs that the cloned proteins
recognize.

6. Cool LB-plates at 4°C for 10 min. Mark position of filter on each plate.

Proceed either to step 7 or 7A

7. Lift nitrocellulose filters and immerse in a deep dish containing BLOTTO [5% Carnation non-fat milk powder, 50 mM Tris HCl, pH 7.5, 50 mM NaCl, 1 mM EDTA, 1 mM DTT]. Incubate for 60 min at room temperature with gentle swirling on an orbital platform shaker.

8. Transfer filters to a dish containing binding buffer [10 mM Tris pH 7.5, 50 mM Nacl, 1 mM EDTA, 1 mM DTT] and incubate as for step 7 for 5 min. Repeat this wash step twice with fresh binding buffer. Filters can be stored in binding buffer at 4°C for up to 24 hr prior to screening.

7A. Lift nitrocellulose filters and air dry them for 15 min at room temperature.

8A. Immerse filters in HEPES binding buffer [25 mM HEPES pH 7.9, 25 mM NaCl, 5 mM MgCl$_2$, 0.5 mM DTT] supplemented with 6 M guanidine hydrochloride (GuHCl). Incubate with gentle shaking at 4°C for 10 min. Repeat this step with fresh HEPES binding buffer containing 6 M GuHCl.

9A. Incubate the filters in HEPES binding buffer containing 3 M GuHCl for 5 min at 4°C (a 1:1 dilution of the 6 M GuHCl from the previous step). Repeat this step four times. Each time use HEPES binding buffer that contains a 2-fold dilution of the GuHCl from the previous step.

10A. Incubate the filters in HEPES binding buffer for 5 min at 4°C. Repeat this step and then block the filters by incubating in BLOTTO (see step 7) at 4°C for 30 min.

11A. Immerse filters in HEPES binding buffer supplemented with 0.25% Carnation non-fat milk powder for 1 min at 4°C. Screen filters as below.

SCREENING OF PROTEIN REPLICA FILTERS

Recognition Site DNA Probe

The highest affinity site among a set of related sequences should be chosen for the synthesis of an oligonucleotide probe. It has been demonstrated that DNA probes containing a single recognition site can be used to isolate the relevant transcription factor clones (MBP-1, XBP, YB-1, see Table 1). However, in a number of cases (Oct-2, Oct-1, E12, see Table 1) the signal was appreciably enhanced with DNA probes containing several copies of the appropriate binding site. Enhancement of the signal with a multi-site probe may be due to the fact that such a probe can simultaneously interact with two or more immobilized protein molecules, thereby increasing the overall stability of the protein-DNA complexes. This type of probe is particularly suitable for the isolation of clones encoding transcription factors with low affinity for their recognition sites.

Multi-site probes can be generated by initially cloning multiple copies (3-10) of the binding site oligomer in a pUC vector (26). The DNA fragment containing tandem binding sites is then end-labeled using α-^{32}P-dNTP and Klenow (36). The labeled binding site DNA fragment is purified by electrophoresis in a 6% non-denaturing polyacrylamide gel. Probe DNA is eluted from the gel and purified on an Elutip-D disposable column (Schleicher and Schuell). With 20 µg of a recombinant pUC plasmid DNA and 200 µCi of α^{32}P-dNTP (5000 Ci/mmol), the probe yield should be 10^8 to 2×10^8 cpm with a specific activity of 2×10^7 to 4×10^7 cpm/pmol. Multi-site probes can also be prepared for screening simply be catenation of a binding site oligonucleotide with DNA ligase, followed by nick translation (25).

Nonspecific Competitor DNA

The addition of an excess of nonspecific competitor DNA in the probe solution reduces background as well as minimizes the detection of recombinant phage encoding non-sequence-specific DNA binding proteins. Several different competitor DNAs have been used to screen expression libraries successfully including poly d(I-C) and denatured calf thymus DNA (19,26,33). The latter DNA is preferred since its inclusion in the probe solution reduces background as well as eliminates signal from recombinant phage encoding single-stranded DNA binding proteins.

Binding and Wash Conditions

The association constants of transcription factors are dependent on ionic strength, temperature and pH. Therefore, these parameters can be manipulated in the binding and wash steps to optimize the detection of a relevant transcription factor clone. If the transcription factor being cloned has an exogenous metal ion requirement (e.g., Mg^{2+}), the binding and wash buffers should be appropriately supplemented.

1. Incubate replica-filters as a stack in binding buffer containing 2.5×10^6 cpm/ml of ^{32}P-labeled recognition site DNA and 5 µg/ml sonicated and heat denatured calf thymus DNA. Shake dish containing the filters gently for 60 min at room temperature.
2. Wash filters in batches, 4 times (7.5 min each wash, 30 min total) at room temperature with aliquots of the binding buffer.

3. Dry filters on blotting paper and perform autoradio-
    graphy with a tungstate intensifying screen at -70°C
    for 12 to 24 hr.
4. Identify the presumptive positive phage plaques by
    aligning the autoradiographs with the LB-plates. To
    reduce the number of false positives, generate auto-
    radiography exposures of varying times with the primary
    filters. Short versus long exposures help to dis-
    tinguish spots with intense centers (likely to be
    artifacts) from those with a diffuse halo-like appear-
    ance (likely to represent true positives).
5. Pick presumptive positives for secondary screens. True
    positives will show enrichment of 10- to 100-fold.
6. Screen secondary platings of true positives with the
    wild-type recognition site probe as well as with
    control DNA probes that either lack the binding site or
    contain mutant versions.
7. Plaque purify phage which are specifically detected
    with the wild-type recognition site probe but not with
    control DNA probes.

CHARACTERIZATION OF RECOMBINANT DNA BINDING PROTEINS

After the isolation of a recombinant phage that is specifically detected with a given binding site probe, but not with
control DNAs, it is necessary to demonstrate that this clone
encodes a recombinant protein of the expected DNA binding
specificity. In the case of a λgt11 recombinant, this is simply
achieved by isolating lysogenized E. coli clones and assaying
extracts of induced lysogens for a β-galactosidase fusion protein
that specifically binds the recognition site probe used in the
screen (see below and Figure 3). Figure 3 shows an analysis of
lysogen extracts with the use of the gel electrophoresis DNA
binding assay. Recombinant phage λh3 and λh4 encode βgal-MBP1
fusion proteins which bind specifically a regulatory element of a
major histocompatibility complex (MHC) class I gene promoter
(19). More rigorous characterization of the DNA binding specificity of the recombinant protein is achieved by chemical and
enzymatic footprinting along with the analysis of mutant binding
sites. It is important to note that even though the above
analyses allow the DNA binding specificity of the recombinant
protein to be thoroughly compared with the native protein, they
do not prove that the cloned protein is the desired transcription
factor. Eventually, direct structural analyses are necessary to
resolve this issue. Antibodies generated against the cloned
protein permit the detection of shared antigenic determinants
(28). Peptide mapping performed on analytical amounts of the
native and cloned proteins constitutes a definitive structural
comparison (38).

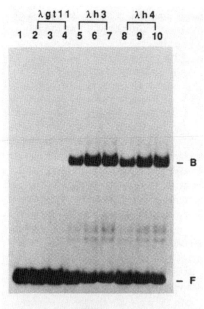

Figure 3. Extracts prepared from induced cultures of λgt11, λh3 and λH4 lysogens were incubated with MHC DNA (25,000 cpm, 1.25 fmoles) and 4 μg poly d(I-C) and the reactions resolved by electrophoresis in a nondenaturing polyacrylamide gel. Free and bound probes are indicated by F and B, respectively. The total protein concentrations in lanes 2, 3 and 4 were 9, 18 and 27 μgs, lanes 5, 6 and 7 were 6, 12 and 18 μgs and lanes 8, 9 and 10 were 6, 12 and 18 μgs.

1. Recombinant phage lysogens are isolated by infecting E. coli strain Y1089 and screening for temperature-sensitive clones (24).

2. Grow 2 ml cultures of the recombinant lysogens at 32°C with good aeration.

3. When cultures reach an O.D.$_{600}$=0.5, shift them to 44°C for 20 min.

4. Add IPTG to 10 mM to induce expression of the β-galactosidase fusion protein and shift the cultures to 37°C for 60 min.

5. Spin 1 ml aliquots of induced cultures in a microfuge for 1 min at room temperature.

6. Discard supernatants and resuspend pellets in 100 μl extract buffer [50 mM Tris HCl pH 7.5, 1 mM EDTA, 1 mM DTT, 1 mM PMSF].

7. Quickly freeze the resuspended cells in liquid nitrogen.

8. Thaw frozen cell suspensions, adjust to 0.5 mg/ml lysozyme and place on ice for 15 min.

9. Adjust cell suspensions to 1 M NaCl, mix thoroughly and incubate on a rotator for 15 min at 4°C.
10. Centrifuge the lysates in microfuge for 30 min at 4°C.
11. Dialyze the supernatants on Millipore filters (type VS, 0.025 μm pore size) against 100 ml extract buffer for 60 min at 4°C. Millipore filters should be floated on dialysis buffer in a 150 mm petri-dish before samples are applied.
12. Freeze the dialyzed extracts immediately and store at -70°C till needed.
13. The DNA binding properties of the recombinant protein can be tested in various ways including the gel mobility shift assay (19,37, see Figure 3) and DNase I footprinting (1,8).

CONCLUSIONS

A strategy for cloning transcription factors by screening cDNA expression libraries with recognition site probes has been described. This strategy circumvents purification of a transcription factor for the purpose of isolating its gene. It simply requires a cDNA library constructed in the phage vector λgt11 and a DNA recognition site probe. As a result of its simplicity and its generality, this strategy has greatly facilitated the analysis of proteins that regulate transcription. Table 1 lists the various transcription factor clones that have been isolated with this approach.

The DNA binding domains of a large number of transcription factors contain either a helix-turn-helix motif or the "zinc finger" motif (39-42). Clones encoding proteins with either of these structural motifs can be detected by in situ screening with the relevant recognition site DNAs. The protein encoded by the MBP-1 cDNA (Table 1) contains "zinc finger motifs". In contrast, the Oct-2 and Oct-1 cDNA clones (Table 1) encode proteins with a predicted helix-turn-helix motif. Thus, the screening method appears not to be restricted to a sub-class of DNA binding domains.

Many transcription factors are functional homodimers. The binding sites of these proteins exhibit two-fold rotational symmetry. In these cases the affinity of the monomer for the complete binding site is significantly lower than that of the dimer (42,43). Clones encoding such homodimeric proteins can be detected by in situ screening. The mammalian transcription factor C/EBP binds its recognition site with high affinity as a dimer (44). A λgt11 recombinant encoding this protein can be detected by screening plaque lifts with the corresponding binding site probe (25). Interestingly, the region of C/EBP required for dimerization, the "leucine zipper," is shared by a number of regulatory proteins including GCN4, Fos, myc and Jun (5). Murre et al. (29) have used the screening strategy described herein to isolate cDNAs encoding a mammalian enhancer binding protein (E12, Table 1) that requires a new type of dimerization domain. These examples clearly show that clones encoding proteins that bind DNA as homodimers, using different dimerization domains, can be successfully screened as a consequence of their functional expression in E. coli.

Most functional DNA binding domains, including elements required for dimerization, are contained within relatively small protein segments, approximately 60 to 200 amino acids, e.g., the DNA binding domains of EBNA-1 (18), GAL-4 (45), GCN4 (46) and Sp-1 (8); therefore successful screening is not dependent on full length cDNAs. It simply requires that a given expression library contain partial cDNA clones spanning the DNA binding domain of the desired transcription factor.

Acknowledgments: I thank Kathy Cobb for expert help in the preparation of this chapter.

REFERENCES

1   Galas, D. and Schmitz, A. (1978) Nucl. Acids Res. 5, 3157-3170.

2   Fried, M. and Crothers, D. (1981) Nucl. Acids Res. 9, 6505-6525.

3   Garner, M. and Revzin, A. (1981) Nucl. Acids Res. 9, 3047-3060.

4   Maniatis, T., Goodbourn, S. and Fischer, J.A. (1987) Science 236, 1237-1245.

5   Landschulz, W.H., Johnson, P.F., Adashi, E.Y., Graves, B.J. and McKnight, S.L. (1988) Genes and Development 2, 786-800.

6   Walter, P., Green, S., Green, G., Krust, A., Bornert, J.-M., Jeltsch, J.-M., Staub, A., Jensen, E., Scrace, G., Waterfield, M. and Chambon, P. (1985) Proc. Nat. Acad. Sci. U.S.A. 82, 7889-7893.

7   Weinberger, C., Hollenberg, S.M., Ong, E.S., Harmon, J.M., Brower, S.T., Cidlowski, J., Thompson, E.B., Rosenfeld, M.G. and Evans, R.M. (1985) Science 228, 740-742.

8   Kadonaga, J.T., Carner, K.R., Masiarz, F.R. and Tjian, R. (1987) Cell 51, 1079-1090.

9   Bodner, M., Castrillo, J.L., Theill, L.E., Deerinck, T., Ellisman, M. and Karin, M. (1988) Cell 55, 505-518.

10 Chodosh, L.A., Carthew, R.W. and Sharp, P.A. (1986) Mol. Cell. Biol. 6, 4723-4733.

11 Kadonaga, J.T. and Tjian, R. (1986) Proc. Nat. Acad. Sci. U.S.A. 83, 5889-5893.

12 Rosenfeld, P.J. and Kelly, T.J. (1986) J. Biol. Chem. 261, 1398-1408.

13   Helfman, D.M., Feramisco, J.R., Fiddes, J.C., Thomas, G.P. and Hughes, S.H. (1983) Proc. Nat. Acad. Sci. U.S.A. 80, 31-35.

14   Young, R.A. and Davis, R.W. (1983) Proc. Nat. Acad. Sci. U.S.A. 80, 1194-1198.

15   Young, R.A. and Davis, R.W. (1983) Science 222, 778-782.

16   Broome, S. and Gilbert, W. (1978) Proc. Nat. Acad. Sci. U.S.A. 75, 2746-2749.

17   Learly, J.J., Brigati, D.J. and Ward, D.C. (1983) Proc. Nat. Acad. Sci. U.S.A. 80, 4045-4049.

18   Rawlins, D.R., Milman, G., Hayward, S.D. and Hayward, G.S. (1985) Cell 42, 859-868.

19   Singh, H., LeBowitz, J.H., Baldwin, A.S. and Sharp, P.A. (1988) Cell 52, 415-423.

20   Singh, H., Clerc, R.G. and LeBowitz, J.H. (1989) BioTechniques 7, 252-261.

21   Haymerle, H., Herz, J., Bressan, G.M., Frank, R. and Stanley, K.K. (1986) Nucl. Acids Res. 21, 8615-8624.

22   Wu, R., Wu, T. and Ray, A. (1987) Methods Enzymol. 152, 343-349.

23   Short, J.M., Fernandez, J.M., Sorge, J.A. and Huse, W.D. (1988) Nucl. Acids Res. 16, 7583-7600.

24   Huynh, T.V., Young, R.A. and Davis, R.W. (1985) in DNA Cloning, Vol. 1: A Practical Approach (Glover, D.M. ed.), pp. 49-78, IRL Press, Oxford.

25   Vinson, C.R., LaMarco, K.L., Johnson, P.F., Landschulz, W.H. and McKnight, S.L. (1988) Genes and Development 2, 801-806.

26   Staudt, L.M., Clerc, R.G., Singh, H., LeBowitz, J.H., Sharp, P.A. and Baltimore, D. (1988) Science 241, 577-580.

27   Müller, M.M., Ruppert, S., Schaffner, W. and Matthias, P. (1988) Nature 336, 544-551.

28   Sturm, R.A., Das, G. and Herr, W. (1988) Genes and Development 2, 1582-1599.

29   Murre, C., Schonleber-McCaw, P. and Baltimore, D. (1989) Cell 56, 777-783.

30   Hsiou-Chi, L., Boothby, M.R. and Glimcher, L.H. (1988) Science 242, 69-71.

31   Reith, W.E., Barras, E., Satola, S., Kobr, M., Reinhart, D., Herrero Sanchez, C. and Mach, B. (1989) Proc. Nat. Acad. Sci. U.S.A. 86, 4200-4204.

32   Didier, D.K., Schiffenbauer, J., Woulfe, S.L., Zacheis, M. and Schwartz, B.D. (1988) Proc. Nat. Acad. Sci. U.S.A. 85, 7322-7326.

33   Miyamoto, M., Fujita, T., Kimura, Y., Maruyama, M., Harada, H., Sudo, Y., Miyata, T. and Taniguchi, T. (1988) Cell 54, 903-913.

34   Ingraham, H.A., Chen, R., Mangalam, H.J., Eisholtz, H.P., Flynn, S.E., Lin, C.R., Simmons, D.M., Swanson, L. and Rosenfeld, M.G. (1988) Cell 55, 519-529.

35   Hoeffler, J.P., Meyer, T.E., Yun, Y., Jameson, J.L. and Habener, J.F. (1988) Science 242, 1430-1433.

36   Maniatis, T., Fritsch, E.F. and Sambrook, J. (1982) Molec-
     ular Cloning: A Laboratory Manual, pp. 113-114, Cold Spring
     Harbor Laboratory Press, Cold Spring Harbor, NY.
37   Chodosh, L.A. (1988) in Current Protocols in Molecular
     Biology (Ausubel, F.M., Brent, R., Kingston, R.E., Moore,
     D.D., Seidman, J.G., Smith, J.A. and Struhl, K., eds.), pp.
     12.2.1-12.2.7, John Wiley and Sons, New York, NY.
38   Clerc, R.G., Corcoran, L.M., LeBowitz, J.H., Baltimore, D.
     and Sharp, P.A. (1988) Genes and Development 2, 1570-1581.
39   Pabo, C.O. and Sauer, R.T. (1984) Annu. Rev. Biochem. 53,
     293-321.
40   Gehring, W.J. (1987) Science 236, 1245-1252.
41   Evans, R.M. and Hollenberg, S.M. (1988) Cell 52, 1-3.
42   Schleif, R. (1988) Science 241, 1182-1187.
43   Ptashne, M. (1986) A Genetic Switch. Cell Press and
     Blackwell Scientific Publications, Cambridge, MA.
44   Landschulz, W.H., Johnson, P.F. and McKnight, S.L. (1989)
     Science 243, 1681-1688.
45   Keegan, L., Gill, G. and Ptashne, M. (1986) Sciences 231,
     699-704.
46   Hope, I.A. and Struhl, K. (1986) Cell 46, 885-894.
47   Rajavashisth, T.B., Taylor, A.K., Andalibi, A., Svenson,
     K.L. and Lusis, A.L. Science (1989) 245, 640-643.
48   Tabata, T., Takase, H., Takamama, S., Mikami, K., Nakatsuka,
     A., Kawata, T., Nakayoma, T. and Iwabuchi, M. (1989) Science
     245, 965-967.

ACG, 106
Actinorhodin, 57,60
Adenovirus, 108-110,252
    EIA protein, 193
African green monkey kidney
    cells see CV-1
afsR, 61
Agrobacterium tumefaciens T-
    DNA, 77
    transformation, 73,79,231
Aldosterone receptor, 185,194
Alfalfa mosaic virus RNA, 101
Alkaline lysis for plasmid
    preparations, 300,301
Alkaline phosphatase, 61
Alternative splicing, 139-
    181,203,218
Aminocyclitol acetyltrans-
    ferase, 61
Aminoglycoside 3'-phospho-
    transferase, 63
2-Aminopurine, 253
Amycolatopsis orientalis, 55
α-Amylase, 61
Anchored PCR, 123,124
Androgene receptor, 185
Anti-sense RNA, 75,76,204,
    209,212,238
aphE, 61,63
Apotyrosinase, 55
Arabidopsis, 84
as-1, 77,78,80-82
as-2, 77,80-82
ASF-1, 77-83
ASF-2, 77,78,80-83
AsnC, 62
Asymmetric PCR, 126
AUG, 100,105-108,244,245,247,
    249,251,257,260
Auxin overproduction, 74

Avian myeloblastosis virus, 46
Avidin, 307

Bacillus cereus electropora-
    tion, 282
Bacillus licheniformis, 64
BamHI, 56,91
bap, 62,63
bar, 62
Barley, 228,230,232,236
B cells, 168
Benzamidine Sepharose, 10
Beta-galactosidase see galac-
    tosidase
Beta-globin see globin
Beta-glucuronidase see GUS
BglII, 55,91
Bialaphos, 60,62
bicoid, 80
Biotin-avidin, 307
Biotinylated substrates,
    211,212
BLOTTO, 323
Bovine growth hormone in
    S. lividans, 64
Bovine pancreatic trypsin
    inhibitor see BPTI
BPTI, 5,17
Branch point sequence, 141
Bread-making wheat quality,
    225-241
Breast cancer, 28
Bromoacetic acid, 88
brpA, 63

ØC31, 57,59,65
Cab see chlorophyll binding
    protein
c-Abl, 245
Calcitonin/CGRP, 150,172-175

Calcitonin gene related
    peptide see calcitonin/CGRP
Campylobacter jejuni electro-
    poration, 275,278,280,290
CaMV, 74,77-79
    35S promoter, 232
Cap, 101-104,208,209,217
    binding protein complex,
        257,258
    structure, 244-250,256,
        258-260
Capping, 201
Carboxypeptidase A, 95-97
CAT, 64,106,187,188,196
Cauliflower mosaic virus see
    CaMV
CD45, 168,169
cDNA gene isolation, 297-316
CD spectrum of t-PA kringle-2,
    14,15
Cell surface proteins, 297,299-
    301,314
293 cells, 252
Cephalosporium acremonium, 63
Chloramphenicol acetyltrans-
    ferase see CAT
Chloramphenicol acetyltrans-
    ferase mRNA, 104,105
Chlorophyll binding protein,
    74,77,80
Chloroquine, 301
CHO cells, 251
Chromosome walking, 23,24
Citraconic anhydride, 10
c-Jun, 245
ClaI, 30,55
CM-CSF see granulocyte-
    macrophage colony
    stimulating factor
c-myc, 247
Collagen, 245
Colony stimulating factors,
    310-313
Consensus sequence, 107,127,
    140,141,208,217,247,259,
    260
Contigs, 24
Copia retroposon, 155
Cortisol receptor, 185,189
cos, 54

COS cells, 171,261,298,299,301
    303-314
COS-1, 197
COUP-TF, 194,195
Couscous, 225
Cowpox virus, 254
CPA see carboxypeptidase A
CREB, 83
Crotalus adamenteus venom, 45
CUG, 106
CV-1, 196,299
Cyanogen bromide in cleavage
    of fusion protein, 9
Cycloheximide, 262,263
CYH2, 155
Cytokine, 301,307-310,313,314

DAI see double-stranded RNA
    activated inhibitor
Deacetoxycephalosporin C
    synthase, 61
DEAD box, 259
Deglycosylation and folding,
    3,4
2'-deoxy-β-D-ribofuranosyl
    nucleosides, 37
Deoxy-2'-β-ribose, 38
2'-deoxy-L-ribonucleosides, 38
DHFR, 106,107,110,248,263
DHFR-(His)$_2$, 95
DHFR-(His)$_6$, 90,91,94
Dihydrofolate reductase see
    DHFR
Disulfide bonds and folding, 3
Disulfide shuffling, 3,7,9
Dithiothreitol in disulfide
    shuffling, 7
DMS and footprinting, 132
α-DNA, 37-52
DNA-binding factors, 74-82,324
    binding protein, 297,299,
        313,314
    factor-binding sites, 78-
        81,319,320
DNase I footprinting, 74,76,
    317
doublesex, 159,160,174,175,178
Double-stranded RNA activated
    inhibitor, 108,252-255
Dreiding models, 38

Drosophila gap gene, 194
  homeotic genes, 76
  orphan receptors, 195
  P-element, 154,156
  sex determination, 158-
    162,174
  splicing, 203,218
  steroid hormone receptors,
    194,195
dsx see doublesex
Durum wheat, 225-227
Dwarfism, 1

E1A protein, 193
EBNA-1, 319,320
E. coli y1090, 321,326
  electroporation, 275,278,
    280,281,284,286,288,291-
    293
  genetic exchange with
    Saccharomyces cerevisiae,
    63
EcoRI, 23,30,64,91
EcoRV, 55
Elastin, 236
Electroporation, 275-295
  of streptomycetes, 54
Electro-transformation see
  electroporation
EMC virus see encephalomyo-
  carditis virus
Encephalomyocarditis virus,
  102,103,246-248
Endosperm of wheat, 226,231
Epibromohydrine, 88
Epidermal growth factor, t-PA
  homology, 11
erbA, 194,195
ERE see estrogen-responsive
  element
eryG, 61
Erythromycin, 60
Erythropoietin receptor,
  308,309
Escherichia coli see E. coli
Esterase D, 23,24
Estradiol receptor, 185,194
Estrogen receptor, 185,189,191
  human, 192,193
  responsive element, 190,192

Exon, 140,201
  skipping, 141,142,149,150,
    157,164-169

F9 see teratocarcinoma
FACS see fluorescence-
  activated cell sorting
Factor VIII, 309
Factor IX, 298
FcRI and FcRII, 307
Ferritin, 245,264
Fibronectin type I in t-PA, 11
Finger domain in t-PA, 11,12
Fluorescence-activated cell
  sorting, 307
Folding intermediates, 2,3
Folding of proteins, 1-19
Fos, 327
Fusion protein, 4,5
  by PCR, 128-130

G418, 31,32,102
GAL4, 192
β-galactosidase fusion,
  184,208,319,326
Gap gene, 194
GCN4, 83,327,328
  mRNA, 245,251
GEF see guanine-nucleotide
  exchange factor
Gel retardation, 132
Gene expression in transgenic
  plants, 73-86
Genomic footprinting with PCR,
  132
Genomic libraries, 289
GF-1, 313
Gliadin, 226-230,233-236
β-Globin gene, human, 263
  mRNA, inhibition of
    translation, 48
Glucocorticoid receptor,
  human, 183,184,189,191-194
  rat, 189,193
  responsive element, 190-
    192,196
β-glucuronidase see GUS
Glutathione in protein
  folding, 7
Glutenin, 226,229,230,232-238

Glycerol as cryoprotectant,
    280,281
Glycerol gradient centrifuga-
    tion, 210
Glycerol kinase, 60
Glycerol-3-phosphate
    dehydrogenase, 60
GM-CSF see granulocyte-
    macrophage colony
    stimulating factor
Gram negative cells and
    electroporation, 285
Granulocyte-macrophage colony
    stimulating factor, 263,310-
    313
GRE see glucocorticoid-
    responsive element
Growth factor, 297,299,301,302,
    307,309,313,314
    hormone, 322
    receptor, 297
GRP78 mRNA, 247
GT-1, 76-78
Guanine-nucleotide exchange
    factor, 109,251
GUS, 73,79
    gene fusion, 232
gylA, 60
gylABX, 62
gylB, 60
gylpl, 62
gylR, 62
gylX, 60,61

Hairpin structure, 245
HBP-1, 81-83
HCR see heme controlled
    repressor
Heat shock protein 189,256
HeLa cells, 172,173,196,247
Helicase, RNA, 259
    in splicing, 217
Heme controlled repressor,
    251,256
Hepadnavirus, 124
Heparin, 211
hER see estrogen receptor,
    human
Herpes simplex virus, 162

Heterogeneous nuclear
    ribonucleoprotein see hnRNP
hex-1 77,78,80,81
hGR see glucocorticoid
    receptor, human
Highly repetitive sequences,
    23,24
HILDA/DIA, 313
HindIII, 30,79,91
His4, 251
Histocompatibility complex,
    127
Histone mRNA, 263
HIV, 103,254,255,261
    gene expression, 218
πH3M, 306
HMG-CoA reductase, 322
hnRNP, 202,203,213,216,217
Homeotic genes, Drosophila, 76
Hordein, 228,229,233,236
Hormone responsive element,
    188,190,191
Housekeeping gene, 28
H-ras splicing, 166,167
HRE see hormone responsive
    element
H.R.S. see highly repetitive
    sequence
HSBF, 77,78,81
HSP see heat shock protein
Human growth hormone, 1
Human plasminogen, 15
hunchback, 80
Hygromycin phosphotransferase,
    64
    resistance, 57

IBP see intron binding protein
IDX see intron derived exon
IL-1, 307
Iminodiacetic acid, 87,88
Immunoglobulin gene, 263
Immunoglobulin κ light-chain
    splicing, 162,163
Inclusion bodies, 1,2,4-6,9,
    92,97
Influenza virus, 252
Initiation of mRNA, 100
Initiator codon, 244,245
Intensifying Screen, 325

Interferon, 252,254,255,307
    in S. lividans, 64
    receptor, 308
Interleukin, 314
    in S. lividans, 64
    receptor, 303
Intron, 140,141,201,217
    binding protein,
        210,213,216
    derived exon, 166,167
    exclusion, 143,149,152-156
    inclusion, 143,152-156
Inverse PCR, 119-121,128,130
Invertase folding, 3
IPNS see isopenicillin N
    synthase
IPTG see isopropyl-β-D-
    thiogalactopyranoside
IRE see iron-responsive element
IRE-BP see IRE binding protein
Iron-responsive element, 246
    binding protein, 246
IS110, 65
Isopenicillin N synthase, 61,63
Isopropyl-β-D-thiogalacto-
    pyranoside, 90,321,323

JUN, 82,83,327

Kanamycin resistance see nptII
KC515, 59
Kinase, 252-255,260
Kinase on retinoblastoma
    protein, 27
Klebsiella pneumoniae
    electroporation, 285
Klenow fragment, 116-118,126,
    187,324
kni, 194
Kringle, 2,11-17

β-Lactamase in S. lividans, 64
Lactobacilli, 290
Lactococcus lactis electropora-
    tion, 281-283
lacZ, 106,125,128,130,319,321
Lambda phage vectors, 288
λgt11, 318-322,325-327
λZAP, 321
Lariat, 141,164,201,217

LDL receptor, 322
Leucine zipper, 263,327
    domain in retinoblastoma,
        25
    proteins, 83
Leukemia, 28
LexA, 192
Ligand binding protein, 300
LUC see luciferase
Luciferase, 29-32,187,188,
    196,197
    in streptomycetes, 58
Lung carcinoma, 28
Lux see luciferase
luxA, 58
luxB, 58
Lymphokine, 263

M13, 125
MT4 cells, HIV-infected, inhi-
    bition of translation, 47-
    49
Maize zein, 232
Mammalian mRNA translation,
    99-113
mel, 55
MEL cells, 313
β-mercaptoethanol in disulfide
    shuffling, 7
Methionyl-tRNA$_i$, 108
Methylation interference,
    76,77
Methylphosphonates, 37
met-tRNA binding, 250
Micromonospora echinospora, 57
    purpurea, 55
Mineralocorticoid receptor,
    185
Minipreparation of plasmids,
    300
Mitosis and dephosphorylation,
    258
Mo, 311
mRNA translation, 99-113
Mutagenesis by PCR, 127
myc, 193,327
Mycobacterium bovis, 64
Myocardial infarction, t-PA
    in, 11
Myoglobin folding, 4

N^ϵ-benzyloxycarbonyl-L-lysine, 88
Negative feedback regulation, 148
Neo see neomycin phosphotransferase
Neomycin phosphotransferase, 29,248
NcoI, 64
Ni(II)-nitrilotriacetic acid, 88,95,96
NMR spectrum, t-PA kringle-2, 15,16
Nocardia orientalis, 55
Nopaline synthase, 77
Northwestern blotting, 216
nptII, 74
NTA-Ni(II) see Ni(II)-nitrilotriacetic acid

Octopine synthase, 77
α-Oligodeoxynucleotides, 37-52
Oligo(dT), 320
opague-2, 81,83
Open reading frame, 245
ORF see open reading frame
Orphan receptors, 185,193-196
Osteosarcoma, 28
Ovalbumin gene, chicken, 194
Oxytetracycline, 60,61

PABP see polyA binding protein
Panning, 306
Pantoyl lactone and electroporation, 286
Pasta, 225,234,235
pATH vectors, 26,27
PCR, 28,115-137
PDGF see platelet-derived growth factor
PDI see protein disulfide isomerase
P-element, 154,156,167
Periplasmic space, 4,5
Phosphatase, 255
Phosphodiesterase, calf spleen, 45
Phosphoramidites, 40
Phosphorothioates, 37
Phosphotriester, 38,39

Photolyase, 61
Phytochrome gene of rice, 76,77
Picornavirus, 246-248,260,264
Picornavirus RNA, 100,102
pIJ702,55,60,62,64
Piperidine and footprinting, 132
Plasminogen, human, 15
Platelet-derived growth factor, 245
Poliovirus, 100-103,246,247, 260-262
PolyA binding protein, 249,260
    sites, 150
    tail, 123,124,248,249,263
    translation activation, 248
Polyadenylation, 155,173,201
    choice, 172,174
poly d(I-C), 324,326
Poly-His fusion proteins, 90-97
Polymerase chain reaction see PCR
Polynucleotide kinase, T4, 44
Polyposis, 21
Polysomes, 264
Preproinsulin mRNA, 103
Progesterone receptor, 185, 189,192,194
Prolactin, 322
Prolamin, 226-234,236,238
Prostate carcinoma, 28
Protease, 260
Protein A of S. aureus, 4,17
Protein denaturation, 2
    disulfide isomerase, 3
    folding, 1-19
    recombinants purified with metal chelates, 87-98
Prothrombin, t-PA homology, 12
Pseudogenes in cereal, 228
Pseudospliceosome, 211
PstI, 56,64
pUC vector, 324
Puromycin, 60
PvuII, 55
pXM, 306

13q14 and retinoblastoma,
    22,23

Rabbit reticulocyte extract,
    degradative properties, 47
    inhibition of translation,
        47-49
RACE see single-sided PCR
RAR see retinoic acid receptor
RARE see retinoic acid-
    response element
RB see retinoblastoma
RbcS, 74,76-79
recA E. coli electroporation,
    284,285,288
Refractile bodies see inclu-
    sion bodies
Reovirus, 252,254,258
Reovirus mRNA translation, 107
Reticulocyte lysate, 251,252,
    255
Retinoblastoma, 21-35
    gene, 24,25
Retinoic acid, 196
    receptor, 183-200
    response element, 197
Retroposon, 155
Rev, 218
Reverse transcriptase, 122,
    124,125
    M-MLV, 46
Ribonuclease folding, 2,3,7
Ribosome binding, 256-258,
    260,262,264
    binding sites, 100,102,103
    dissociation, 250
    initiation, 244,245
    recognition, 260
Ribulose 1,5-bisphosphate
    carboxylase see RbcS
RNA helicase, 259
    in splicing, 217
RNA polymerase III, 131
RNase H, avian myeloblastosis
    virus, 46
    cleavage, 203,204,209,210
    Drosophila, 45
    E. coli, 45,46
RNase protection, 211,216
RNP consensus sequence, 208,217

RPL32, 153,154,167,175,176
rRNA, 61
RSRDRE, 178
RT see reverse transcriptase
Rye, 228

Saccharomyces cerevisiae,
    genetic exchange with E.
    coli, 63
    introns, 153
    PABP, 249
    suppressor mutants, 251
sak, 64
Salmonella electroporation,
    285
ScaI, 30
Scanning model for initiation
    of translation, 100,244,
    245,248,251
SCF see splicing complementing
    factor
SCP2*, 57
Secretion signal, 5
Self splicing, 203
Sense mRNA, 75,76
Sepharose CL-6B, 88,89
Sex determination
    Drosophila, 158-162,174
Sex-lethal, 159-162,165,167,
    174-177
Sicco-5, 234
Single-sided PCR, 122-124
SLP1, 56
SmaI, 30
Smaller nuclear ribonucleo-
    protein see snRNP
Small nuclear RNA see snRNA
snRNA, 141,205,216
snRNP, 141,202-216,218
    Hela, 205
Somatostatin, 322
SphI, 55
Spliceosomes, 141,202,210,
    212,218
Splicing complementing factor,
    213,214
    factors, 201-224
    RNA, 139-181
    regulators, 175,176
    site competition, 151-152

Spm, 81
SstI, 55
STII, 5,13,16,17
Staphylococcus aureus phage
    42D, 64
Staphylokinase in S. lividans,
    64
Stem and loop structure of
    mRNA, 245-247,254,255,263
Steroid receptor, 183-200
sti, 56
Streptavidin agarose, 211
Streptococci, 290
Streptomyces alboniger, 57
    albus G, 57
    ambofaciens, 58,61
    antibioticus, 55
    avermitilis, 55,65
    cacaoi, 56
    clavuligerus, 55,57,61,63
    coelicolor, 54,56,57,60,
        64,65
    erythreus, 57
    fradiae, 63,65
    glaucescens, 57,66
    granaticolor, 55
    griseofuscus, 55,59
    griseus, 61,63
    hygroscopicus, 55,62
    jumonjinensis, 61
    lincolnensis, 65
    lipmanii, 61
    lividans, 54,56,57,61,64,65
    marcescens, 64
    peucetius, 57
    polysporaerythraea, 61
    rimosus, 57,61
    veolaceoniger, 55
Streptomycetes in cloning, 53-
    72
Streptomycin, 60
Streptomycin-3"-phosphotrans-
    ferase, 61,63
Sucrose gradient centrifuga-
    tion, 215
Sulfitolysis in protein
    folding, 9,10
su(w^a), 154-156,167,175,178

SV40 replication origin-
    containing vector, 299,300,
        306,314
    T antigen splicing, 164,165
    translation, 106
Sxl see Sex lethal
Synovial sarcoma, 28
Synthons, 37-40

T antigen in COS cells, 299
    splicing, 164,165
T cells, 168,169
3T3 mouse fibroblast extract,
    degradative properties, 47
T4 DNA polymerase, 126
Taq polymerase, 117-119,125,
    126,128
TAR, 254,255,261
tat see trans-activating
    protein
TAT see tyrosine aminotrans-
    ferase
TATA box, 232,233
TdT see terminal transferase
TER, 154,166
Teratocarcinoma cells, 172-
    174,197
Terminal transferase, 124
Testosterone receptors, 185
Tetracenomycin, 60
TFIIIA, 131,191
TGA1a, 83
TGA1b, 83
TGFα see transforming growth
    factor α
Thermus aquaticus, 117,118
Thiostrepton resistance,
    55,57,58,61
Thymidine kinase gene, 162
Thyroid hormone receptor, 183-
    200
    responsive element, 190,197
tipA, 61
Tissue plasminogen activator
    see t-PA
TM1, 169,170
tms2, 74
Tn5, 57
Tn4560, 65

Tobacco transformation,
   231,232
t-PA, 2,11-17,310
tra see transformer
Trans-activating protein, 254
Transcription factor, 297
   cloning, 317-330
Transesterification, 141
Transferrin receptor, 245,246
Transformation of strepto-
   mycetes, 54,55
transformer, 158-162,165,167,
   172-178
Transforming growth factor α,
   8-10
Transgenic plants, 73-86
Translation activation, 248
Translation initiation, 243-
   273
Transposase, 156
Transposons in streptomycetes,
   64,65
Transposon tagging, 74,81
TRE see thyroid hormone-
   responsive element
Trimethylguanosine cap, 209
Triose phosphate isomerase
   genes, 263
Triticum aestivum see bread
   wheat
Tropomyosin, 169-171
trpE fusion protein, 26,27
Trypanosoma brucei mRNA, inhi-
   bition of translation, 48
Trypsin inhibitor, 5
Tubulin mRNA, 262
Tyrosinase, 55,57
Tyrosine aminotransferase,
   184,185

Urokinase folding, 10

VA$_I$, 108-110
Vaccinia virus, 252,254
Vesicular stomatitis virus N
   protein, inhibition of
   translation, 48
Vibrio harveyi, 57
Viomycin resistance, 57

Vitamin D3 receptor, 183,
   185,194
VP2, 106
VP3, 106
vph, 65

Wheat germ extract, degradative properties, 47
   inhibition of translation,
      47-49
   storage proteins, 225-241
Wheat germ histone H3 gene,
   322
Wilms' tumor, 21

XbaI, 91
Xenopus oocyte, degradative
   properties, 47
   inhibition of translation,
      47-49
   splicing, 163,203,204,206,
      207,209,210
   translation, 301,311
X-glucuronide, 232
XhoI, 79,91

Zein, 232
Zinc fingers, 190,192,194,263